T0145232

Methods in Molecular Biology

Series Editor
John M. Walker
School of Life and Medical Sciences
University of Hertfordshire
Hatfield, Hertfordshire, UK

For further volumes:
http://www.springer.com/series/7651

For over 35 years, biological scientists have come to rely on the research protocols and methodologies in the critically acclaimed *Methods in Molecular Biology* series. The series was the first to introduce the step-by-step protocols approach that has become the standard in all biomedical protocol publishing. Each protocol is provided in readily-reproducible step-by-step fashion, opening with an introductory overview, a list of the materials and reagents needed to complete the experiment, and followed by a detailed procedure that is supported with a helpful notes section offering tips and tricks of the trade as well as troubleshooting advice. These hallmark features were introduced by series editor Dr. John Walker and constitute the key ingredient in each and every volume of the *Methods in Molecular Biology* series. Tested and trusted, comprehensive and reliable, all protocols from the series are indexed in PubMed.

Post-Transcriptional Gene Regulation

Third Edition

Edited by

Erik Dassi

Department of Cellular, Computational and Integrative Biology (CIBIO), University of Trento, Trento, Italy

Editor
Erik Dassi
Department of Cellular, Computational
and Integrative Biology (CIBIO)
University of Trento
Trento, Italy

ISSN 1064-3745 ISSN 1940-6029 (electronic)
Methods in Molecular Biology
ISBN 978-1-0716-1853-0 ISBN 978-1-0716-1851-6 (eBook)
https://doi.org/10.1007/978-1-0716-1851-6

This Humana imprint is published by the registered company Springer Science+Business Media, LLC, part of Springer Nature.
The registered company address is: 1 New York Plaza, New York, NY 10004, U.S.A.

Preface

Post-transcriptional regulation of gene expression (PTR) has emerged, in the last few years, as a pervasive and complex facet of gene expression control. The advent of high-throughput techniques empowered by next-generation sequencing, such as ribosome footprinting and protein-RNA interaction determination methods such as CLIP and its variants, has allowed us to start unraveling the contribution of RNA-binding proteins, noncoding RNAs, and, most recently, the RNA modifications shaping cell physiology and pathology. This volume presents the most recent advances in techniques for studying this critical level of gene expression regulation. With sections on bioinformatics approaches, expression profiling, the protein and RNA interactome, the mRNA lifecycle, and RNA modifications, it aims at guiding molecular biologists to harness the power of this new generation of techniques, while also introducing the data analysis skills that these high-throughput techniques bring along. This book in the *Methods in Molecular Biology* series is organized into five parts. First of all, Part I presents bioinformatics approaches for studying post-transcriptional regulation (Chapters 1–3). Readers are then introduced to techniques for expression profiling and perturbation in Part II (Chapters 4–7). Part III presents protein-RNA and ncRNA-RNA interaction analysis techniques (Chapters 8–12), while Part IV presents methods for studying various aspects of RNA biogenesis and processing (Chapters 13–18). Finally, Part V introduces a series of methods for profiling RNA modifications (Chapters 19–22). Recognizing the ever-increasing contribution of bioinformatics in empowering the effectiveness of these techniques, many chapters include hints for data analysis alongside the much-needed tips for benchwork. The final aim of this volume is thus to offer a versatile resource to researchers studying post-transcriptional regulation, both introducing the most recent techniques and providing a comprehensive guide to their implementation.

Part I: Bioinformatics

Given the ever-increasing amount of data generated by high-throughput techniques as well as the possibility to more efficiently drive and select the experimental work to be performed through analyses and predictions, ultimately enabling the reduction of the hypothesis space to be explored, bioinformatics is more and more regarded as an invaluable tool for any researcher studying post-transcriptional regulation of gene expression. Recognizing this fundamental contribution, the first part includes three chapters that illustrate the possibilities offered by current bioinformatics approaches for PTR. Chapter 1 introduces tools and databases dealing with aspects ranging from processing PTR omics datasets, current knowledge about regulatory factors and their interactions with RNA, to interaction prediction and motif search; such an introduction should allow researchers to start implementing pipelines able to answer to their data analysis needs. Chapter 2, by Delli Ponti and Tartaglia, presents their CROSS and CROSSalive algorithms to predict the RNA secondary structure propensity both in vitro and in vivo. Chapter 3, by Paz and colleagues, deals instead with another critical problem for PTR, that of predicting binding sites for RNA-binding proteins (RBPs). Their approach, named RBPmap, leverages the motif environment to improve prediction accuracy. Ultimately, an additional bioinformatics chapter (Chapter 10), presenting a tool for the analysis of CLIP datasets, found its logical place alongside the chapters describing these techniques and is thus included in Part III.

Part II: Expression Studies

Understanding changes in the molecular output (such as protein synthesis levels, alternative splicing isoforms usage, mRNA degradation, etc.) produced by post-transcriptional regulatory mechanisms due to the impact of a stimulus, a treatment, or more generally to the difference between two conditions is one of the main avenues of research in PTR. Indeed, by such studies, one can formulate hypotheses on acting trans-factors and the regulatory networks they give rise to. Thus, this part aims at presenting tools and techniques allowing the study of this aspect, both on a genome-wide and single-transcript scale. First of all, Chapter 4, by Cammas and colleagues, introduces us to the polysomal profiling technique, allowing the identification of polysomes-bound mRNAs and the study of the protein machinery associated with the mRNAs being translated. Then, in Chapter 5, Cope and colleagues describe how ribosome positioning on translating transcripts can be investigated by ribosome profiling, thus allowing translation mechanics in different systems and under various conditions to be systematically analyzed. They also introduce RiboViz, a tool for the analysis and visualization of ribosome profiling data. This part is then concluded by two chapters describing methods to characterize different aspects of CRISPR-Cas proteins, now widely used to perform transcriptional and post-transcriptional regulation of genes of interest in a highly specific manner. Chapter 6, by Sharma and Sharma, presents a method to globally identify the direct RNA-binding partners of CRISPR-Cas RNPs, allowing us to infer design principles for nucleic acid-targeting applications and fitted to each nuclease of interest. Finally, Chapter 7, by Wandera and Beisel, describes a cell-free transcription-translation system to characterize Cas13 nucleases from Type VI CRISPR-Cas systems. These nucleases employ guide RNAs to recognize complementary RNA targets and are exploited for numerous applications, including programmable gene silencing. The method thus allows measuring on-target and collateral RNA cleavage by these nucleases.

Part III: Interactomics

RNA-binding proteins and noncoding RNA are among the most prominent players in controlling the fate of RNA molecules at the post-transcriptional level. However, the lack of a complete catalog of their targets is hampering our ability to reconstruct the PTR networks concurring to shape the cell phenotype. Fortunately, this issue has recently received a great deal of attention. Techniques have been developed to tackle the identification of RBPs binding preferences, mode of action (e.g., controlling mRNA stability rather than its splicing), and processes and functions affected by the regulatory activity they exert. This part therefore begins by presenting the most recent techniques approaching this task. Chapter 8, by Rossi and Inga, describes the RNA immunoprecipitation (RIP) technique, allowing us to immunoprecipitate ribonucleoprotein complexes (RNPs) and identify targets for the RBP of interest in an efficient way. Chapter 9, by Danan and colleagues, presents PAR-CLIP, coupling UV-crosslinking and photoactivatable ribonucleosides to the immunoprecipitation to precisely map binding sites for the RBP of interest on its RNA targets. Chapter 10 by Sahadevan and colleagues introduces the reader to the htseq-clip and

DEWSeq tools, realizing a workflow for the analysis of eCLIP and iCLIP datasets and providing a practical usage tutorial. Identifying RBP targets is of paramount importance to trace PTR networks and mechanisms, but so is determining the interactions of specific target RNAs with noncoding RNAs and proteins. Along this line, Chapter 11, by Zeni and colleagues, describes miR-CATCH, a method using affinity capture biotinylated antisense oligonucleotides to co-purify a target transcript of interest together with all its endogenously bound miRNAs. MicroRNAs (miRNAs) are potent players in many PTR processes, involved both in normal physiology and diseases such as cancer, neurodevelopmental syndromes, and others. Lastly, Chapter 12 by Henke and colleagues presents HyPR-MS, a method for probing the in vivo protein interactomes of one or more target RNAs based on hybridization purification of RNA-protein complexes followed by mass spectrometry.

Part IV: The RNA Lifecycle

A host of different post-transcriptional processes shapes the lifecycle of RNA molecules at various stages, turning this layer of gene expression regulation into complex, finely tuned machinery. Among these are alternative polyadenylation, mRNA localization and degradation, and many others; of course, not all of them can be addressed here. Thus, this part presents a selection of the most recent techniques dealing with several aspects of the RNA lifecycle.

The first two chapters focus on employing imaging tools to investigate processes related to the metabolism of RNA molecules. Chapter 13, by Arora and colleagues, describes a protocol for single-molecule fluorescent in situ hybridization, which uses multiple approaches for synthesizing the fluorescent probes and enables the visualization and quantification of subcellular RNA localization. Chapter 14, by Cawte and colleagues, details a method using the fluorogenic RNA aptamer Mango II to tag and image RNAs with single-molecule sensitivity in both fixed and live cells, thus allowing the visualization of RNA dynamics. Then Chapter 15, by Kwak and Kwak, presents TED-seq, a technique sequencing size-selected 3′ RNA fragments including the poly(A) tail pieces to provide an accurate measure of transcriptome-wide poly(A)-tail lengths at high resolution. Chapter 16, by Gupta and Rouskin, describes instead a dimethyl sulfate-based chemical probing method coupled with high-throughput sequencing, called DMS-MaPseq, to study RNA secondary structure in vivo. Then Chapter 17 by Fasching and colleagues presents SLAMseq, a technique based on metabolic RNA labeling that provides accurate information on transcript half-lives across annotated features in the genome, including by-products of transcription such as introns. This part is eventually concluded by Chapter 18. There, May and McManus present FACS-uORF, a massively parallel reporter assay to simultaneously test thousands of yeast uORFs to evaluate the impact of codon usage on their functions.

Part V: RNA Modifications

A recently emerged aspect of PTR deals with the more than 150 different post-transcriptional modifications to which an RNA can be subjected. These modifications, ranging from the A-to-I editing and pseudouridylation to m^5C, m^6A, and 2′-*O* RNA methylation, have been found to be much more pervasive than previously thought. While many modifications target noncoding RNA species such as tRNAs, mRNAs are considerably modified as well. Furthermore, processes affected by these modifications and their impact on RNA stability, secondary structure, and translation are just beginning to be elucidated, as is their role in human pathology. Thus, the four chapters composing this part aim to present techniques to profile three of these modifications: Chapter 19, by Bhattarai and Aguilo, thus describes MeRIP-seq, a powerful technique to map m^6A sites at the transcriptome-wide level and thus generate comprehensive methylation profiles. Chapter 20 by Tegowski and colleagues presents in vitro DART-seq, an antibody-free method for m^6A profiling from low-input RNA samples. Chapter 21, by Sibbritt and colleagues, introduces a protocol for the positional profiling of the 5-methylcytosine RNA modification. Based on RNA bisulfite conversion, it employs locus-specific PCR amplification and detection of candidate sites by sequencing on the Illumina MiSeq platform. Last of all, Chapter 22 by Zhu and colleagues concludes this part by describing RibOxi-seq, a high-throughput approach to detect 2′-*O*-methylation sites in mRNA and noncoding RNAs using low-input material and sequencing depth.

Finally, I would like to thank all the authors for their invaluable contribution. Their work has shaped this book into a passionate and hopefully useful guide to current methods in such a dynamic and constantly expanding field. Sharing our knowledge, thus allowing others to build on it, is the ultimate essence of who we are as scientists. It is indeed by "*standing on the shoulders of giants*" that we nourish our fascination for the not yet understood and the unexpected.

Trento, Italy ***Erik Dassi***

The updated online version of the FM was revised. The correction to this chapter is available at https://doi.org/10.1007/978-1-0716-1851-6_23

Contents

Contributors

FRANCESCA AGUILO • *Department of Molecular Biology, Umeå University, Umeå, Sweden; Wallenberg Center for Molecular Medicine, Umeå University, Umeå, Sweden*

STEFAN L. AMERES • *IMBA—Institute of Molecular Biotechnology, Vienna Biocenter (VBC), Vienna, Austria; Max Perutz Labs, University of Vienna, Vienna Biocenter (VBC), Vienna, Austria*

AMIR ARGOETTI • *Faculty of Biology, Technion—Israel Institute of Technology, Haifa, Israel*

ANKITA ARORA • *Department of Biochemistry and Molecular Genetics, University of Colorado Anschutz Medical Campus, Aurora, CO, USA*

CHASE L. BEISEL • *Helmholtz Institute for RNA-based Infection Research (HIRI), Helmholtz-Centre for Infection Research (HZI), Würzburg, Germany; Medical Faculty, University of Würzburg, Würzburg, Germany*

DEVI PRASAD BHATTARAI • *Department of Molecular Biology, Umeå University, Umeå, Sweden; Wallenberg Center for Molecular Medicine, Umeå University, Umeå, Sweden*

ANNE CAMMAS • *Cancer Research Center of Toulouse, INSERM UMR 1037, Toulouse, France; Université Toulouse III Paul Sabatier, Toulouse, France*

GORDON G. CARMICHAEL • *Department of Genetics and Genome Sciences, UCONN Health, Farmington, CT, USA*

ADAM D. CAWTE • *Single Molecule Imaging Group, MRC London Institute of Medical Sciences, London, UK; Section of Virology, Department of Infectious Disease, Faculty of Medicine, Imperial College London, London, UK; Developmental Epigenetics, Department of Biochemistry, University of Oxford, Oxford, UK*

SUSAN J. CLARK • *Epigenetics Research Laboratory, Genomics and Epigenetics Division, Garvan Institute of Medical Research, Sydney, NSW, Australia; St. Vincent Clinical School, University of New South Wales, Sydney, NSW, Australia*

NOA COHEN • *Faculty of Biology, Technion—Israel Institute of Technology, Haifa, Israel; Department of Computer Sciences, Technion—Israel Institute of Technology, Haifa, Israel*

ALEXANDER L. COPE • *Department of Genetics, Rutgers University, Piscataway, NJ, USA; Human Genetics Institute of New Jersey, Piscataway, NJ, USA*

CHARLES DANAN • *RNA Molecular Biology Group, NIAMS, Bethesda, MD, USA*

ERIK DASSI • *Department of Cellular, Computational and Integrative Biology (CIBIO), University of Trento, Trento, Italy; Laboratory of RNA Regulatory Networks, Department of Cellular, Computational and Integrative Biology (CIBIO), University of Trento, Trento, Italy*

RICCARDO DELLI PONTI • *School of Biological Sciences, Nanyang Technological University, Singapore, Singapore*

MICHELA A. DENTI • *Department of Cellular, Computational and Integrative Biology—CIBIO, University of Trento, Trento, Italy*

ELIANA DESTEFANIS • *Department of Cellular, Computational and Integrative Biology (CIBIO), University of Trento, Trento, Italy*

LEÏLA DUMAS • *Cancer Research Center of Toulouse, INSERM UMR 1037, Toulouse, France; Université Toulouse III Paul Sabatier, Toulouse, France*

NIV EVEN • *Faculty of Biology, Technion—Israel Institute of Technology, Haifa, Israel; Department of Computer Sciences, Technion—Israel Institute of Technology, Haifa, Israel*

NINA FASCHING • *IMBA—Institute of Molecular Biotechnology, Vienna Biocenter (VBC), Vienna, Austria*
JOHN S. FAVATE • *Department of Genetics, Rutgers University, Piscataway, NJ, USA; Human Genetics Institute of New Jersey, Piscataway, NJ, USA*
RAEANN GOERING • *Department of Biochemistry and Molecular Genetics, University of Colorado Anschutz Medical Campus, Aurora, CO, USA; RNA Bioscience Initiative, University of Colorado Anschutz Medical Campus, Aurora, CO, USA*
MARGHERITA GRASSO • *Department of Cellular, Computational and Integrative Biology—CIBIO, University of Trento, Trento, Italy; L.N.Age Srl—Link Neuroscience and Healthcare, Pomezia (RM), Italy*
MARCO GUARNACCI • *Department of Genome Sciences, John Curtin School of Medical Research, Australian National University, Canberra, ACT, Australia*
PAROMITA GUPTA • *Whitehead Institute for Biomedical Research, Cambridge, MA, USA*
MARKUS HAFNER • *RNA Molecular Biology Group, NIAMS, Bethesda, MD, USA*
KATHERINE B. HENKE • *Department of Chemistry, University of Wisconsin-Madison, Madison, WI, USA*
PAULINE HERVIOU • *Cancer Research Center of Toulouse, INSERM UMR 1037, Toulouse, France; Université Toulouse III Paul Sabatier, Toulouse, France*
VERONIKA A. HERZOG • *IMBA—Institute of Molecular Biotechnology, Vienna Biocenter (VBC), Vienna, Austria*
CHRISTOPHER L. HOLLEY • *Department of Medicine (Cardiology Division), Duke University School of Medicine, Durham, NC, USA; Department of Molecular Genetics and Microbiology, Duke University School of Medicine, Durham, NC, USA*
HARUKI IINO • *Single Molecule Imaging Group, MRC London Institute of Medical Sciences, London, UK; Section of Virology, Department of Infectious Disease, Faculty of Medicine, Imperial College London, London, UK*
ALBERTO INGA • *Department of Cellular, Computational and Integrative Biology (CIBIO), University of Trento, Trento, Italy*
RACHEL A. KNOENER • *Department of Chemistry, University of Wisconsin-Madison, Madison, WI, USA*
HOJOONG KWAK • *Department of Molecular Biology and Genetics, Cornell University, Ithaca, NY, USA*
YEONUI KWAK • *Department of Molecular Biology and Genetics, Cornell University, Ithaca, NY, USA; Graduate Field of Genetics, Genomics, and Developmental Biology, Cornell University, Ithaca, NY, USA*
YAEL MANDEL-GUTFREUND • *Faculty of Biology, Technion—Israel Institute of Technology, Haifa, Israel; Department of Computer Sciences, Technion—Israel Institute of Technology, Haifa, Israel*
SUDHIR MANICKAVEL • *RNA Molecular Biology Group, NIAMS, Bethesda, MD, USA*
GEMMA E. MAY • *Department of Biological Sciences, Carnegie Mellon University, Pittsburgh, PA, USA*
C. JOEL MCMANUS • *Department of Biological Sciences, Carnegie Mellon University, Pittsburgh, PA, USA; Computational Biology Department, Carnegie Mellon University, Pittsburgh, PA, USA*
KATE D. MEYER • *Department of Biochemistry, Duke University School of Medicine, Durham, NC, USA; Department of Neurobiology, Duke University School of Medicine, Durham, NC, USA*

RACHEL M. MILLER • *Department of Chemistry, University of Wisconsin-Madison, Madison, WI, USA*
STEFANIA MILLEVOI • *Cancer Research Center of Toulouse, INSERM UMR 1037, Toulouse, France; Université Toulouse III Paul Sabatier, Toulouse, France*
INBAL PAZ • *Faculty of Biology, Technion—Israel Institute of Technology, Haifa, Israel*
JAN PETRŽÍLEK • *IMBA—Institute of Molecular Biotechnology, Vienna Biocenter (VBC), Vienna, Austria; Vienna BioCenter PhD Program, Doctoral School of the University of Vienna and Medical University of Vienna, Vienna, Austria*
NIKO POPITSCH • *IMBA—Institute of Molecular Biotechnology, Vienna Biocenter (VBC), Vienna, Austria*
THOMAS PREISS • *Department of Genome Sciences, John Curtin School of Medical Research, Australian National University, Canberra, ACT, Australia; Victor Chang Cardiac Research Institute, Sydney, NSW, Australia*
ANNALISA ROSSI • *Department of Cellular, Computational and Integrative Biology (CIBIO), University of Trento, Trento, Italy*
SILVIA ROUSKIN • *Whitehead Institute for Biomedical Research, Cambridge, MA, USA*
DAVID S. RUEDA • *Single Molecule Imaging Group, MRC London Institute of Medical Sciences, London, UK; Section of Virology, Department of Infectious Disease, Faculty of Medicine, Imperial College London, London, UK*
SUDEEP SAHADEVAN • *European Molecular Biology Laboratory (EMBL), Heidelberg, Germany*
MARK SCALF • *Department of Chemistry, University of Wisconsin-Madison, Madison, WI, USA*
ULRIKE SCHUMANN • *Department of Genome Sciences, John Curtin School of Medical Research, Australian National University, Canberra, ACT, Australia*
THOMAS SCHWARZL • *European Molecular Biology Laboratory (EMBL), Heidelberg, Germany*
THILEEPAN SEKARAN • *European Molecular Biology Laboratory (EMBL), Heidelberg, Germany*
ANDREW SHAFIK • *Department of Genome Sciences, John Curtin School of Medical Research, Australian National University, Canberra, ACT, Australia*
PREMAL SHAH • *Department of Genetics, Rutgers University, Piscataway, NJ, USA; Human Genetics Institute of New Jersey, Piscataway, NJ, USA*
CYNTHIA M. SHARMA • *Molecular Infection Biology II, Institute of Molecular Infection Biology (IMIB), University of Würzburg, Würzburg, Germany*
SAHIL SHARMA • *Molecular Infection Biology II, Institute of Molecular Infection Biology (IMIB), University of Würzburg, Würzburg, Germany*
TENNILLE SIBBRITT • *Department of Genome Sciences, John Curtin School of Medical Research, Australian National University, Canberra, ACT, Australia*
KYLE S. SKALENKO • *Department of Genetics, Rutgers University, Piscataway, NJ, USA; Waksman Institute, Rutgers University, Piscataway, NJ, USA*
LLOYD M. SMITH • *Department of Chemistry, University of Wisconsin-Madison, Madison, WI, USA*
MICHELE SPINIELLO • *Department of Chemistry, University of Wisconsin-Madison, Madison, WI, USA; Immuno-Hematology and Transfusion Medicine, Cardarelli Hospital, Naples, Italy*

J. MATTHEW TALIAFERRO • *Department of Biochemistry and Molecular Genetics, University of Colorado Anschutz Medical Campus, Aurora, CO, USA; RNA Bioscience Initiative, University of Colorado Anschutz Medical Campus, Aurora, CO, USA*
GIAN GAETANO TARTAGLIA • *Center for Human Technologies, Istituto Italiano di Tecnologia, Genoa, Italy*
MATTHEW TEGOWSKI • *Department of Biochemistry, Duke University School of Medicine, Durham, NC, USA*
PETER J. UNRAU • *Department of Molecular Biology and Biochemistry, Simon Fraser University, Burnaby, BC, Canada*
PEDRO TIRADO VELEZ • *RNA Bioscience Initiative, University of Colorado Anschutz Medical Campus, Aurora, CO, USA*
SANGEEVAN VELLAPPAN • *Department of Genetics, Rutgers University, Piscataway, NJ, USA; Waksman Institute, Rutgers University, Piscataway, NJ, USA*
KATHARINA G. WANDERA • *Helmholtz Institute for RNA-based Infection Research (HIRI), Helmholtz-Centre for Infection Research (HZI), Würzburg, Germany*
SRUJANA S. YADAVALLI • *Department of Genetics, Rutgers University, Piscataway, NJ, USA; Waksman Institute, Rutgers University, Piscataway, NJ, USA*
ANDREA ZENI • *Department of Cellular, Computational and Integrative Biology—CIBIO, University of Trento, Trento, Italy; Istituto Rainerum—SDB, Bolzano, Italy*
HUANYU ZHU • *Department of Biochemistry, Duke University School of Medicine, Durham, NC, USA*
YINZHOU ZHU • *Department of Medicine (Cardiology Division), Duke University School of Medicine, Durham, NC, USA*

Part I

Bioinformatics

Chapter 1

Introduction to Bioinformatics Resources for Post-transcriptional Regulation of Gene Expression

Eliana Destefanis and Erik Dassi

Abstract

Untranslated regions of mRNA (UTRs) are involved in defining the fate of the transcript through processes such as mRNA localization, degradation, translation initiation regulation, and several others: the action of trans-factors such as RNA-binding proteins and non-coding RNAs, combined with the presence of defined sequence and structural cis-elements, ultimately determines protein synthesis levels. Identifying functional regions in UTRs and uncovering post-transcriptional regulators acting upon these is thus of paramount importance to understand this regulatory layer: these tasks can now be approached computationally to reduce the testable hypothesis space and drive the experimental validation in a more effective way.

This chapter will focus on presenting databases and tools allowing to study the various aspects of post-transcriptional regulation, including the profiling of actively translated mRNAs, regulatory network analysis (e.g., RBP and ncRNA binding sites), trans-factor binding sites prediction, motif search (sequence and secondary structure), and other aspects of this regulatory layer: two potential analysis pipelines are also presented as practical examples of how these tools could be integrated and effectively employed.

Key words Bioinformatics, UTR, Database, Prediction, Data analysis, Pipeline, Omics, Polysomal profiling, RBP, ncRNA, Binding site, Secondary structure, Motif

1 Introduction

Post-transcriptional regulation of gene expression (PTR) has been object, in recent years, of an ever-increasing interest. This led to the development, also thanks to high-throughput techniques such as next-generation sequencing, of a whole new set of experimental assays aiming at profiling and unraveling these mechanisms on a genome-wide scale [1–3]. The advent of next-generation sequencing has indeed provided the possibility to address tasks that were previously out of reach, such as determining an RBP binding specificity. However, to fully exploit the huge amounts of data produced by such experiments, one needs a well-built toolkit of analysis tools and databases, which should ultimately allow tracing

Erik Dassi (ed.), *Post-Transcriptional Gene Regulation*, Methods in Molecular Biology, vol. 2404,
https://doi.org/10.1007/978-1-0716-1851-6_1,

the regulatory networks underlying the system and conditions under study, thus deriving meaningful functional insights.

A common example of task one could perform is to profile and quantify PTR under different stimuli or treatments by genome-wide profiling of translating transcripts: this is usually addressed with ribosome or polysome profiling techniques [2, 4]. Resulting datasets need, first of all, specific tools to properly identify differences in translated and untranslated transcripts sets. A second class of tools, i.e., regulatory elements and binding sites databases/predictors, is then needed to study the mRNA of the identified interesting genes set, trying to understand which trans-factors (RBPs or ncRNA) may be playing a role in producing the observed differences, and thus in uncovering related regulatory mechanisms.

More generally, one may need to study PTR events, as mediated by UTRs, on a single gene or a set of interesting ones, even if not derived from genome-wide approaches. Sequence and structure features of such UTRs must thus be investigated by exploiting existing knowledge (using databases) and prediction algorithms capabilities.

Another example of a frequent task, from the opposite point of view, consists in determining binding specificities for an RNA-binding protein: this is usually approached by the CLIP family of techniques [5–13] or by methods such as RNAcompete [14], RNAcompete+S [15], Bind-n-Seq [16], and SEQRS [17]. In addition, alternative methods have been developed, including the use of protein purification and mass spectrometry (comprehensively reviewed by Lin and Miles [18]). Data analysis for those high-throughput assays is far from being trivial and requires dedicated tools and statistical approaches (as Chap. 10 of this book will describe in detail). Results (i.e., identified RNA targets and binding sites) will then be amenable to further analyses, such as sequence and secondary structure motif search, using yet another class of tools to eventually characterize the RBP specificity.

These two tasks are, of course, only two representative examples of the many possible data analysis workflows one may need to implement while studying PTR. However, these already give a clear indication that one should invest time in learning about available tools, what they are best suited for, and how to combine them; this knowledge will allow a researcher to be effective in such data analysis tasks.

This chapter will first introduce the reader to analysis tools, databases, prediction algorithms, and other software which are currently available to address such tasks, with a particular emphasis on practical considerations and integration of multiple resources: by means of two example pipelines, we will then aim at providing guidance in setting up and performing the analysis of typical PTR datasets.

2 Tools for Omics Datasets Analysis

The first class of tools we describe deals with omic datasets: these resources, listed in Table 1, can be divided into three broad groups according to the kind of data they deal with. *Expression profiling* datasets are derived by polysome and ribosome profiling and allow to quantify the mRNAs translational efficiency; *small RNA* datasets are obtained by RNA-seq performed on the short RNAs fraction and permit to identify and quantify RNAs such as miRNAs and piRNA. Eventually, *binding* datasets aim at identifying targets and binding sites for an RNA-binding protein and can be produced by techniques such as RIP, CLIP and its variants. We will now proceed to describe and compare all tools related to these experimental approaches.

2.1 Expression Profiling

Polysomal profiling consists in the isolation of polysome-bound, and thus actively translating, transcripts by means of a sucrose gradient separation. This assay is usually complemented with a total mRNA profiling of the same sample, in order to compute translational efficiency (TE) and identify regulatory events affecting this value.

anota [19] is an algorithm, available as an R package, stemming from the consideration that polysomal mRNA levels are dependent on the total (cytoplasmic) mRNA amounts in the samples. The method thus performs an analysis of partial variance in combination with a random variance model, to avoid considering false buffering events, produced by the commonly used log-ratio approach, as true differential translation phenomena. *tRanslatome* [20] is also an R package, but adopting a wider perspective: it allows to identify differentially expressed genes with several methods (limma, *t*-test, TE, RankProd, anota, DEseq, and edgeR), and to perform Gene Ontology and regulatory enrichment analyses on the differentially translated genes; every step is then illustrated through a variety of plot types (scatterplot, MA, SD and identity plot, histogram, heatmap, similarity, and radar plot).

Ribosome profiling, a much more recent technique, is instead based on a nuclease protection assay and provides a snapshot of ribosomal "footprints" on the mRNAs, allowing the quantification of their translational levels; in the last years, several tools were developed to analyze this kind of data. *Babel* [21] is a statistical framework to assess the significance of translational control differences between conditions. Available as an R package, it infers an expected ribosome occupancy level (based on mRNA expression) by means of an errors-in-variables regression model; the significance of the deviation of the actual ribosome occupancy from this estimate is then assessed by a parametric bootstrap procedure, thus obtaining *p*-values for all genes. *Xtail* [22] represents another R

Table 1
Data analysis tools for PTR omics datasets. The table lists tools available to process PTR omics datasets: listed are the application of each tool (either expression, i.e., translation levels determination, small RNA, i.e., miRNA and other ncRNAs profiling, or binding, i.e., RBP targets identification), the type of supported assays, their specific approach, and where to obtain the software from (URL and reference)

Name	Application	Supported data types	Approach	URL	References no.
Anota	Expression	Polysomal profiling	analysis of partial variance tuned to avoid "false buffering" associated to cytosolic mRNA levels correction	bioconductor.org/packages/release/bioc/html/anota.html	[19]
tRanslatome	Expression	Polysomal profiling	implements several DEGs identification methods, plot types, and functional analyses	bioconductor.org/packages/release/bioc/html/tRanslatome.html	[20]
Babel	Expression	Ribo-seq	based on an errors-in-variables regression model to identify translational control changes	cran.r-project.org/web/packages/babel/	[21]
Xtail	Expression	Ribo-seq	quantify the magnitudes and statistical significances of differential translations	github.com/xryanglab/xtail	[22]
Riborex	Expression	Ribo-seq	uses existing model-based frameworks for gene expression analysis to estimate all parameters of the differential translation analysis model	github.com/smithlabcode/riborex	[24]
RiboDiff	Expression	Ribo-seq	performs differential translation analysis by using a GLM and accounting for different sequencing protocols	bioweb.me/ribodiff	[23]
RiboToolkit	Expression	Ribo-seq	Web-based service allowing unbiased mRNA translation efficiency and	rnabioinfor.tch.harvard.edu/RiboToolkit	[53]

(continued)

Table 1 (continued)

Name	Application	Supported data types	Approach	URL	References no.
			differential translation analysis		
scikit-ribo	Expression	Ribo-seq	performs translation efficiency inference using a codon-level GLM with ridge penalty	github.com/schatzlab/scikit-ribo	[54]
anota2seq	Expression	Polysomal profiling, Ribo-seq	analyze changes in translational efficiencies affecting protein levels and buffering and not affected by spurious correlations	bioconductor.org/packages/release/bioc/html/anota2seq.html	[25]
RIVET	Expression	Polysomal profiling, Ribo-seq	allows translatomics data exploration and differential analysis	ruggleslab.shinyapps.io/RIVET	[26]
CPSS	small RNA	RNA-seq	Webserver integrating available analysis tools for small RNA prediction and quantification	114.214.166.79/cpss2.0/	[28]
iMir	small RNA	RNA-seq	pipeline performing miRNA prediction and quantification by several available tools	www.labmedmolge.unisa.it/italiano/home/imir	[29]
UEA sRNA workbench	small RNA	RNA-seq	pipeline performing small RNA prediction, quantification and visualization by ad hoc implemented tools	srna workbench.cmp.uea.ac.uk	[27]
CAP-miRSeq	small RNA	RNA-seq	pipeline performing miRNA quantification and prediction, SNV detection, data visualization, and differential expression	bioinformaticstools.mayo.edu/research/cap-mirseq/	[30]
Chimira	small RNA	RNA-seq	Webserver allowing miRNA quantification, epitranscriptomics modifications identifications, data visualization and	www.ebi.ac.uk/research/enright/software/chimira	[31]

(continued)

Table 1
(continued)

Name	Application	Supported data types	Approach	URL	References no.
			differential expression analysis		
mirPRo	small RNA	RNA-seq	pipeline for miRNAs prediction and quantification	sourceforge.net/projects/mirpro/	[32]
miRMaster	small RNA	RNA-seq	Webserver for quantification and identification of miRNAs, isomiRs, mutations, exogenous RNAs, and motifs prediction	www.ccb.uni-saarland.de/mirmaster	[35]
sRNAnalyzer	small RNA	RNA-seq	pipeline for ncRNAs quantification	srnanalyzer.systemsbiology.net/.	[36]
SPAR	small RNA	RNA-seq	Webserver for the analysis, visualization, and annotation with reference small RNA datasets from ENCODE and DASHR	www.lisanwanglab.org/SPAR	[38]
sRNAPipe	small RNA	RNA-seq	Galaxy-based pipeline for ncRNAs quantification	github.com/brassetjensen/sRNAPipe	[37]
miRge 2.0	small RNA	RNA-seq	pipeline for miRNA quantification and prediction, A-to-I editing analysis, and isomiR detection	github.com/mhalushka/miRge.	[39]
Prost!	small RNA	RNA-seq	pipeline for miRNAs quantification and identification with user-defined annotation database and post-transcriptional modifications frequency analysis	prost.readthedocs.io/en/latest/	[40]
COMPSRA	small RNA	RNA-seq	small RNAs identification, quantification and basic differential analysis	regepi.bwh.harvard.edu/circurna/	[41]

(continued)

Table 1
(continued)

Name	Application	Supported data types	Approach	URL	References no.
sRNAtoolbox	small RNA	RNA-seq	Webserver for small RNAs quantification and prediction, RNA targets identification and differential analysis by several available tools	bioinfo2.ugr.es/srnatoolbox/	[34]
PIRANHA	Binding	RIP-seq, HITS-CLIP, PAR-CLIP, iCLIP	allows external covariates, such as expression data, to guide binding sites identification	smithlab.usc.edu	[42]
ASPeak	Binding	RIP-seq, HITS-CLIP	computes expression-sensitive backgrounds to improve peak calling	sourceforge.net/projects/as-peak	[43]
RIPSeeker	Binding	RIP-seq, PAR-CLIP	infers significant binding regions by EM and HMM modeling	bioconductor.org/packages/release/bioc/html/RIPSeeker.html	[44]
PARalyzer	Binding	PAR-CLIP	employs a non-parametric kernel density estimate classifier to identify binding sites from T > C conversions and read density	ohlerlab.mdc-berlin.de/software/PARalyzer_85/	[46]
PIPE-CLIP	Binding	HITS-CLIP, PAR-CLIP, iCLIP	customizable Galaxy pipeline based on zero-truncated negative binomial likelihoods	github.com/QBRC/PIPE-CLIP	[50]
wavClusteR	Binding	PAR-CLIP	uses a mixture model and a wavelet-based peak calling procedure exploiting coverage function geometry at binding sites regions	bioconductor.org/packages/devel/bioc/html/wavClusteR.html	[47]
CLIPper	Binding	CLIP	models background read counts using a Poisson distribution and identifies regions that	github.com/YeoLab/clipper	[45]

(continued)

Table 1
(continued)

Name	Application	Supported data types	Approach	URL	References no.
			are higher than expected by chance		
omniCLIP	Binding	CLIPs	identifies target sites via an unsupervised segmentation of the genome and performs peaks calling by learning diagnostic events from the data	github.com/philippdre/omniCLIP	[52]
PureCLIP	Binding	iCLIP, eCLIP	calls crosslinking sites and models possible sources of bias	github.com/skrakau/PureCLIP	[51]
MiClip	Binding	HITS-CLIP, PAR-CLIP	infers binding sites and CLIP clusters through two rounds of HMM	cran.r-project.org/web/packages/MiClip/index.html	[48]

package computing differential translation. Here, two pipelines have been implemented for the quantification of the translational changes between two conditions. The dissimilarity is calculated, on one side, between the changes in the total mRNA and the ribosome-derived mRNA expressions, and on the other side between the total-to-ribosome mRNA ratios. Thus, the statistical significance of the differential translation is inferred from the probability distribution of these translational changes. *RiboDiff* [23] was developed to analyze the effect of the treatment on mRNAs translation efficiency. It uses a generalized linear model (GLM) and produces robust dispersion estimates accounting for the different characteristics and statistical properties between total mRNAs and ribosome-derived mRNAs counts. As a last tool specifically implemented for the ribosome profiling data analysis, *Riborex* [24] identifies differentially translated genes by applying a GLM with parameters estimated from existing model-based frameworks for gene expression analysis. Thus it needs shorter computation times compared to the other described methods.

Recently, a few methods able to analyze data coming from both polysome profiling and ribosome profiling have been proposed. The *anota2seq* [25] R package was introduced to fill the need to analyze translation efficiencies changes, not only from continuous (e.g., DNA-microarray) but also from count data (i.e., RNA-seq). This algorithm allows statistics-based separate identification of changes in translational efficiency affecting protein levels and buffering, and as anota [19], it is not affected by spurious correlations. *RIVET* [26] is instead an R shiny-based graphical user interface. It performs differential expression analysis by providing several statistical modeling methods and offering comprehensive data visualization alternatives. Here, the user can decide to perform the analysis separately at the transcription or translation levels or to take both levels into account (translational efficiency).

2.2 Small RNA Profiling

Identification and quantification of short RNAs such as, for instance, microRNAs and piwi-interacting RNAs (piRNAs) is performed by means of an RNA-seq assay in which the RNAs selected for sequencing are approximately 20–30 nucleotides long. *UEA sRNA workbench* [27] provides a set of tools able to address various recurrent tasks (novel ncRNA identification, differential expression analysis, target prediction, etc.) in small ncRNA data analysis. Furthermore, it offers several visualization options, such as secondary structure plots and annotation and alignments display tools. *CPSS 2.0* [28] is a webserver which combines existing tools to analyze NGS small RNA data, including quantification, differential expression, target predictions, functional analysis, and novel miRNA identification: the output is presented as a convenient graphic summary in the browser, with detailed results as downloadable files; on the same line, *iMir* [29] and *CAP-miRSeq* [30] are pipelines for the processing and analysis of miRNA-seq data. The first integrates many pre-existing open-source tools, providing an easy to use graphical user interface and allowing to analyze NGS data to identify small ncRNAs (existing and novel), analyze their differential expression, and predict their targets. It is a stand-alone tool and can be installed on Unix systems only. The second instead focuses on the pre-processing, quantification, and prediction steps as well as SNV detection, differentially expression and data visualization of small RNA-seq data. It can be run locally or through a virtual machine for small-scale study samples; an Amazon cloud is instead provided to run a large number of samples. Similarly, *Chimira* [31] has been developed for fast and user-friendly analysis of small RNA datasets. In addition, it identifies 3′UTR, 5′UTR, and internal epitranscriptomic modifications allowing the extraction of global modification profiles across and within each of the input samples. On the same line, *mirPRO* [32] performs the pre-processing of miRNA-seq data and miRNA prediction and quantification with different approaches. It also offers other functions such as the miRNA variant detection, including isomiRs and “arm switching,” and the seed region check.

sRNAbench [33] performs the analysis of small RNA high-throughput datasets, and represents the core of the *sRNAtoolbox* [34]. The latter is a webserver collecting small RNA research tools. Apart from the small RNA expression profiling, it provides other downstream analysis capabilities such as differential expression, target prediction, and analysis of unmapped reads. *miRMaster* [35] is a web-based NGS data analysis tool allowing the processing, detection, and quantification of miRNA and miRNAs variants and isoforms; furthermore, application programming interfaces (APIs) are available for downstream analyses such as target and functional enrichment. By mapping non-human small RNA reads, it further allows the detection of contaminants, infections, or exogenous miRNAs as done by *sRNAnalyzer* [36]. sRNAnalyzer is divided into a data pre-processing, sequence mapping/alignment, and a result summarization module. In particular, in the second module, both endogenous RNA profiling and identification of RNA sequences from exogenous species are performed. *sRNAPipe* [37] is a Galaxy tool composed of several approaches wrapped together, which allows the processing, identification, and analysis of different categories of small genomic sequences from multiple small RNA libraries in parallel. Similarly, *SPAR* [38] allows the analysis of data from a variety of sncRNA experimental protocols for the identification of several small RNA types. It further includes DASHR and ENCODE reference sncRNA databases, thus enabling the comparison of the input data with publicly available experimental data. Two Python tools have also been developed recently. The first is *miRge 2.0* [39] which is a fast tool for the analysis of single or multiple small RNA-seq samples. It focuses on maximizing the capture of true miRNAs and isomiRs and on measuring the miRNAs entropy. With the latest update the authors implemented a new miRNA detection method based on a machine learning algorithm (SVM), improved alignment parameters and a module for the identification of A-to-I changes. The second is *Prost!* [40]. By allowing the analysis of many samples in parallel, it aligns the sequencing reads on user-provided genomic reference and annotates them with user-defined databases, thus enabling the study of miRNAs of any species and of potentially novel miRNAs. The last tool is *COMPSRA* [41], a recent stand-alone platform for small RNA-seq data analysis. It includes RNA databases and sequencing processing tools for the identification and quantification of small RNA molecules, also including fragmental microbial RNAs.

2.3 Binding Sites Identification

Identifying the targets of an RNA-binding protein and the related binding sites is currently performed mostly by RIP-seq (RNA-immunoprecipitation coupled with NGS), HITS-CLIP, PAR-CLIP, iCLIP, and eCLIP (collectively referred to as CLIPs, i.e., cross-linking and immunoprecipitation). While RIP is mostly used to identify RNA targets only, the CLIPs can also provide precise binding sites localization on these, eventually allowing the definition of the studied RBP binding specificity (*see* Subheading 4). As these techniques have appeared only in the last few years, analysis methods are still evolving and increasing in terms of approaches variety. The first two tools we describe, *PIRANHA* [42] and *ASPeak* [43], share the principle of exploiting a coupled expression dataset to help in true binding sites identification. The former, applicable to RIP-seq and all CLIPs, expands this concept to allow any covariate other than expression data, such as genome mappability information or the position of relevant sites (e.g., splice sites). It then applies a zero-truncated negative binomial (NB) regression to extract binding sites from the dataset and statistically score them. The latter instead requires a coupled expression dataset, used to compute an expression-sensitive background for binding data. It then runs a NB test over each nucleotide to produce a precise site definition; furthermore, it can exploit multiple processors to speed up the computation. Another tool able to deal with RIP-seq data is *RIPSeeker* [44], which can also handle PAR-CLIP data. Its approach is based on stratifying the genome in bins of equal size and applying a two-state Hidden Markov Model (HMM) with NB emission and a Viterbi algorithm yielding the peak calls, eventually tested for statistical significance. Similar to PIRANHA, the *CLIPper* [45] algorithm was designed to analyze CLIP-seq datasets. It identifies clusters representing RBPs binding sites by employing a three-pass filter on the detected peaks to reduce false positives. At first, it determines the minimum heights of CLIP-seq reads which satisfy the FDR threshold, then it removes peaks with few reads and interpolates reads height using cubic splines; the clusters are then identified from the fitted curve. *PARalyzer* [46] and *wavClusteR* [47] are specifically designed to deal with PAR-CLIP datasets. The former generates two smoothed kernel density estimates, one for T > C transitions and one for non-transitions. Nucleotides with a minimum read depth and a conversion likelihood higher than non-conversion ones are considered interaction sites, which are then extended to define the full binding site. A motif search can also be performed to define the RBP binding motif. The latter tool is instead based on a mixture model, defined on the observed T > C substitutions and the read coverage of the related nucleotides; a continuous wavelet transform is then applied to the model, exploiting the coverage function geometry to detect significant discontinuities in coverage, ideally representing true binding sites boundaries. To mitigate the underlying model assumptions and increase the types of CLIP-seq datasets that can be handled, the *MiClip* [48] algorithm was introduced. Able to analyze PAR-CLIP and HITS-CLIP datasets, it is implemented as both R package and Galaxy plugin [49] and identifies protein-RNA binding sites with two rounds of a Hidden Markov model (HMM). Another Galaxy-based pipeline is *PIPE-clip* [50]: it exploits a zero-truncated NB likelihood model to identify enriched clusters, then selecting interesting ones by exploiting the assay properties (e.g., mutations for PAR-CLIP).

Since no method could address peak calling and crosslinking site detection simultaneously, the *PureCLIP* [51] tool was developed to detect protein-RNA interaction in iCLIP/eCLIP datasets, with the possibility to be extended to other similar single-nucleotide resolutions protocols. It applies the HMM strategy and enables the identification of crosslinking sites by considering both regions hosting a significantly greater number of reads (peaks) and the sites originating from truncated cDNAs. Furthermore, it corrects for biases, such as transcript abundances, background binding, and crosslinking sequence preferences. Eventually, *omniCLIP* [52] is a probabilistic method whose parameters are learned directly from the data, thus making it applicable to all the CLIP-seq protocols-derived data. It employs a negative binomial (NB)-based GLM to model the coverage, which accounts for gene expression, local effects, and excess variance. Then, a Dirichlet-multinomial mixture (DMM) models the CLIP-specific crosslinking site, and a Non-Homogeneous Hidden Markov Model (NHMM) is used for peak calling. Thus, it accounts for replicates and background information as well as technical and biological variance.

3 PTR Databases

We now proceed in our exploration of PTR resources by describing the many databases storing and presenting the current knowledge about this class of processes: these tools, listed in Table 2, can be broadly classified according to the kind of data they hold, namely binding sites for RNA-binding proteins (*RBP*) or *miRNA*, *cis-elements* location or *ncRNAs* identity and features in general; a few of these resources can be termed *integrative*, in the sense that they present multiple types of information in the same setting, thus allowing for the integration of several PTR facets.

3.1 RBPs Binding Sites

These resources collect and present either binding sites for RNA-binding proteins on mRNAs, derived both by low- and high-throughput experimental approaches, possibly in multiple organisms, or focus on RBP binding specificities, presenting binding motifs and related information. *CLIPZ* [55] is a database and analysis environment for CLIPs datasets. Aside from visualizing the included datasets (amounting to ~100 CLIPs including replicates) at the genome level, users can upload and analyze their data down to the identification of enriched binding motifs and binding sites on the mRNAs; furthermore, miRNA-specific tools are provided to handle Argonaute CLIPs used to identify targets for these ncRNAs. *CISBP-RNA* [56] focuses instead on RBP motifs, collecting many experiments both collected from the literature or performed by the database maintainers. As multiple techniques and organisms are included, one can obtain a precise picture of the binding

Table 2
Databases presenting available PTR data. The table lists databases containing data about PTR, such as RBP and miRNA binding sites and cis-elements regions. Shown are the data types (i.e., RBP and non-coding RNA binding sites, all cis-elements or a specific type of these) provided by each database, the species for which data is available, and how these databases can be reached (URL and reference)

Name	Data types	Species	URL	References no.
AURA 2	RBP, miRNA, cis-elements	Human and mouse	aura.science.unitn.it	[62]
CISBP-RNA	RBP	Human, mouse, and 22 more	cisbp-rna.ccbr.utoronto.ca	[56]
CLIPZ	RBP	Human, mouse, and *C. elegans*	edoc.unibas.ch/19104/	[55]
RBPDB	RBP	Human, mouse, *D. melanogaster*, and *C. elegans*	rbpdb.ccbr.utoronto.ca/	[57]
ATtRACT	RBP	Human, mouse, and 21 more	attract.cnic.es	[58]
oRNAment	RBP	Human, mouse, *D. melanogaster*, *D. rerio*, and *C. elegans*	rnabiology.ircm.qc.ca/oRNAment/	[61]
POSTAR2	RBP	Human, mouse, and 4 more	lulab.life.tsinghua.edu.cn/postar	[59]
MotifMap-RNA	RBP	Human, mouse	motifmap-rna.ics.uci.edu/#	[92]
doRiNa2	RBP, miRNA	Human, mouse, *D. melanogaster*, and *C. elegans*	dorina.mdc-berlin.de	[63]
UTRdb/UTRsite	RBP, miRNA, cis-elements	Human, mouse, and 77 more	utrdb.ba.itb.cnr.it / utrsite.ba.itb.cnr.it	[65]
lncRNAdb	lncRNA	Human, mouse, and 58 more	rnacentral.org/	[76]
lncRNA2Target	lncRNA	Human, mouse	123.59.132.21/lncrna2target	[78]
miRGator	miRNA	Human	mirgator.kobic.re.kr	[70]
miRTarBase	miRNA	Human, mouse, and 30 more	miRTarBase.cuhk.edu.cn/	[69]
miRWalk2	miRNA	Human, mouse, and rat	zmf.umm.uni-heidelberg.de/mirwalk2	[72]

(continued)

Table 2
(continued)

Name	Data types	Species	URL	References no.
miRNAMap2	miRNA	Human, mouse, and 11 more	miRNAMap.mbc.nctu.edu.tw/	[68]
DIANA-TarBAse	miRNA	Human, mouse, and 16 more	www.microrna.gr/tarbase	[74]
miRBase	miRNA	Human, mouse, and 269 more	mirbase.org/	[66]
miRGate	miRNA	Human, mouse, and rat	mirgate.bioinfo.cnio.es	[71]
STarMirDB	miRNA	Human, mouse, and *C. elegans*	sfold.wadsworth.org/starmirDB.php	[73]
DIANA-LncBase	miRNA	Human, mouse	microrna.gr/LncBase	[75]
miRDB	miRNA	Human, mouse, and 3 more	mirdb.org	[67]
NONCODE	ncRNA	Any	noncode.org	[77]
NPInter	ncRNA	Human, mouse, and 32 more	bigdata.ibp.ac.cn/npinter	[80]
RAID	ncRNA, RBP	Human	rna-society.org/raid	[79]
Rfam	ncRNA, cis-elements	Any	rfam.sanger.ac.uk	[81]
starBase 2	ncRNA, RBP	Human, mouse, and 4 more	starbase.sysu.edu.cn/starbase2/index.php	[64]
APADB	Alternative polyadenylation sites	Human, mouse, and chicken	tools.genxpro.net/apadb	[83]
ARED-Plus	AU-rich elements	Human	brp.kfshrc.edu.sa/ared	[84]
AREsite2	AU, GU and U-rich elements, RBP	Human, mouse, *D. melanogaster*, *D. rerio*, and *C. elegans*	rna.tbi.univie.ac.at/AREsite	[85]
IRESite	Internal ribosome entry sites	Human, mouse, and 7 more	iresite.org	[86]
IRESbase	Internal ribosome entry sites	Human, mouse and 13 more and 19 virus	reprod.njmu.edu.cn/cgi-bin/iresbase/index.php	[87]
SelenoDB	SECIS element	Human, mouse, and 6 more	selenodb.crg.eu/	[89]

(continued)

Table 2
(continued)

Name	Data types	Species	URL	References no.
SIREs	Iron-responsive element	Any	ccbg.imppc.org/sires	[90]
Transterm	Cis-elements	Any	crispr.otago.ac.nz/TT/index.php	[91]
RegRNA 2.0	Cis-elements, AU-rich elements, polyadenylation sites, miRNA	Any	regrna2.mbc.nctu.edu.tw/	[82]

specificities for the RBP of interest; furthermore, tools are provided to predict instances of an RBP binding motif in RNA sequences or to compare custom motifs with the ones stored in the database. *RBPDB* [57] consists of a database of experimentally observed RNA-RBP interactions. It collects all the RBPs with a known RNA-binding domain from in vitro and in vivo experiments in four metazoan species. The database can be searched, on one side, for RNA-binding data by RBP, species, or experiment type; on the other side, the user can submit nucleotide sequences to scan for matches with RBP motifs associated with full Positional Weight Matrices (PWMs). The RBP-RNA interactions and RBP binding motifs collected in CISBP-RNA, RBPDB, and other databases were gathered in the *ATtRACT* database [58]. It combines already characterized data with in-silico analysis of protein-RNA structures available in the Protein Data Bank (PDB), to allow searching for RBPs or motifs. Furthermore, it implements a scan sequence and a de novo motif discovery interface, which enables the user to scan a file of RNA/DNA sequences searching for the presence of motifs. *POSTAR2* [59] includes the CLIPdb [60] data and collects evidence of post-transcriptional regulatory logics by assembling CLIP-seq and Ribo-seq experimental results. It provides three modules of exploration: the first consists of several annotations of the RBPs, while the second link the query gene to RBP binding sites and to their experimental or predicted miRNA bindings as well as RNA modifications, editings, SNVs, and associated disorders. The third and last module to characterize the RNAs translational landscape from predicted ORFs. The most recent database is *oRNAment* [61], which collects the putative motif instances of 223 RBPs covering coding and non-coding RNAs of four model organisms. RBP binding specificities are obtained through RNA-compete and RNA Bind-n-Seq methodologies [14, 16], and then transcriptome-wide scanned by a custom pipeline. The user can query the database by RBP, transcript, or by a specific combination of transcript attributes. In the first case, the user can specify the PWM specificity threshold, which allows the inclusion of motifs with different degrees of similarity to the most probable in vitro defined consensus motif.

The other resources presenting RBP binding sites, namely *AURA2* [62], *doRiNa2* [63], *starBase2* [64], *x*\ and *UTRdb/UTRsite* [65], also provide other types of data (e.g., miRNA binding sites) and will thus be described in the integrative resources section (Subheading 3.4).

3.2 miRNAs Binding Sites and ncRNAs

These databases aim at presenting data about non-coding RNAs such as their identity, role, and molecular targets. The majority of these is focused on collecting miRNA-mRNA interactions, both derived experimentally and computationally through the plethora of available predictors; the functions of individual miRNAs are also often presented, derived by analyzing processes and pathways in which their targets are involved.

miRBase [66] is the primary public repository and online resource for microRNAs sequences and annotations of 271 organisms, which further includes links to external information such as predicted and experimentally validated miRNAs targets. *miRDB* [67] includes data on 3.5 million predicted targets regulated by 7000 miRNAs in five species and allows custom target prediction with user-provided sequences. The recently updated database also enables the prediction of miRNA functions and of cell-specific miRNA targets based on the expression profile of over 1000 cell lines. *miRNAMap* [68] and *miRTarBase* [69] provide a curated collection of several thousand experimentally verified and/or computationally predicted miRNA-mRNA interactions (in miRNAMap compiled by three different algorithms). miRNAMap and miRTarBase also provide miRNA and target expression profiles, with the last collectioning also upstream and downstream miRNAs regulators, miRNAs associations to diseases and miRNA-targets networks. On the same line but heavily focused on deep sequencing data, *miRGator* [70] also provides a catalog of NGS-derived miRNAs for various tissues and organs, a dedicated browser concurrently displaying miRNA sequencing data and secondary structure and miRNA-target expression correlations, all combined with the goal of identifying true regulatory, functional and pathological associations. *miRGate* [71] tries to overcome the lack of overlap between the different target prediction methods by developing a tool based on a common miRNAs sequence and a complete 3′UTR sequences dataset. In addition, it also stores experimentally validated miRNA-mRNA targets. *miRWalk* [72] is a comprehensive archive of experimentally verified and predicted (13 prediction datasets) miRNA-mRNA/lncRNA interactions. It further allows to generate customized lists of predicted putative targets of miRNAs of interest and to obtain information on miRNA target genes' pathways, ontologies and associated disorders.

STarMirDB [73] is a database that collects miRNA binding sites in 3′UTR, 5′UTR, and CDS mRNAs regions, predicted from CLIP studies on human, mouse, and worm. It can be searched by miRNAs or mRNAs separately or in combination, by entering the gene ID or sequence. *DIANA-TarBase* and *DIANA-LncBase* [74, 75] catalog experimental-supported miRNA-gene and miRNA-lncRNA interactions, respectively, through a manual curation of the available publications and the analysis of high-throughput datasets. The first allows the retrieval of the interactions through miRNA and gene name queries. In the second, the user can query the database through miRNA and/or lncRNA names and inspect the lncRNAs expression profiles.

Among resources dedicated to ncRNAs features, or to a specific class of these, *lncRNAdb* [76] and *NONCODE* [77] are dedicated to long non-coding RNAs (lncRNAs), including annotations such as sequence and structure information, expression profiles, conservation and function for multiple organisms. *lncRNA2Target* [78] provides a comprehensive resource of all literature-based lncRNA–target associations in human and mouse from high-throughput and low-throughput experiments.

RAID [79] focuses instead on providing a literature-curated catalog of RNA-associated interactions, including both RNA-RNA and RNA-protein interactions. This data potentially helps understanding the role, based on their binding properties, of various ncRNA molecules still not fully characterized. The interactions are classified by molecule type of the participants (e.g., lncRNA-associated or snoRNA-associated), and annotations such as binding site, tissue type, experimental technique, and others are included. Similarly, *NPInter* [80] collects ncRNAs interaction data through manual literature mining and processing of high-throughput sequencing data for 35 organisms. It provides interactions information on ncRNAs, including lncRNA, miRNA, circRNA, snoRNA, snRNA at the RNA-RNA, RNA-protein, RNA-DNA, and RNA-TF levels. Each interaction entry is associated with a detailed annotation and prediction score; furthermore, links to ncRNAs interactions and diseases or biological processes are also provided.

Eventually, *Rfam* [81] presents an extensive set of RNA families describing the various RNA genes types (including miRNAs, lncRNAs, snoRNAs, and many others) and mRNA cis-elements. Families are defined through covariate models by sequence alignments and primarily having a conserved structure. Entries are extensively described in a Wikipedia-like format with description, figures, and references. Furthermore, a tool allows user RNA sequences to be scanned, thus identifying matches with Rfam families.

The other resources presenting miRNA binding sites or ncRNAs data, namely *AURA2* [62], *doRiNa2* [63], *starBase2* [64], *RegRNA2* [82], and *UTRdb/UTRsite* [65], also provide other types of data (e.g., RBP sites or cis-elements) and will thus be described in the integrative resources section (Subheading 3.4).

3.3 Cis-Elements

These resources usually focus on one or multiple types of cis-elements and aim at presenting related features such as their instances on mRNAs, factors binding to and mediating the role of the element. Several databases also provide predictive tools to help in identifying previously undetected instances of these elements.

APADB [83] provides 3′end sequencing-derived information about alternative polyadenylation sites in 3′UTRs, including both coding and non-coding genes. The data is displayed through a genome browser, organized by tissue/organ and available for human and other two organisms. Potential losses of miRNA binding sites are also highlighted to help in understanding regulatory changes due to alternative polyadenylation events. *ARED-Plus* [84] and *AREsite2* [85] are two databases focusing on AU-rich Elements (ARE), a well-characterized cis-element commonly found in 3′UTRs of mRNAs and involved in their stability (through the action of several RBPs termed ARE-binding proteins, or ARE-BPs). Besides AREs motifs, AREsite2 allows the investigation of GU and U-rich elements (GRE, URE) in four species aside from human. Both resources are based on a computationally mapped catalog, obtained by matching one or more sequence patterns. The analysis focuses not only on 3′UTR regions but also on introns and pre-mRNAs (ARED-Plus) or the whole gene body and ncRNAs (AREsite2). Furthermore, they offer some degree of annotation, including graphical representations of found AREs, structural information, phylogenetic conservation, and supporting evidence extracted from the literature. In addition, AREsites2 includes data from CLIP-Seq experiments in order to highlight motifs with validated protein interaction. *IRESite* [86] and *IRESbase* [87] aim at producing a curated catalog of cellular and viral internal ribosome entry sites (IRES): these elements mediate translation initiation in the absence of a 5′ cap structure, thus allowing protein synthesis in stress conditions and of viral mRNAs. IRESite provides detailed information about each IRES (sequences, translation efficiency, condition, etc.), extracted from the literature, along with tools to search custom sequences against known IRESs to detect potentially novel instances of this element. IRESbase collects both eukaryotic (mRNAs, lncRNAs, circRNAs) and experimentally validated viral IRESes and provides other potential eukaryotic host transcripts by sequence identity prediction. It expands the annotations, compared to IRESite, including targeting miRNAs for human IRESes, and providing a BLAST search option where users can search for a similar IRES in a query RNA sequence. The last tool concerning IRES is the *IRESPred* [88] algorithm and webserver. It has been developed for the prediction of both cellular and viral IRES using a Support Vector Machine (SVM) which takes into consideration both UTRs sequences, structure properties and the probabilities of interactions between UTR and small subunit ribosomal proteins (SSRPs). As the next two resources underscore, mRNA cis-elements are considerably involved in the metabolism of several chemical elements. *SelenoDB* [89] is devoted to the description of selenogenes and selenocysteine insertion sequence (SECIS) elements. This element, found in the 3′UTRs, recruits proteins involved in selenium metabolism to the mRNA through its characteristic stem-loop structure. Instances of this element were computationally predicted and then manually curated: these are graphically displayed and correlated with several annotations. *SIREs* [90] is instead a webserver for the prediction of iron-responsive elements (IREs), specific cis-elements found in the mRNA of proteins involved in iron metabolism; this element is well characterized in both its sequence and secondary structure. This resource allows users to input their own sequence and, based on patterns derived from this characterization, will predict IREs position, features, and binding specificity for the iron proteins (IRP1 or IRP2).

Transterm [91] is a database of regions affecting translation, including both experimentally validated regulatory elements and the ability to scan user-provided sequences to identify instances of the many cis-elements classes for which a searchable pattern could be defined (extracted from the literature). A detailed description of the various elements classes is provided, as are several basic annotations (e.g., initiation codon context).

The other resources presenting data about cis-elements, namely *AURA2* [62], *UTRdb/UTRsite* [65], and *RegRNA2* [82], also provide other types of data (e.g., RBP and miRNA binding sites) and will thus be described in the integrative resources section (Subheading 3.4).

3.4 Integrative

We define a PTR database as *integrative* if it collects data about multiple aspects of post-transcriptional regulation such as, for instance, RBP- and miRNA-mediated regulation. The principle behind these resources is to allow a more precise and complete definition of an mRNA potential for PTR, through the parallel observation of many factors possibly mediating this potential. This approach allows to make the most of publicly available data. However, collecting such amount and variety of data is a daunting task, often requiring manual literature search. Indeed, just a few resources following this approach are currently available.

AURA2 [62] is a meta-database focused on the UTRs and providing data regarding the multiple aspects of PTR these regions mediate. It includes experimentally derived RBP and miRNA binding sites, cis-elements, RNA methylation, and SNP data, complemented with multiple annotations such as phylogenetic conservation, secondary structures, and functional descriptions. A custom UTR browser, along with several additional views and batch tools, allows the simultaneous display of all the various datasets, thus helping in obtaining a complete understanding of the regulation to which a UTR is subjected. Also *UTRdb/UTRsite* [65] focus on 5′ and 3′UTRs, providing annotations for these regions in many different organisms. It includes instances of cis-elements (including polyadenylation signals), phylogenetic conservation, SNPs, and experimentally determined miRNA targets. RBP binding sites are however absent from this resource. Furthermore, the UTRsite section allows to provide a custom sequence and predict the occurrences of many cis-elements types.

RegRNA2 [82] is a webserver for RNA motifs and site identification. It collects data from several databases and websites and incorporates different prediction models. From an input RNA sequence, it allows the retrieval of many RNA regulatory sites (e.g., polyadenylation sites, RNA editing), AU-rich and cis-elements, as well as splicing, UTR and DNA motifs. Furthermore, miRNA target sites can also be extracted in a computationally predicted form.

doRiNa2 [63] is a database dedicated to RNA interactions, including both RBP and miRNA binding sites as derived by CLIP approaches or custom data uploads; miRNA targets are also predicted by several algorithms. Searches can be performed by selecting a trans-factor and an mRNA region of interest: detailed results can then be displayed with the help of a genome browser. Furthermore, the combinatorial search tool allows the identification of mRNAs regulated by multiple factors of interest. Also *starBase2* [64] exploits exclusively CLIP data; however, it aims at identifying interactions between miRNAs and several other types of RNA, namely mRNA, lncRNA, ceRNA, circRNA, and other non-coding RNAs, also including protein-RNA interactions derived from the same sources. Interactions are displayed by category and annotated with expression data; furthermore, miRNA, and ceRNA functions can be predicted through dedicated tools leveraging functional ontologies terms.

4 Prediction Tools

When PTR data about genes of interest are not available, prediction tools can help in formulating a biological hypothesis concerning these genes. While some problems are relatively easy to address, and thus many tools are available (e.g., microRNA targets identification), others are more complex (e.g., RBP binding sites). We will now describe the various methods allowing to predict the presence of regulatory elements and their role in PTR. These tools are listed in Table 3 and can be broadly grouped by the type of prediction they provide, namely identifying *RBP targets*, *miRNA targets* or the effects of *SNPs on RNA secondary structures*.

4.1 RBP Targets

Predicting the targets of an RBP and the location of its binding sites on RNA molecules is a challenging task. Both identifying which residues of a protein may bind RNA and which sequence or structure specificities they confer are complex problems for which no precisely defined rules exist. Indeed, only a few tools have tried addressing this problem so far.

catRapid [93] is a webserver offering an algorithm performing de novo prediction of protein-RNA interactions based on physicochemical properties of polypeptides and nucleotide chains. The interaction propensity of these molecules is thus calculated solely based on their sequence: in particular, secondary structure, hydrogen bonding, and Van der Waals propensities are computed and combined together to yield an interaction profile; eventually, an interaction propensity score and an evaluation of the interaction statistical significance derived by a discriminative power calculation are returned. *omiXcore* [94] and *GlobalScore* [95] were developed on top of *catRapid*. The first is trained on eCLIP data and predicts RBP interactions with human long intergenic RNAs (lincRNAs); the second instead predicts protein interactions with large RNAs and is trained on PAR-CLIP and HITS-CLIP data. *RPISeq* [96] is a family of classifiers to predict RNA-protein interactions based exclusively on sequence-derived information. The prediction model is generated by normalized k-mers frequency distribution of RNA and protein sequences and applying the support vector machines and random forest methods. These classifiers were trained only on experimental RNA-protein interaction data. *RBPmap* [97] is instead based on a weighted-rank approach, accepting any RNA sequence as input and exploiting currently available RBP binding motifs extracted from the literature. It is optimized for human, mouse, and fruit fly, although other organisms are supported too. The algorithm matches the motifs matrices to the user input sequence; it then takes into account the propensity of binding sites for clustering, and the overall conservation of the region, in order to guide the identification of true binding sites. *ScanForMotifs* [98] is a webserver enabling the prediction of RBP binding sites, miRNA targets, and cis-elements by means of a set of 3′UTR alignments, known RBP binding motifs, miRNA seeds, and Transterm [91] elements. Users can provide a gene symbol or a sequence alignment. The tool will then run three parallel jobs to deal with each prediction type; results will eventually graphically show each identified site on the input alignment.

Table 3
Prediction tools for post-transcriptional regulation events. The table lists tools for predicting post-transcriptional phenomena such as RBP or miRNA binding and SNP effect on RNA secondary structure. Listed are the prediction type, the approach adopted by the tool, and how to retrieve it (URL and reference)

Name	Prediction type	Approach	URL	References no
catRapid	RBP targets	computes interaction propensity distribution and ranks the results	service.tartaglialab.com/page/catrapid_group	[93]
RBPmap	RBP targets	uses a weighted-rank algorithm considering sites clustering propensity and conservation	rbpmap.technion.ac.il	[97]
ScanForMotifs	RBP and miRNA targets, cis-elements	combines multiple data sources to predict regulatory elements in 3′UTRs	bioanalysis.otago.ac.nz/sfm	[98]
iDeepS	RBP targets	identifies sequence and structure motifs using CNN and a BLSTM	github.com/xypan1232/iDeepS	[99]
RPISeq	RBP targets	employees SVM and RF classifiers to predict whether or not a RNA-protein pair interact	pridb.gdcb.iastate.edu/RPISeq/	[96]
omiXcore	RBP targets	predicts interaction with large RNAs by evaluating local physicochemical properties of polypeptide and nucleotide sequences	service.tartaglialab.com/grant_submission/omixcore	[94]
GlobalScore	RBP targets	predicts interaction with large RNAs	service.tartaglialab.com/grant_submission/omixcore	[95]
beRBP	RBP targets	predicts RBP-RNA interaction based on random forest for RBPs with both known or unknown targets	bioinfo.vanderbilt.edu/beRBP/	[100]
ProNA2020	RBP targets	sequence-based system for the identification of protein-protein/DNA/RNA interaction and prediction of binding residues	github.com/Rostlab/ProNA2020.git	[104]
RPI-Net	RBP targets	secondary structure-based approach for RNA-protein interaction prediction with graph neural network	github.com/HarveyYan/RNAonGraph	[101]

(continued)

Table 3 (continued)

Name	Prediction type	Approach	URL	References no
PredPRBA	RBP targets	predicts the quantitative binding affinities of protein-RNA complexes through a gradient boosted regression trees	PredPRBA.denglab.org/	[102]
NucleicNet	RBP targets	Structure-based framework to predict binding preference of RNA constituents	github.com/NucleicNet/NucleicNet	[103]
ComiR	Combinatorial miRNA targeting	uses thermodynamic modeling and machine learning techniques coupled to expression data	benoslab.pitt.edu/comir	[116]
DIANA-microT	miRNA targets	predicts targets by exploiting positive and negative recognition element sets as defined by PAR-CLIP data	microrna.gr/webServer	[113]
miRanda	miRNA targets	employs sequence position-specific and conservation rules	mirnablog.com/microrna-target-prediction-tools/	[118]
miRmap	miRNA targets	ranks by repression strength through the use of multiple features	mirmap.ezlab.org	[106]
PicTar	miRNA targets	employs vertebrates alignment to identify targets	pictar.mdc-berlin.de	[120]
PITA	miRNA targets	exploits energy changes in miRNA-target duplex formation to predict targeting	genie.weizmann.ac.il/pubs/mir07/	[121]
TargetProfiler	miRNA targets	uses an HMM trained on experimentally verified miRNA targets	mirna.imbb.forth.gr/Targetprofiler.html	[115]
TargetScan	miRNA targets	predict targets by identifying conserved seed complementarity	targetscan.org	[105]
TargetSpy	miRNA targets	combines machine learning techniques with deep sequencing data for miRNA target prediction	webclu.bio.wzw.tum.de/targetspy/index.php?search=true	[110]
RNAhybrid	miRNA targets	predicts miRNA target by determining the most favorable hybridization site between two sequences	bibiserv.techfak.uni-bielefeld.de/rnahybrid	[107]

(continued)

Table 3
(continued)

Name	Prediction type	Approach	URL	References no
MultiMiTar	miRNA targets	combines SVM with multi-objective metaheuristic based feature selection technique for miRNA target prediction	www.isical.ac.in/~bioinfo_miu/multimitar.htm	[111]
MirTarget2	miRNA targets	SVM-based machine learning algorithm for miRNA target prediction	mirdb.org	[108]
TarPmiR	miRNA targets	random-forest-based approach to predict miRNA target sites	hulab.ucf.edu/research/projects/miRNA/TarPmiR/	[117]
RNAsnp	SNP effects on RNA structure	computes SNP effect on ensemble of structures and deriving *p*-values	rth.dk/resources/rnasnp/	[122]
SNPfold	SNP effects on RNA structure	use a partition function calculation that considers the ensemble of possible RNA conformations	ribosnitch.bio.unc.edu/snpfold	[123]
SNIPER	SNP effects on RNA structure	detects SNVs that alter ncRNA secondary structure and identify those enriched or depleted in tumors	github.com/suzhixi/SNIPER/	[124]

Recently, a series of machine learning-based approaches have been implemented for the prediction of RBP-RNA related interactions. *iDeepS* [99] algorithm aims at the prediction of RBPs binding sites through the identification of sequence and structure motifs. It is a deep-learning based method implemented in Python, where one-hot encodings for the sequences and secondary structure are initially loaded in a convolutional neural network (CNN) and then in a bidirectional long short-term memory (BLSTM), which detects the dependencies between binding sequence and structure motifs identified by the previous step. The *beRBP* [100] webserver applies a random forest classifier to analyze four standardized RNA sequence/structure features scores (motif matching, clustering, accessibility, and conservation) and thus, to predict RNA sequences bound by an RBP characterized with a specific PWM. In particular, it builds a specific model for each RBP with known targets, and a general model by pooling known targets of all the different RBPs. This last model can be applied to any RBPs with known PWMs, user-provided PWMs, or even RBP sequences, from which PWMs are inferred. *RPI-Net* [101] uses a graph neural network (GNN) to learn the RNA secondary structure directly at the molecular level from the RNA sequences and structures. Furthermore, it also implements a method for RBPs sequences and structures binding sites extraction. Other approaches consists in *PredPRBA* [102] which predicts protein-RNA binding affinity using gradient boosted regression trees, and *NucleicNet* [103] which by retrieving physicochemical characteristics on protein surface, assesses whether an RNA fits with a binding pocket and predict new RBPs and their binding pockets/preferences.

Finally, *ProNA2020* [104] identifies proteins interacting with DNA/RNA/protein and predicts the residues involved in the binding. It requires only the protein sequence as input and detects the involved residues by applying a combination of homology-based inference, motif-based profile-kernel, and word-based approaches.

4.2 miRNA Targets

Contrary to predicting an RBP binding site, identifying targets for a miRNA may seem a quite straightforward task: once the miRNA seed sequence is known, the problem consists in finding matching complementary sequences in the mRNAs. However, this intuitive procedure has been proven to produce many false positives and miss non-canonical miRNA binding sites. Nevertheless, the accessible nature of this task has led to the development of many tools, exploiting different principles to attempt at discriminating true sites from the bulk.

TargetScan [105] is an algorithm predicting miRNA targets by exploiting phylogenetic conservation information over 46 vertebrate species: under the assumption that conservation often implies function, conserved sites (7-mer and 8-mer) that match the seed region of miRNAs are extracted and used to associate a miRNA to its target mRNAs. Predictions are then ranked by a targeting efficacy score, determined by keeping into account various features of the site context. Going beyond seed match identification alone, *miRmap* [106] is a webserver based on a Python library employing thermodynamic, evolutionary, and sequence-based features. These features are then combined using a linear model to yield the "*miRmap score*," representing the predicted miRNA repression strength. The library can be integrated into other applications through a REST service, and precomputed predictions can be downloaded in full from the website. *RNAhybrid* [107] is an online tool and web service whose algorithm determines the most favorable hybridization sites between two sequences, where the short sequence (miRNA) is hybridized to the best fitting part of the long one (target). The already discussed miRDB [67] database includes a custom prediction and a target mining module, both based on the *MirTarget2* [108] algorithm. This prediction tool was trained on a large microarray transcriptional profiling dataset [109], from which the SVM extracts 131 features. The target prediction model was then constructed by non-linearly integrating the known ones, such as seed pairing, target site conservation and structural accessibility, and novel features. On the same line, *TargetSpy* [110] and *MultiMiTar* [111]. The first algorithm is based on a machine learning and automatic feature selection approach, which does not take into consideration the seed match and the evolutionary conservation, and allows the prediction of species-specific and 3′ compensatory target sites. MultiMiTar algorithm combines SVM with Archived Multi-Objective Simulated Annealing [112] as a novel feature selection and classification tool. Furthermore, it computes a prediction score which ranks the miRNAs that target a specific gene or genes targeted by a single miRNA, thus allowing combinatorial interaction study of the most favorable miRNAs. *DIANA-microT* [113] is a webserver detecting miRNA sites both in 3′UTRs and in coding sequences: the algorithm is trained on a positive and negative microRNA-recognition element sets defined by an Argonaute PAR-CLIP assay. Potential 3′UTR and CDS sites are considered separately, and specific features (including conservation, flanking AU content and others) are computed for every candidate; these are then combined and eventually scored by generalized linear models. Furthermore, this resource offers a useful plugin for the Taverna [114] workflow platform, allowing the inclusion of miRNA target prediction into an analysis pipeline. Also *TargetProfiler* [115] exploits a small set of experimentally derived miRNA targets to train a model, an Hidden Markov Model (HMM) in this case, then used to probabilistically learn miRNA-target associations. Predicted targets are then filtered according to the HMM score, the miRNA-mRNA hybrid free energy, and the eight-species phylogenetic conservation of the site region. The webserver provides precomputed predictions for all human genes and miRNAs. Rather than focusing on a single miRNA, *ComiR* [116] aims at computing the potential of mRNAs to be regulated by a set of miRNAs, directly provided by the user or derived by input expression levels. By employing four different methods complementing each other and integrating expression levels, the tool first computes miRNA-mRNA interaction probabilities for each miRNA, then integrating these probabilities by an SVM model. The output is a list of genes, ranked by the probability of being targets of the miRNA set. Eventually, *TarPmiR* [117] tool has been shown to outperform miRanda [118], TargetScan [105], and miRmap [106] in the miRNA target sites prediction. It is a Python package relying on a random forest approach which integrates 13 prediction features extracted from the public crosslinking ligation and sequencing of hybrids (CLASH) data [119].

A few other tools devoted to miRNA target prediction, namely *miRanda* [118], *PicTar* [120], and *PITA* [121], are not described in detail here but listed in Table 3 along with all other tools.

4.3 SNPs Impact on RNA Secondary Structure

The effects of genetic variation are most often studied on protein-coding sequences only (e.g., exome sequencing), thus focusing on changes in protein domains and related features. However, variants in the non-coding portions of an mRNA may heavily affect the regulation of these transcripts, thus altering the abundance of an otherwise functional peptide. To study the impact of such variants a few tools are now available, investigating structural consequences connected to the presence of SNPs.

The first tool, *RNAsnp* [122], is a webserver based on computing the structural differences between wild-type and mutated sequence using an RNA folding method and a dynamic programming algorithm over windows of fixed length. It includes three algorithms; one tuned for short sequences (<1000 nt), one for long ones, and a last method consisting in the combination of the other two approaches. All three algorithms report as output the window of maximum base pair distance and the related *p*-value. *SNPfold* [123] first computes a partition function (i.e., a matrix representing the probability of base pairing for all possible pairs in the sequence) over the wild-type and the mutated sequence; then, it determines how much the two structure differs (by means of a correlation coefficient) and also where this difference is the greatest. Through this partition function, the ensemble of all possible structural conformations for both wild-type and mutated sequences are considered, thus reporting results with more confidence and a more reliable *p*-value. The last available tool is *SNIPER* [124] which detects single-nucleotide variations (SNVs) that alter the secondary structure (riboSNitches) of UTRs and ncRNAs employing RNAplfold algorithm [125] and MeanDiff and EucDiff methods and identifies riboSNitches enriched or depleted in cancer genomes.

5 Tools for RNA Sequence and Structure Motif Search

We conclude our tools presentation by a set of resources, listed in Table 4, aimed at *sequence* and/or *structural* RNA motifs identification. This task is particularly frequent and is needed to, for instance, determine binding preferences for an RBP or identify what regulatory element may be responsible for a shared translational control pattern observed in a group of genes (e.g., as identified by polysome profiling). While the majority of tools yield motifs focused on one of the two aspects (either sequence or structure), most algorithms take a step further by integrating both aspects to define the motifs.

CMfinder [126] is a tool based on an expectation-maximization (EM) algorithm, including RNA secondary structure information by using covariance models (CM). While the EM algorithm drives the search, motifs distribution in sequences is

Table 4
Tools for RNA motif search. The table lists algorithms for RNA motifs identification, both at the sequence and secondary structure levels. Searched motif type, the adopted approach, and how to retrieve each tool (URL and reference) are indicated

Name	Motif type	Approach	URL	References no.
CMfinder	Sequence	Expectation-maximization algorithm using covariate models to describe motifs	bio.cs.washington.edu/yzizhen/CMfinder	[126]
MEMERIS	Sequence/structure	Expectation-maximization algorithm which uses secondary structure to guide search towards single-stranded RNA regions	bioinf.uni-freiburg.de/~hiller/MEMERIS	[127]
RBPmotif	Sequence/structure	uses binding data to predict RBP binding specificities and sequence motifs to predict structural preferences	rnamotif.org	[138]
RNAcontext	Sequence/structure	computes structural context distributions on bound sequences to train a model	www.cs.toronto.edu/~hilal/rnacontext/	[139]
RNAmotifs	Multivalent RNA motifs	identifies multivalent RNA motifs from analysis of differentially regulated exons	github.com/matteocereda/RNAmotifs	[140]
SSMART	Sequence/structure	simultaneously models the primary sequence and the secondary structure of the RNA	ohlerlab.mdc-berlin.de/software/SSMART_137/	[137]
GraphProt	Sequence/structure	uses bound and unbound sites to predict RBP binding specificities and calculates the secondary structure	www.bioinf.uni-freiburg.de/Software/GraphProt	[134]
SMARTIV	Sequence/structure	Webserver performing RNA sequence and secondary structure prediction	smartiv.technion.ac.il/	[136]
Zagros	Sequence/structure	combines sequence/structure-specificity with RBP-specific crosslink signals from CLIP-seq data	smithlabresearch.org	[135]
RNApromo	Structure	uses a stochastic context-free grammar and a probabilistic inference algorithm	genie.weizmann.ac.il/pubs/rnamotifs08/rnamotifs08_predict.html	[129]
TEISER	Structure	finds enriched/depleted structures that explain patterns in genome-wide data	tavazoielab.c2b2.columbia.edu/TEISER	[130]

(continued)

Table 4
(continued)

Name	Motif type	Approach	URL	References no.
CapR	Structure	computes the RNA structural profiles by enumerating all the possible secondary structures	sites.google.com/site/fukunagatsu/software/capr	[131]
BEAM	Structure	Webserver allowing RNA secondary structure identification by prediction and BEAR encoding	beam.uniroma2.it/	[133]
RNAProfile	Structure	identifies similar regions considering both sequence similarity and associated secondary structure	www.beaconlab.it/modtools/	[132]

described by a mixture model, and motifs themselves are modeled by a CM; given the complexity of the so-defined search space, the algorithm uses heuristics to select interesting candidates. Only motifs with stable secondary structures are considered and then aligned to define the motif consensus. Results are eventually refined by a second EM algorithm to yield the final predicted motifs. Exploiting EM algorithms as well, *MEMERIS* [127] is based on the popular MEME motif search software [128], identifying sequence motifs by guiding the search towards single-stranded regions; this criterion is justified by the preference of several RBPs for binding to such regions. This guidance is made possible by replacing the MEME uniform motif start probability distribution by a single-strandedness distribution computed on the input RNA sequence; furthermore, maximum flexibility is granted by allowing to tune the weight of the single-strandedness assumption. *RNA-promo* [129] is an algorithm based instead on stochastic context-free grammars (SCFGs), devoted to the identification of short secondary structure motifs in RNA sequences. To reduce the search space, the algorithm requires a suggested structure as input, along with the set of sequences supposingly sharing such motif. The algorithm first identifies a set of structures that appear in as many sequences as possible; these are then refined and statistically evaluated by means of a probabilistic inference algorithm. Also based on context-free grammars, *TEISER* [130] is aimed at discovering structural motifs that can be correlated with genome-wide measurements such as, to cite one, mRNA stability. TEISER uses mutual information to understand the impact of presence/absence of many possible structural elements on the provided measurements. It is thus possible, for instance, to deduce the dependency of mRNA stability on the presence of a specific hairpin in the mRNA 5′ or 3′UTRs. Candidate motifs are then refined by selecting the ones with the greatest impact on such measurements, which are eventually statistically assessed to yield truly relevant motifs. In addition, other secondary structure motif discovery tools have been developed. *CapR* [131] calculates the secondary structural contexts of RNA molecules retrieved from CLIP-seq data of various RBPs. It computes the probability that each RNA base position is located within one of the six secondary structural contexts taken into account, accordingly with the Turner energy model of RNAs. Thus, a structural profile is defined for each RNA base and consists of the six probabilities that the base belongs to each context. *RNAProfile* [132] detects conserved secondary structures motifs in functionally related RNA sequences. From a set of input RNA sequences, it identifies potential motifs and compares these candidate regions by computing a pairwise alignment with a scoring function that considers similarity at both sequence and structure level. The *BEAM* [133] webserver allows the secondary structure prediction by applying two alternative methods and converts the RNA structures in a string of characters (BEAR encoding). Thus, it handles the motif discovery of big datasets along their entire length.

GraphProt [134] learns models of RBP binding preferences from both CLIP-seq and RNAcompete data. The training step comprised the collection of RBP bound and unbound sites from the high-throughput experiments and the calculation of highly probable secondary structures. Subsequently, the application phase allows the motif detection and visualization with sequence logos, and the prediction of novel RBPs target sites. Also, Bahrami-Samani et al. developed the *Zagros* [135] algorithm, specifically designed for CLIP-seq derived RNA binding sites, which is based not only on RNA sequence and secondary structure information but also on technology-specific crosslinking events. In the same direction, the *SMARTIV* [136] algorithm allows the discovery of combined sequence and structure motifs in RNA from CLIP experiments. Here, however, the predicted secondary structures are combined with the sequence information in a single representation. From this joint information, the Hyper Geometric statistics extracts the enriched k-mers that are further clustered to generate the PWMs. Similarly, also the *SSMART* [137] tool simultaneously models the primary and secondary structure of RNAs from high-throughput in vivo and in vitro data.

The last tools we describe are devoted to a specific motif identification problem, rather than being generally applicable to any set of RNA sequences. *RBPmotif* [138] is a webserver focused on discovering the sequence and structural binding preferences of RNA-binding proteins. If no such preference is known, the tool will run an algorithm (*RNAcontext* [139], requiring a set of bound and unbound sequences as input) to investigate this aspect. On the other end, if a sequence motif is already available for the RBP, an additional analysis will be run to identify potential structural contributions to the RBP sequence binding preference. Statistically evaluated motifs are eventually returned and can be compared either by considering motif instances in bound and unbound sequences or by looking at similar binding motifs of other RBPs. Eventually, *RNAmotifs* [140] stands out of the pack because of its particular application: indeed, it is aimed at identifying motifs involved in splicing regulation of a set of differentially regulated alternative exons. In particular, motifs are defined as either degenerate or non-degenerate tetramers found around enhanced or repressed exons: these tetramers are tested for enrichment in sequences surrounding these exons and statistically evaluated by a Fisher test and a bootstrap procedure. Furthermore, splicing maps derived by the enriched tetramers score profiles can also be visualized.

6 Pipelines for PTR: Two Case Studies

The tools and databases we described, which can be used to analyze various types of PTR data, all are individually useful and serve a purpose on their own relating to a single analysis aspect. Nevertheless, to get the most out of the data deluge coming out from such genome-wide approaches, and reach the highest possible resolution and accuracy, these tools must be combined and integrated into full-blown analysis workflows. Towards this goal, we will thus now conclude the chapter by presenting two tentative pipelines, combining several of these resources to address the analytic needs of two different usage cases. Through these examples, illustrated in Fig. 1, this section thus aims at providing initial practical guidance to researchers approaching common PTR data analysis tasks for the first time.

6.1 Case Study 1: Impact of a Treatment on Translation

Our first case study deals with a particularly common experiment in PTR, consisting in profiling the effects of a treatment/stimuli on the translational control behavior of a suitable cellular model system. The ultimate goal would be understanding which regulatory factors/elements are influencing the translatability of the transcripts following the treatment. This task is often addressed by means of coupled total and polysomal RNA extraction followed by an RNA-seq assay on the poly(A+) fraction, eventually allowing transcripts quantification. In such an experimental setting, the analysis can be subdivided in five phases, illustrated by Fig. 1a. First of all, one needs to identify genes (called DEGs for differentially expressed genes) which are significantly changing their expression solely at the translation level and not concordantly at the transcription and translation levels, following the treatment: this task can be performed by *tRanslatome* [20], by providing a table

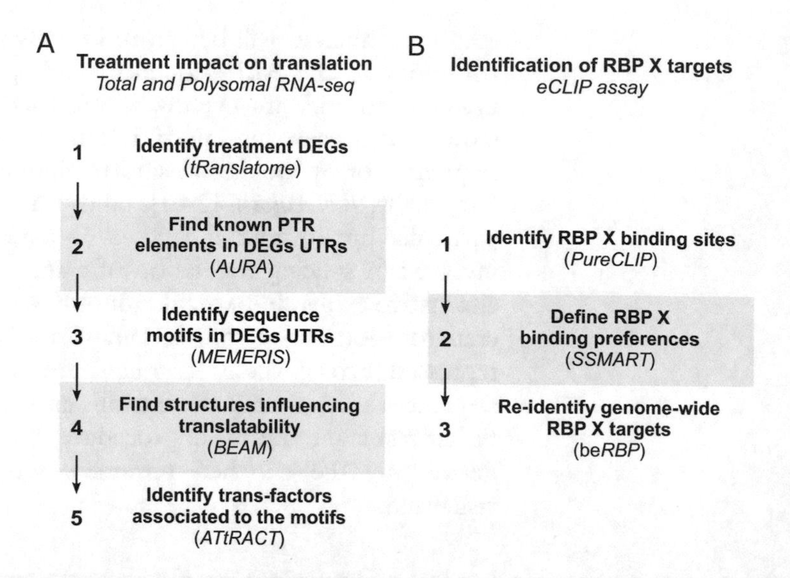

Fig. 1 Representative data analysis pipelines to address two PTR case studies. The figure displays potential data analysis pipelines for two common types of experiment in post-transcriptional regulation of gene expression. (**a**) describes a potential pipeline for the analysis of a study to determine the impact of a treatment/stimuli on translation. This can be performed by total and polysomal profiling coupled with RNA-seq. Differentially translated genes (DEGs) can first of all be identified by *tRanslatome*; known PTR elements influencing translation (RBP binding sites, cis-elements, etc.) can then be identified by *AURA* in DEGs UTRs. Next, two motif search analyses can be performed to identify potential determinants of the observed translational changes at the sequence (*MEMERIS*) and structure (*BEAM*) levels. Eventually, these motifs may be characterized by attempting at identifying trans-factors binding to them by means of *ATtRACT*. (**b**) describes a workflow aimed at the identification of targets and binding preferences for a generic RBP, named X. Starting from an eCLIP assay in the system of interest, binding sites for the RBP are first identified using *PureCLIP*. Subsequently, its sequence and structure binding specificities are determined through the *SSMART* tool. Eventually, the targets list is re-assessed by predicting binding sites (using the motif computed in the previous step) with *beRBP*, to further confirm eCLIP results and identify additional targets which may be non-expressed in the studied model (and thus potentially missed by the assay)

with per-gene read counts (steps required to obtain these counts from raw reads are related to the RNA-seq technique and will thus not be described here; readers can refer to this review [141]) as input and choosing one of the available methods for DEGs calling. The output will consist of a list of genes with significant changes, representing the treatment-induced phenotype. The next step aims at obtaining a first overview of which trans-factors binding sites and cis-elements are already known to be present in the mRNAs untranslated regions of the significant DEGs: this task can be performed by inputting the gene list to the batch tools (the

Regulatory Element Enrichment one in particular) of *AURA2* [62]; these results could also be complemented by predicting the presence of RBPs binding sites through a tool such as *beRBP* [100]. At the end of this step, one should already be able to understand whether one or a few trans-factors/cis-elements are significantly shared by many DEGs, and thus decide to focus on these as interesting candidates to explain the effect of the treatment.

If this is not the case, the next step consists in looking for shared sequence motifs in DEGs UTRs by exploiting also secondary structure information: this analysis can be efficiently performed by providing DEGs 5′ and 3′ UTRs sequences (in two distinct executions) to *MEMERIS* [127]; a further step in the same direction, which can be addressed by means of *BEAM* [133], consists in detecting structural motifs which can explain DEGs changes in translational levels. These two steps will eventually yield sequence and structure motifs shared by a consistent number of DEGs UTRs and could represent the distinctive post-transcriptional features acted upon to obtain the treatment effects.

As a last step, these motifs should be analyzed to attempt at understanding which trans-factors are targeting them to mediate the observed effect: to do so, one can try to find matches with known binding motifs by means of *ATtRACT* [58]; obviously, this analysis should be coupled to a thorough literature search, especially for structure motifs, to maximize the chances of identifying the factors at play.

6.2 Case Study 2: Identification of Targets and Binding Preference for an RBP

Our second case study focuses instead on an increasingly common kind of PTR experiment, namely identifying targets and binding preferences for an RNA-binding protein: this kind of experiment has recently been made possible by the advent of the CLIP family of techniques. In particular, our case study will focus on the most recent eCLIP technique, which exploits the truncation induced by the termination of the retrotranscription at the RNA-protein crosslinking site to precisely pinpoint the binding sites location. The analysis workflow for such an experiment, illustrated in Fig. 1b, can be subdivided in three phases. The first phase consists of processing aligned reads to detect "peaks" (i.e., RNA regions hosting a significantly greater number of reads with respect to the genome-wide background) and intersect these with crosslinking sites, originating from truncated cDNAs. This processing, done by means of *PureCLIP* [51], produces a list of binding sites for our RBP of interest, along with an estimation of their statistical significance. From this list, the set of targets for this RBP can be implicitly inferred, along with processes and pathways in which these are involved: we can thus obtain an overall view of the role that this RBP is playing in the system under study. While this information is extremely useful, we still lack, at this point, a definition of how this RBP chooses its binding sites on target RNAs. To address this issue, the next step, performed by means of the *SSMART* [137] tool, exploits the set of bound sequences to learn the binding preference of the RBP and derive the probability of each position in the resulting motif to be paired or unpaired. That done, we are still left with a last potential issue: our assay was performed in a single model system, expressing a specific set of genes (and thus RNAs), most likely considerably smaller than the total number of transcribed loci in the genome. We are thus potentially missing a fraction of all targets due to them not being expressed in our system. To alleviate this issue, we could exploit the binding preferences we derived in the previous step to predict, over the whole transcriptome, potential binding sites for our RBP: this task can be performed using *beRBP* [100], providing the sequence motif (either its consensus sequence or its probability matrix) as input to the tool. This analysis will also allow us to compute the goodness of fit between experimentally derived binding sites and predicted motif matches, thus eventually enabling the evaluation of the motif quality and, possibly, its refinement.

References

1. Metzker ML (2010) Sequencing technologies—the next generation. Nat Rev Genet 11:31–46
2. Ingolia NT (2014) Ribosome profiling: new views of translation, from single codons to genome scale. Nat Rev Genet 15:205–213
3. Milek M, Wyler E, Landthaler M (2012) Transcriptome-wide analysis of protein-RNA interactions using high-throughput sequencing. Semin Cell Dev Biol 23:206–212
4. Arava Y (2003) Isolation of polysomal RNA for microarray analysis. Methods Mol Biol 224:79–87
5. Ule J, Jensen KB, Ruggiu M et al (2003) CLIP identifies Nova-regulated RNA networks in the brain. Science 302:1212–1215
6. Chi SW, Zang JB, Mele A, Darnell RB (2009) Argonaute HITS-CLIP decodes microRNA–mRNA interaction maps. Nature 460:479–486
7. Hafner M, Landthaler M, Burger L et al (2010) Transcriptome-wide identification of RNA-binding protein and microRNA target sites by PAR-CLIP. Cell 141:129–141
8. König J, Zarnack K, Rot G et al (2010) iCLIP reveals the function of hnRNP particles in splicing at individual nucleotide resolution. Nat Struct Mol Biol 17:909–915
9. Linder B, Grozhik AV, Olarerin-George AO et al (2015) Single-nucleotide-resolution mapping of m6A and m6Am throughout the transcriptome. Nat Methods 12:767–772
10. Van Nostrand EL, Pratt GA, Shishkin AA et al (2016) Robust transcriptome-wide discovery of RNA-binding protein binding sites with enhanced CLIP (eCLIP). Nat Methods 13:508–514
11. Zarnegar BJ, Flynn RA, Shen Y et al (2016) irCLIP platform for efficient characterization of protein–RNA interactions. Nat Methods 13:489–492
12. Kargapolova Y, Levin M, Lackner K, Danckwardt S (2017) sCLIP—an integrated platform to study RNA–protein interactomes in biomedical research: identification of CSTF2tau in alternative processing of small nuclear RNAs. Nucleic Acids Res 45:6074–6086
13. George H, Ule J, Hussain S (2017) Illustrating the epitranscriptome at nucleotide resolution using methylation-iCLIP (miCLIP). Methods Mol Biol 1562:91–106
14. Ray D, Kazan H, Chan ET et al (2009) Rapid and systematic analysis of the RNA recognition specificities of RNA-binding proteins. Nat Biotechnol 27:667–670
15. Cook KB, Vembu S, Ha KCH et al (2017) RNAcompete-S: combined RNA sequence/structure preferences for RNA binding proteins derived from a single-step in vitro selection. Methods 126:18–28

16. Lambert N, Robertson A, Jangi M et al (2014) RNA Bind-n-Seq: quantitative assessment of the sequence and structural binding specificity of RNA binding proteins. Mol Cell 54:887–900
17. Campbell ZT, Bhimsaria D, Valley CT et al (2012) Cooperativity in RNA-protein interactions: global analysis of RNA binding specificity. Cell Rep 1:570–581
18. Lin C, Miles WO (2019) Beyond CLIP: advances and opportunities to measure RBP-RNA and RNA-RNA interactions. Nucleic Acids Res 47:5490–5501
19. Larsson O, Sonenberg N, Nadon R (2011) anota: analysis of differential translation in genome-wide studies. Bioinformatics 27:1440–1441
20. Tebaldi T, Dassi E, Kostoska G et al (2014) tRanslatome: an R/Bioconductor package to portray translational control. Bioinformatics 30:289–291
21. Olshen AB, Hsieh AC, Stumpf CR et al (2013) Assessing gene-level translational control from ribosome profiling. Bioinformatics 29:2995–3002
22. Xiao Z, Zou Q, Liu Y, Yang X (2016) Genome-wide assessment of differential translations with ribosome profiling data. Nat Commun 7:11194
23. Zhong Y, Karaletsos T, Drewe P et al (2017) RiboDiff: detecting changes of mRNA translation efficiency from ribosome footprints. Bioinformatics 33:139–141
24. Li W, Wang W, Uren PJ et al (2017) Riborex: fast and flexible identification of differential translation from Ribo-seq data. Bioinformatics 33:1735–1737
25. Oertlin C, Lorent J, Murie C et al (2019) Generally applicable transcriptome-wide analysis of translation using anota2seq. Nucleic Acids Res 47:e70–e70
26. Ernlund AW, Schneider RJ, Ruggles KV (2018) RIVET: comprehensive graphic user interface for analysis and exploration of genome-wide translatomics data. BMC Genomics 19:809
27. Stocks MB, Moxon S, Mapleson D et al (2012) The UEA sRNA workbench: a suite of tools for analysing and visualizing next generation sequencing microRNA and small RNA datasets. Bioinformatics 28:2059–2061
28. Wan C, Gao J, Zhang H et al (2017) CPSS 2.0: a computational platform update for the analysis of small RNA sequencing data. Bioinformatics 33:3289–3291
29. Giurato G, De Filippo MR, Rinaldi A et al (2013) iMir: an integrated pipeline for high-throughput analysis of small non-coding RNA data obtained by smallRNA-Seq. BMC Bioinformatics 14:362
30. Sun Z, Evans J, Bhagwate A et al (2014) CAP-miRSeq: a comprehensive analysis pipeline for microRNA sequencing data. BMC Genomics 15:423
31. Vitsios DM, Enright AJ (2015) Chimira: analysis of small RNA sequencing data and microRNA modifications. Bioinformatics 31:3365–3367
32. Shi J, Dong M, Li L et al (2015) mirPRo—a novel standalone program for differential expression and variation analysis of miRNAs. Sci Rep 5:14617
33. Barturen G, Rueda A, Hamberg M et al (2014) sRNAbench: profiling of small RNAs and its sequence variants in single or multi-species high-throughput experiments. Methods Next Gen Seq 1:21–31
34. Aparicio-Puerta E, Lebrón R, Rueda A et al (2019) sRNAbench and sRNAtoolbox 2019: intuitive fast small RNA profiling and differential expression. Nucleic Acids Res 47: W530–W535
35. Fehlmann T, Backes C, Kahraman M et al (2017) Web-based NGS data analysis using miRMaster: a large-scale meta-analysis of human miRNAs. Nucleic Acids Res 45:8731–8744
36. Wu X, Kim T-K, Baxter D et al (2017) sRNAnalyzer—a flexible and customizable small RNA sequencing data analysis pipeline. Nucleic Acids Res 45:12140–12151
37. Pogorelcnik R, Vaury C, Pouchin P et al (2018) sRNAPipe: a Galaxy-based pipeline for bioinformatic in-depth exploration of small RNAseq data. Mob DNA 9:25
38. Kuksa PP, Amlie-Wolf A, Katanic Ž et al (2018) SPAR: small RNA-seq portal for analysis of sequencing experiments. Nucleic Acids Res 46:W36–W42
39. Lu Y, Baras AS, Halushka MK (2018) miRge 2.0 for comprehensive analysis of microRNA sequencing data. BMC Bioinformatics 19:275
40. Desvignes T, Batzel P, Sydes J et al (2019) miRNA analysis with Prost! Reveals evolutionary conservation of organ-enriched expression and post-transcriptional modifications in three-spined stickleback and zebrafish. Sci Rep 9:3913
41. Li J, Kho AT, Chase RP et al (2020) COMPSRA: a COMprehensive Platform for Small RNA-Seq data Analysis. Sci Rep 10:4552
42. Uren PJ, Bahrami-Samani E, Burns SC et al (2012) Site identification in high-throughput

RNA-protein interaction data. Bioinformatics 28:3013–3020

43. Kucukural A, Özadam H, Singh G et al (2013) ASPeak: an abundance sensitive peak detection algorithm for RIP-Seq. Bioinformatics 29:2485–2486
44. Li Y, Zhao DY, Greenblatt JF, Zhang Z (2013) RIPSeeker: a statistical package for identifying protein-associated transcripts from RIP-seq experiments. Nucleic Acids Res 41:e94
45. Lovci MT, Ghanem D, Marr H et al (2013) Rbfox proteins regulate alternative mRNA splicing through evolutionarily conserved RNA bridges. Nat Struct Mol Biol 20:1434–1442
46. Corcoran DL, Georgiev S, Mukherjee N et al (2011) PARalyzer: definition of RNA binding sites from PAR-CLIP short-read sequence data. Genome Biol 12:R79
47. Sievers C, Schlumpf T, Sawarkar R et al (2012) Mixture models and wavelet transforms reveal high confidence RNA-protein interaction sites in MOV10 PAR-CLIP data. Nucleic Acids Res 40:e160
48. Wang T, Chen B, Kim M et al (2014) A model-based approach to identify binding sites in CLIP-Seq data. PLoS One 9:e93248
49. Blankenberg D, Von Kuster G, Coraor N et al (2010) Galaxy: a web-based genome analysis tool for experimentalists. Curr Protoc Mol Biol. https://doi.org/10.1002/0471142727.mb1910s89
50. Chen B, Yun J, Kim MS et al (2014) PIPE-CLIP: a comprehensive online tool for CLIP-seq data analysis. Genome Biol 15:R18
51. Krakau S, Richard H, Marsico A (2017) Pure-CLIP: capturing target-specific protein–RNA interaction footprints from single-nucleotide CLIP-seq data. Genome Biol 18:240
52. Drewe-Boss P, Wessels H-H, Ohler U (2018) omniCLIP: probabilistic identification of protein-RNA interactions from CLIP-seq data. Genome Biol 19:183
53. Liu Q, Shvarts T, Sliz P, Gregory RI (2020) RiboToolkit: an integrated platform for analysis and annotation of ribosome profiling data to decode mRNA translation at codon resolution. Nucleic Acids Res 48:W218–W229
54. Fang H, Huang Y-F, Radhakrishnan A et al (2018) Scikit-ribo enables accurate estimation and robust modeling of translation dynamics at codon resolution. Cell Syst 6:180–191.e4
55. Khorshid M, Rodak C, Zavolan M (2011) CLIPZ: a database and analysis environment for experimentally determined binding sites of RNA-binding proteins. Nucleic Acids Res 39: D245–D252
56. Ray D, Kazan H, Cook KB et al (2013) A compendium of RNA-binding motifs for decoding gene regulation. Nature 499:172–177
57. Cook KB, Kazan H, Zuberi K et al (2011) RBPDB: a database of RNA-binding specificities. Nucleic Acids Res 39:D301–D308
58. Giudice G, Sánchez-Cabo F, Torroja C, Lara-Pezzi E (2016) ATtRACT—a database of RNA-binding proteins and associated motifs. Database. https://doi.org/10.1093/database/baw035
59. Zhu Y, Xu G, Yang YT et al (2019) POSTAR2: deciphering the post-transcriptional regulatory logics. Nucleic Acids Res 47:D203–D211
60. Yang Y-CT, Di C, Hu B et al (2015) CLIPdb: a CLIP-seq database for protein-RNA interactions. BMC Genomics 16:51
61. Benoit Bouvrette LP, Bovaird S, Blanchette M, Lécuyer E (2020) oRNAment: a database of putative RNA binding protein target sites in the transcriptomes of model species. Nucleic Acids Res 48:D166–D173
62. Dassi E, Re A, Leo S et al (2014) AURA 2: empowering discovery of post-transcriptional networks. Translation 2:e27738
63. Blin K, Dieterich C, Wurmus R et al (2014) DoRiNA 2.0—upgrading the doRiNA database of RNA interactions in post-transcriptional regulation. Nucleic Acids Res 43:D160–D167
64. Li J-H, Liu S, Zhou H et al (2014) starBase v2.0: decoding miRNA-ceRNA, miRNA-ncRNA and protein-RNA interaction networks from large-scale CLIP-Seq data. Nucleic Acids Res 42:D92–D97
65. Grillo G, Turi A, Licciulli F et al (2010) UTRdb and UTRsite (RELEASE 2010): a collection of sequences and regulatory motifs of the untranslated regions of eukaryotic mRNAs. Nucleic Acids Res 38:D75–D80
66. Kozomara A, Birgaoanu M, Griffiths-Jones S (2019) miRBase: from microRNA sequences to function. Nucleic Acids Res 47: D155–D162
67. Chen Y, Wang X (2020) miRDB: an online database for prediction of functional microRNA targets. Nucleic Acids Res 48: D127–D131
68. Hsu S-D, Chu C-H, Tsou A-P et al (2008) miRNAMap 2.0: genomic maps of microRNAs in metazoan genomes. Nucleic Acids Res 36:D165–D169

69. Huang H-Y, Lin Y-C-D, Li J et al (2019) miRTarBase 2020: updates to the experimentally validated microRNA–target interaction database. Nucleic Acids Res 48:D148–D154
70. Cho S, Jang I, Jun Y et al (2013) MiRGator v3.0: a microRNA portal for deep sequencing, expression profiling and mRNA targeting. Nucleic Acids Res 41:D252–D257
71. Andrés-León E, González Peña D, Gómez-López G, Pisano DG (2015) miRGate: a curated database of human, mouse and rat miRNA–mRNA targets. Database. https://doi.org/10.1093/database/bav035
72. Dweep H, Gretz N (2015) miRWalk2.0: a comprehensive atlas of microRNA-target interactions. Nat Methods 12:697
73. Rennie W, Kanoria S, Liu C et al (2016) STarMirDB: a database of microRNA binding sites. RNA Biol 13:554–560
74. Karagkouni D, Paraskevopoulou MD, Chatzopoulos S et al (2017) DIANA-TarBase v8: a decade-long collection of experimentally supported miRNA–gene interactions. Nucleic Acids Res 46:D239–D245
75. Karagkouni D, Paraskevopoulou MD, Tastsoglou S et al (2019) DIANA-LncBase v3: indexing experimentally supported miRNA targets on non-coding transcripts. Nucleic Acids Res 48:D101–D110
76. Quek XC, Thomson DW, Maag JLV et al (2015) lncRNAdb v2.0: expanding the reference database for functional long noncoding RNAs. Nucleic Acids Res 43:D168–D173
77. Bu D, Yu K, Sun S et al (2012) NONCODE v3.0: integrative annotation of long noncoding RNAs. Nucleic Acids Res 40:D210–D215
78. Cheng L, Wang P, Tian R et al (2019) LncRNA2Target v2.0: a comprehensive database for target genes of lncRNAs in human and mouse. Nucleic Acids Res 47: D140–D144
79. Zhang X, Wu D, Chen L et al (2014) RAID: a comprehensive resource for human RNA-associated (RNA-RNA/RNA-protein) interaction. RNA 20:989–993
80. Teng X, Chen X, Xue H et al (2020) NPInter v4.0: an integrated database of ncRNA interactions. Nucleic Acids Res 48:D160–D165
81. Burge SW, Daub J, Eberhardt R et al (2013) Rfam 11.0: 10 years of RNA families. Nucleic Acids Res 41:D226–D232
82. Chang T-H, Huang H-Y, Hsu JB-K et al (2013) An enhanced computational platform for investigating the roles of regulatory RNA and for identifying functional RNA motifs. BMC Bioinformatics 14(Suppl 2):S4
83. Müller S, Rycak L, Afonso-Grunz F et al (2014) APADB: a database for alternative polyadenylation and microRNA regulation events. Database. https://doi.org/10.1093/database/bau076
84. Bakheet T, Hitti E, Khabar KSA (2018) ARED-Plus: an updated and expanded database of AU-rich element-containing mRNAs and pre-mRNAs. Nucleic Acids Res 46: D218–D220
85. Fallmann J, Sedlyarov V, Tanzer A et al (2016) AREsite2: an enhanced database for the comprehensive investigation of AU/GU/U-rich elements. Nucleic Acids Res 44: D90–D95
86. Mokrejs M, Masek T, Vopálensky V et al (2010) IRESite—a tool for the examination of viral and cellular internal ribosome entry sites. Nucleic Acids Res 38:D131–D136
87. Zhao J, Li Y, Wang C et al (2020) IRESbase: a comprehensive database of experimentally validated internal ribosome entry sites. Genomics Proteomics Bioinformatics 18:129–139
88. Kolekar P, Pataskar A, Kulkarni-Kale U et al (2016) IRESPred: web server for prediction of cellular and viral internal ribosome entry site (IRES). Sci Rep 6:27436
89. Castellano S, Gladyshev VN, Guigó R, Berry MJ (2008) SelenoDB 1.0 : a database of selenoprotein genes, proteins and SECIS elements. Nucleic Acids Res 36:D332–D338
90. Campillos M, Cases I, Hentze MW, Sanchez M (2010) SIREs: searching for iron-responsive elements. Nucleic Acids Res 38: W360–W367
91. Jacobs GH, Chen A, Stevens SG et al (2009) Transterm: a database to aid the analysis of regulatory sequences in mRNAs. Nucleic Acids Res 37:D72–D76
92. Liu Y, Sun S, Bredy T et al (2017) MotifMap-RNA: a genome-wide map of RBP binding sites. Bioinformatics 33:2029–2031
93. Agostini F, Zanzoni A, Klus P et al (2013) catRAPID omics: a web server for large-scale prediction of protein-RNA interactions. Bioinformatics 29:2928–2930
94. Armaos A, Cirillo D, Gaetano Tartaglia G (2017) omiXcore: a web server for prediction of protein interactions with large RNA. Bioinformatics 33:3104–3106
95. Cirillo D, Blanco M, Armaos A et al (2016) Quantitative predictions of protein interactions with long noncoding RNAs. Nat Methods 14:5–6
96. Muppirala UK, Honavar VG, Dobbs D (2011) Predicting RNA-protein interactions

using only sequence information. BMC Bioinformatics 12:489

97. Paz I, Kosti I, Ares M Jr et al (2014) RBPmap: a web server for mapping binding sites of RNA-binding proteins. Nucleic Acids Res 42:W361–W367
98. Biswas A, Brown CM (2014) Scan for Motifs: a webserver for the analysis of post-transcriptional regulatory elements in the 3′ untranslated regions (3' UTRs) of mRNAs. BMC Bioinformatics 15:174
99. Pan X, Rijnbeek P, Yan J, Shen H-B (2018) Prediction of RNA-protein sequence and structure binding preferences using deep convolutional and recurrent neural networks. BMC Genomics 19:511
100. Yu H, Wang J, Sheng Q et al (2019) beRBP: binding estimation for human RNA-binding proteins. Nucleic Acids Res 47:e26
101. Yan Z, Hamilton WL, Blanchette M (2020) Graph neural representational learning of RNA secondary structures for predicting RNA-protein interactions. Bioinformatics 36:i276–i284
102. Deng L, Yang W, Liu H (2019) PredPRBA: prediction of Protein-RNA binding affinity using gradient boosted regression trees. Front Genet 10:637
103. Lam JH, Li Y, Zhu L et al (2019) A deep learning framework to predict binding preference of RNA constituents on protein surface. Nat Commun 10:4941
104. Qiu J, Bernhofer M, Heinzinger M et al (2020) ProNA2020 predicts protein–DNA, protein–RNA, and protein–protein binding proteins and residues from sequence. J Mol Biol 432:2428–2443
105. Lewis BP, Burge CB, Bartel DP (2005) Conserved seed pairing, often flanked by adenosines, indicates that thousands of human genes are microRNA targets. Cell 120:15–20
106. Vejnar CE, Blum M, Zdobnov EM (2013) miRmap web: comprehensive microRNA target prediction online. Nucleic Acids Res 41: W165–W168
107. Krüger J, Rehmsmeier M (2006) RNAhybrid: microRNA target prediction easy, fast and flexible. Nucleic Acids Res 34:W451–W454
108. Wang X, El Naqa IM (2008) Prediction of both conserved and nonconserved microRNA targets in animals. Bioinformatics 24:325–332
109. Linsley PS, Schelter J, Burchard J et al (2007) Transcripts targeted by the microRNA-16 family cooperatively regulate cell cycle progression. Mol Cell Biol 27:2240–2252
110. Sturm M, Hackenberg M, Langenberger D, Frishman D (2010) TargetSpy: a supervised machine learning approach for microRNA target prediction. BMC Bioinformatics 11:292
111. Mitra R, Bandyopadhyay S (2011) MultiMiTar: a novel multi objective optimization based miRNA-target prediction method. PLoS One 6:e24583
112. Bandyopadhyay S, Saha S, Maulik U, Deb K (2008) A simulated annealing-based multiobjective optimization algorithm: AMOSA. IEEE Trans Evol Comput 12:269–283
113. Paraskevopoulou MD, Georgakilas G, Kostoulas N et al (2013) DIANA-microT web server v5.0: service integration into miRNA functional analysis workflows. Nucleic Acids Res 41:W169–W173
114. Wolstencroft K, Haines R, Fellows D et al (2013) The Taverna workflow suite: designing and executing workflows of Web Services on the desktop, web or in the cloud. Nucleic Acids Res 41:W557–W561
115. Oulas A, Karathanasis N, Louloupi A et al (2012) A new microRNA target prediction tool identifies a novel interaction of a putative miRNA with CCND2. RNA Biol 9:1196–1207
116. Coronnello C, Benos PV (2013) ComiR: combinatorial microRNA target prediction tool. Nucleic Acids Res 41:W159–W164
117. Ding J, Li X, Hu H (2016) TarPmiR: a new approach for microRNA target site prediction. Bioinformatics 32:2768–2775
118. John B, Enright AJ, Aravin A et al (2004) Human MicroRNA targets. PLoS Biol 2:e363
119. Helwak A, Kudla G, Dudnakova T, Tollervey D (2013) Mapping the human miRNA interactome by CLASH reveals frequent noncanonical binding. Cell 153:654–665
120. Krek A, Grün D, Poy MN et al (2005) Combinatorial microRNA target predictions. Nat Genet 37:495–500
121. Kertesz M, Iovino N, Unnerstall U et al (2007) The role of site accessibility in microRNA target recognition. Nat Genet 39:1278–1284
122. Sabarinathan R, Tafer H, Seemann SE et al (2013) The RNAsnp web server: predicting SNP effects on local RNA secondary structure. Nucleic Acids Res 41:W475–W479
123. Halvorsen M, Martin JS, Broadaway S, Laederach A (2010) Disease-associated mutations that alter the RNA structural ensemble. PLoS Genet 6:e1001074
124. He F, Wei R, Zhou Z et al (2019) Integrative analysis of somatic mutations in non-coding

regions altering RNA secondary structures in cancer genomes. Sci Rep 9:8205

125. Corley M, Solem A, Qu K et al (2015) Detecting riboSNitches with RNA folding algorithms: a genome-wide benchmark. Nucleic Acids Res 43:1859–1868

126. Yao Z, Weinberg Z, Ruzzo WL (2006) CMfinder—a covariance model based RNA motif finding algorithm. Bioinformatics 22:445–452

127. Hiller M, Pudimat R, Busch A, Backofen R (2006) Using RNA secondary structures to guide sequence motif finding towards single-stranded regions. Nucleic Acids Res 34:e117

128. Bailey TL, Boden M, Buske FA et al (2009) MEME SUITE: tools for motif discovery and searching. Nucleic Acids Res 37: W202–W208

129. Rabani M, Kertesz M, Segal E (2008) Computational prediction of RNA structural motifs involved in posttranscriptional regulatory processes. Proc Natl Acad Sci U S A 105:14885–14890

130. Goodarzi H, Najafabadi HS, Oikonomou P et al (2012) Systematic discovery of structural elements governing stability of mammalian messenger RNAs. Nature 485:264–268

131. Fukunaga T, Ozaki H, Terai G et al (2014) CapR: revealing structural specificities of RNA-binding protein target recognition using CLIP-seq data. Genome Biol 15:R16

132. Zambelli F, Pavesi G (2015) De novo secondary structure motif discovery using RNAProfile. Methods Mol Biol 1269:49–62

133. Pietrosanto M, Adinolfi M, Casula R et al (2018) BEAM web server: a tool for structural RNA motif discovery. Bioinformatics 34:1058–1060

134. Maticzka D, Lange SJ, Costa F, Backofen R (2014) GraphProt: modeling binding preferences of RNA-binding proteins. Genome Biol 15:R17

135. Bahrami-Samani E, Penalva LOF, Smith AD, Uren PJ (2015) Leveraging cross-link modification events in CLIP-seq for motif discovery. Nucleic Acids Res 43:95–103

136. Polishchuk M, Paz I, Yakhini Z, Mandel-Gutfreund Y (2018) SMARTIV: combined sequence and structure de-novo motif discovery for in-vivo RNA binding data. Nucleic Acids Res 46:W221–W228

137. Munteanu A, Mukherjee N, Ohler U (2018) SSMART: sequence-structure motif identification for RNA-binding proteins. Bioinformatics 34:3990–3998

138. Kazan H, Morris Q (2013) RBPmotif: a web server for the discovery of sequence and structure preferences of RNA-binding proteins. Nucleic Acids Res 41:W180–W186

139. Kazan H, Ray D, Chan ET et al (2010) RNAcontext: a new method for learning the sequence and structure binding preferences of RNA-binding proteins. PLoS Comput Biol 6:e1000832

140. Cereda M, Pozzoli U, Rot G et al (2014) RNAmotifs: prediction of multivalent RNA motifs that control alternative splicing. Genome Biol 15:R20

141. Conesa A, Madrigal P, Tarazona S et al (2016) A survey of best practices for RNA-seq data analysis. Genome Biol 17:13

Chapter 2

Predicting RNA Secondary Structure Using In Vitro and In Vivo Data

Riccardo Delli Ponti and Gian Gaetano Tartaglia

Abstract

The new flow of high-throughput RNA secondary structure data coming from different techniques allowed the further development of machine learning approaches. We developed CROSS and CROSSalive, two algorithms trained on experimental data able to predict the RNA secondary structure propensity both in vitro and in vivo. Since the in vivo folding of RNA molecules depends on multiple factors due to the cellular crowded environment, prediction is a complex problem that needs additional calculations for the interaction with proteins and other molecules. In the following chapter, we will describe the differences in predicting RNA secondary structure propensity using experimental data as input for an Artificial Neural Network (ANN) in vitro and in vivo.

Key words Artificial neural networks, RNA secondary structure, Machine learning, SHAPE, RNA structure in vivo

1 Introduction

The RNA secondary structure (RSS) is a fundamental property to understand several RNA biological functions, from the three-dimensional structure to the interaction with proteins [1]. Crystallographic techniques such as X-ray and NMR can also be applied on RNA structures, but up to date very little RNA crystals are available, especially for specific classes of RNAs, and the technique is still challenging for long and complex RNAs [2]. The advent of high-throughput techniques, enzymatic, such as Parallel Analysis of RNA Structure (PARS), or chemical-based, such as Selective 2-′-hydroxyl Acylation Analyzed By Primer Extension (SHAPE), provided a new flow of data to be employed to train machine learning approaches [3, 4].

Using five datasets of genome-wide experimental techniques we developed Computational Recognition of Secondary Structure (CROSS), a neural network trained on high-propensity in vitro data

Erik Dassi (ed.), *Post-Transcriptional Gene Regulation*, Methods in Molecular Biology, vol. 2404,
https://doi.org/10.1007/978-1-0716-1851-6_2,

to predict the RNA secondary structure propensity [5]. CROSS algorithm was the first trained of high-throughput RNA secondary structure data, without sequence length restriction and at single-nucleotide resolution.

However, while developing CROSS, we realized that predicting RNA secondary structure in vivo data needed a different approach compared with in vitro data. The RNA folding in vivo is more complex than in vitro, especially for the presence of a more crowded cellular environment and for the interactions with proteins such as chaperones [6, 7]. We used this information while building CROSSalive, the first method focused on predicting RSS in vivo data using additional information coming from protein interactions [8].

In the following chapter, we will focus on the differences between predicting the RNA secondary structure using in vitro and in vivo data. While some procedures and methodologies are similar between the two approaches, it is interesting to notice how specific key-steps and filtering procedures were essential to improve the predictive power of both methods. To have a better comparison between the in vitro and the in vivo predictive approach, marked Notes (*) describe in more detail a step that shows crucial differences between the two approaches. Moreover, we focused on describing in detail the functionality and the importance of each step to create our algorithms, since statistics, datasets, and performances of both algorithms were extensively analyzed in the original manuscripts.

2 Materials

2.1 Artificial Neural Network

1. We trained CROSS models using a fully connected artificial neural network with one hidden layer and two adaptive weight matrices ω_k^i and Ω^k that are optimized using backpropagation, and accordingly variated at each iteration to avoid overfitting.
2. The input F_i is propagated to the first hidden layer of k nodes as:

$$h_k = \tanh(\omega_k^i F_i)$$

3. The score Π of the nucleotide in the center of the window is then given by:

$$\Pi = \tanh(\Omega^k h_k)$$

4. The contributions of h_k for the hidden layer are weighted by Ω^k.
5. We used FANN architecture (http://leenissen.dk/fann/wp/) built in C++ to train and test the previously described networks.

2.2 Protein-RNA Interactions Using the catRAPID Algorithm

1. *cat*RAPID algorithm (available at www.tartaglialab.com) employs physico-chemical properties from proteins and RNA molecules [1], including information about the RNA secondary structure using RNAfold [9], to predict the interaction between proteins and RNA.
2. *cat*RAPID *omics* is part of *cat*RAPID suite, able to compute interactions between multiple RNA molecules and the entire set of RNA-binding proteins of a specific organism [10]. In our case, we used a set of >7700 protein regions coming from 640 putative RNA-binding proteins (RBPs).
3. To classify our interactions, we used *cat*RAPID Discriminative Power (DP) as the selective score. The discriminative power is a statistical measure introduced to evaluate the interaction propensity with respect to *cat*RAPID training. It represents the confidence of the prediction. The Discriminative Power (DP) ranges from 0% (unpredictability) to 100% (predictability). DP values above 50% indicate that the interaction is likely to take place, whereas DPs above 75% represent high-confidence predictions [1].

3 Methods

3.1 Predicting In Vitro Data

3.1.1 Selecting Higher-Propensity Nucleotides from Different Experimental Techniques

1. Selecting nucleotides with a high propensity to be in double- or single-stranded conformation for each experimental technique (*see* **Note 1**, Fig. 1).
2. For Parallel Analysis of RNA Structure (PARS) applied on yeast and human transcriptomes [3, 11], we selected five *maxima/minima* for each profile, in order to collect a consistent signal for each transcript (*see* **Note 2**).
3. For Selective 2′ Hydroxyl Acylation analyzed by Primer Extension (SHAPE) data, we used the standard thresholds of confidence (>0.5 for single-stranded, <0.2 for double-stranded) provided by the authors of the SHAPE study [12]. By using the provided threshold, we were able to select high-propensity signals by filtering out the noise of uncertain nucleotides.
4. For in vivo click selective 2-hydroxyl acylation and profiling experiment (icSHAPE) data [13], the entire mouse transcriptomes showed a bimodal distribution (Supplementary Fig. 11 of [5]). To build our predictive approach, we selected the

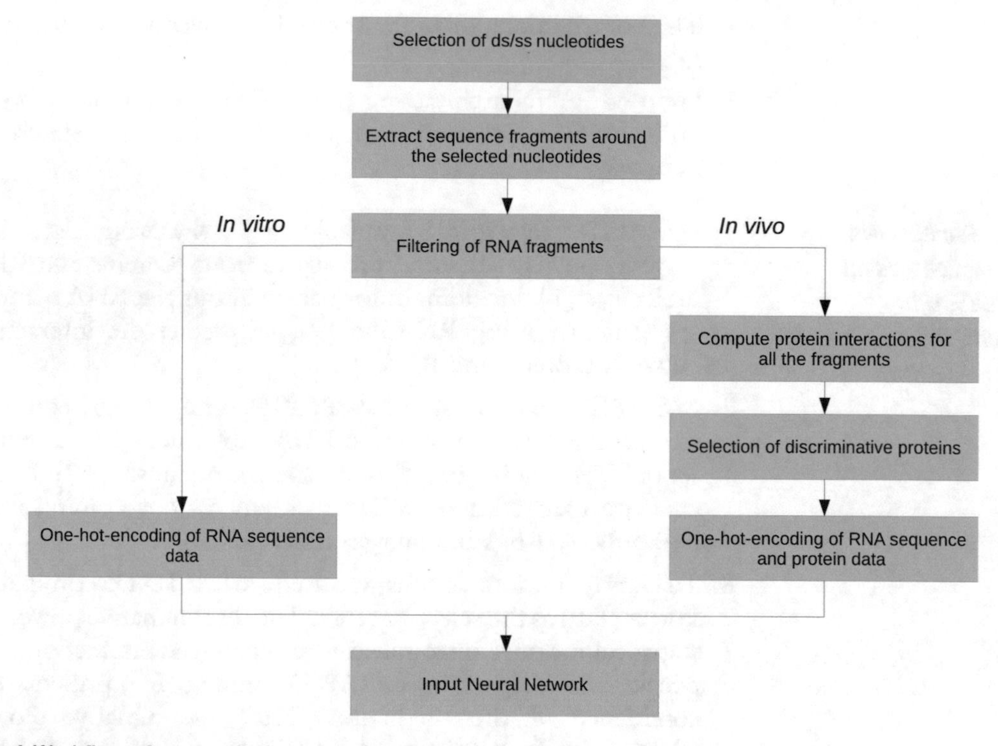

Fig. 1 Workflow of our predictive procedure both for the in vitro and the in vivo algorithms

highest and lowest reactivity scores (1 for single-stranded, 0 for double-stranded; *see* **Note 3***).

3.1.2 Encoding Sequence Information into the Predictive Approach

1. The first step to predict the RNA structure, both in vitro and in vivo, was to encode the information of the RNA sequences associated with higher-propensity nucleotides (*see* **Note 4**).
2. In the previous section, we provided details on how we identified nucleotides encoding for a high signal according to different techniques. However, the input of our method is the vector encoding the information on fragments of specific length centered around the higher-propensity nucleotides previously selected. The fragments encoding the RNA sequence information were extracted by using a sliding window spanning the precedent and subsequent six residues (i.e., 13 nucleotides; *see* **Note 5***).
3. The fragments were further filtered for ambiguity, meaning that fragments both encoding for a higher-propensity to be in single- and double-stranded conformation were removed from our training set. Moreover, each fragment was selected only once for each dataset (positive and negative set during the training) to avoid overfitting of the data.
4. Moreover, for PARS data, we included only fragments in which the experimental score of the central nucleotide was in the same

state (single- or double-stranded) in >90% of the occurrences within the yeast/human transcriptome (*see* **Note 6**).

3.1.3 In Vitro ANN Architecture: Training and Testing

1. To encode the signal coming from the RNA fragments inside our predictive approach, we used a one-hot encoding 4-mer notation to represent each nucleotide: A = (1, 0, 0, 0), C = (0, 1, 0, 0), G = (0, 0, 1, 0), and U = (0, 0, 0, 1). This procedure is commonly applied to use strings as input inside neural networks for biological applications ([14]; *see* **Note 7***).
2. For our approach, we developed a fully connected straight-forward neural network with one hidden layer. We used a tuning approach to define the number of internal variables and epochs (*see* **Note 8**).
3. Both positive (double-stranded fragments) and negative (single-stranded fragments) sets were balanced during the training.
4. The output of the network was associated with the secondary structure propensity, with a continuous score of >0 for high-propensity double-stranded fragments, and <0 for single-stranded nucleotides. Same architecture and tuning was also applied to the in vivo approach.
5. A tenfold cross validation was employed to test each independent network. Ninty percent of each dataset was used for the training, while the unused 10% was selected for testing purposes. Moreover, each model was also tested on the datasets of the other techniques to have an even more independent testing.

3.2 Predicting In Vivo Data

3.2.1 Filtering and Selecting In Vivo Data

1. The first steps of the predictive approach are quite similar between in vitro and in vivo (Fig. 1). We used a similar procedure to select icSHAPE data and to encode RNA sequence information inside an ANN.
2. icSHAPE data were selected using the same approach employed for the in vitro architecture (*see* Subheading 3.1.1). To balance the datasets, we selected RNA fragments carrying the central nucleotide with the highest (single-stranded; 10^5 non-redundant sequences) and lowest icSHAPE reactivities (double-stranded; 10^5 non-redundant sequences) for the three different icSHAPE datasets (in vitro and in vivo with and without adenosine methylation *Mettl3,* which we refer to as m6a+ and m6a−; *see* **Note 9**).
3. As previously explained (*see* Subheading 3.1.2 for the in vitro approach), we used a wider window of 51 nucleotides to extract a fragment around the *maxima/minima* of icSHAPE data (*see* **Note 10**).

4. The fragments of 51 nucleotides were integrated inside the ANN using again the one-hot encoding procedure: A = (1, 0, 0, 0), C = (0, 1, 0, 0), G = (0, 0, 1, 0), and U = (0, 0, 0, 1).
5. The networks trained on icSHAPE in vitro, in vivo m6a+ (with *Mettl3*) and in vivo m6a– were trained in the same conditions, and tested using a tenfold cross validation. The results were quite interesting since the in vitro network was the only one able to achieve good performances (AUC 0.86) using data coming only from the RNA sequence (*see* **Note 11**).

3.2.2 Integrating Protein Data into the In Vivo Predictive Approach

1. Integrating information from the interaction with proteins for each RNA fragment was the key feature to improve the predictive power of in vivo data, and the crucial difference with the in vitro predictive approach (Fig. 2).
2. We used the *cat*RAPID *omics* algorithm [10] to study the interaction of 7797 domains from a library of 640 canonical RNA-binding proteins against our datasets comprising a total of 200,000 single- and double-stranded fragments (*see* **Note 12**).
3. We identified proteins able to discriminate nucleotides in single- and double-stranded states, with the aim of integrating specific and not noisy data in our predictive approach. First, we used catRAPID score to select proteins with accuracies >0.6 to predict the RNA secondary structure propensity.
4. Proteins passing the previous filter and with gene ontology related to RNA structure (double- and single-stranded RNA binding; helicase activity; m6a+: 101 protein domains; m6a–: 81 protein domains) were further selected to be integrated with RNA sequence data.

3.2.3 In Vivo ANN Architecture Including Protein Data: Training and Testing

1. RNA sequence and protein interactions for each selected fragment were integrated together as input for the ANN.
2. As previously explained, sequence data followed the same processing and integration as in the in vitro architecture, where the four nucleotides are converted in arrays of integers. The only difference is the wider window, necessary to predict interactions using the *cat*RAPID algorithm, with an input array of 204 integers ($51 \times 4 = 204$, 1 or 0). This approach better captures properties of the local folding.
3. Integrating protein data was a more complex step. After selecting an independent set of discriminative proteins (for accuracy and relevant biological function) for the in vivo datasets, with and without methylation, we collected all the catRAPIDs interactions for all the proteins for each RNA fragment.

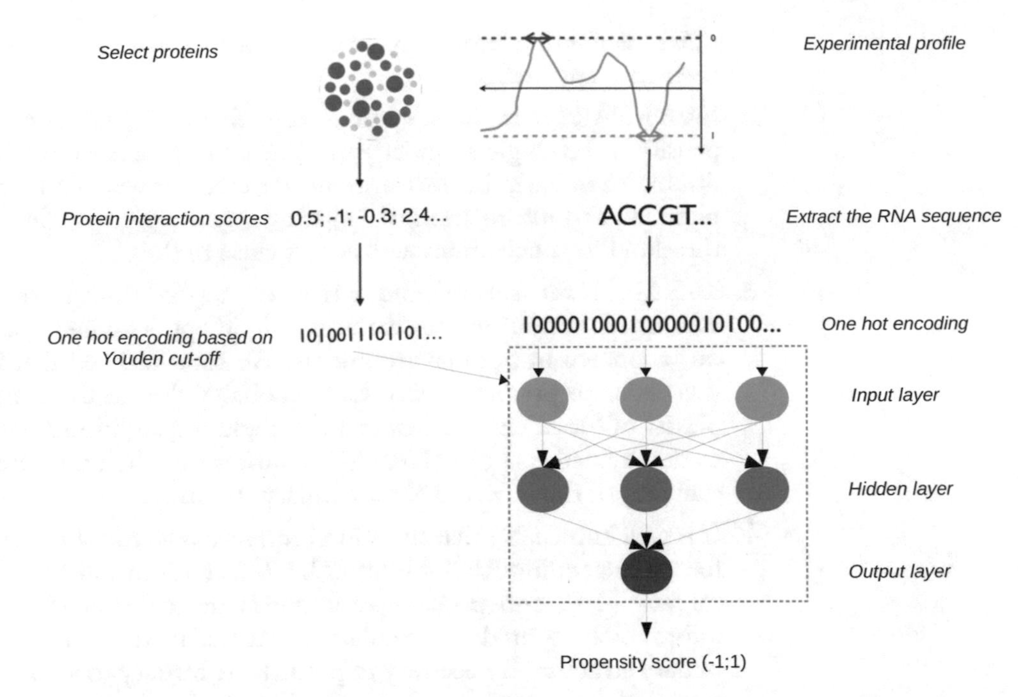

Fig. 2 Graphical summary showing the different approaches to predict the RNA secondary structure using only sequence information (in vitro data) and integrating protein information (in vivo data). The ANN architecture and the input data are also represented in the figure

4. We then used the Youden cut-off computed on *cat*RAPID scores for each protein in the dataset to normalize the data (*see* **Note 13**). Scores above the cut-off were set to 1 (0 otherwise).
5. The arrays encoding sequence information ($n = 204$) and protein interactions ($n = 101$ for m6a+; $n = 81$ for m6a−) were combined in one input layer ($n = 305$ or 285; Fig. 2).
6. The ANN architecture was similar to the one applied for sequence-only ANN, a fully connected neural network with one hidden layer.
7. Tuning and testing also followed similar strategies as before (*see* Subheading 3.1.3). As an independent test set we used in vivo SHAPE-MaP data of the entire lncRNA *Xist* (~18,000 nt; [15]).

4 Notes

1. This was the first and most important step to predict the RNA secondary structure based on experimental techniques. However, each individual technique has a specific scoring system and a different range to identify what could be noise, so we needed

different filtering procedures to implement clean data in our predictive approach.

2. Even if PARS score has a specific range from −7 (higher propensity to be single-stranded) to +7 (higher propensity to be double-stranded), the distributions of the scores were far to be bimodal, and information was missing regarding a confidence threshold for nucleotides with scores close to 0.
3. icSHAPE is the only technique that was applied both in vitro and in vivo conditions, providing a unique opportunity to test our approach in both environments. We soon realized that it was easier to predict in vitro data, probably due to the complexity of the in vivo environment. Integrating additional data to the predictive approach could be the key for a better understanding of the in vivo RNA secondary structure.
4. It is well known [9] that the RNA sequence is a crucial feature for understanding RNA folding. The RNA folds in a hierarchical way [16], and primary structure is the main source of information to predict secondary structure. In turn, the secondary structure is necessary to predict the tertiary structure.
5. We used as a reference the V1 enzyme used in PARS that requires at least six nucleotides around the cutting point [17]. However, a window of seven nucleotides centered on the nucleotide of interest (3 + 1 + 3 nt) is not sufficient to capture the signal as well as a 13 nucleotides window. We realized that longer fragments did not substantially improve the method, while reducing the coverage on the extremities of a molecule. In this step, the in vitro and in vivo approaches diverged, since to use the information coming from protein interactions to predict in vivo data we needed to employ larger fragments of 51 nucleotides (25 + 1 + 25).
6. Since the selection of maxima and minima is transcript-specific for PARS data, the procedure includes ambiguous fragments (i.e., fragments that are not always associated with extreme experimental values outside the transcript of selection). This was a critical step to select unambiguous fragments encoding a higher propensity to be single- or double-stranded for PARS data. This step was not applied for other data in which the scores followed an almost bimodal distribution or if the authors of the technique provided a clear threshold to exclude noise. This is also true for in vivo data, for which only icSHAPE data were available.
7. For the in vitro predictive approach, each 13 nt fragment was converted in an array of 52 integers (1 or 0), which was used as input data for the ANN. However, for the in vivo architecture, a wider window (51 nt = 204 integers) was used to maximize

the interactions with proteins and to correctly use the catRAPID algorithm.

8. For the fine-tuning procedure, we tried several combinations of internal variables for the inner layer and number of epochs, till defining the best combination in terms of performances on a testing set and of computational time. For more details regarding the computational timing of the algorithm, check the General Information of the online Tutorial (http://service.tartaglialab.com/static_files/algorithms/cross/tutorial.html). This architecture was consistent for each model trained on the fragments extracted from each experimental technique.
9. Adenosine methylation (m6a) has also an effect on the structure, promoting the transition toward single strands [13, 18]. For this reason, in vivo icSHAPE experiments were performed with and without *Mettl3* to provide an overall comparison of the effects of the methylation on the structure at single-nucleotide resolution. For the two datasets, we independently selected the RNA fragments and the discriminative proteins, building two independent predictive models: in vivo m6a + (with m6a methylation) and m6a− (without m6a methylation).
10. The *cat*RAPID *omics* algorithm exploits several physicochemical properties, including *RNAfold* data, to predict the interaction between proteins and RNA. The algorithm is able to predict thousands of interactions between the complete set of RNA-binding proteins of a specific organism, in our case mouse, and wide RNA fragments of at least 50 nucleotides.
11. To predict the RNA structure in vivo it was necessary to integrate additional data into our predictive model. The cellular environment is far more complex than the in vitro one, and it was previously shown that the presence of chaperones and a crowded environment have a positive effect on the RNA folding [7], leading us to introduce selective protein data inside our algorithm.
12. *cat*RAPID was used to compute $>12 \times 10^7$ comparisons between proteins and multiple 51-long RNA fragments. Selecting only discriminative proteins and interactions was a crucial step to improve the predictive power of in vivo RNA secondary structure data.
13. The Youden cut-off was generated for each individual protein based on the AUROC generated to discriminate between double- and single-stranded RNA fragments using catRAPID score. By using the Youden cut-off we were able to optimally normalize the catRAPID score and to implement it together with one-hot encoded sequence data inside our in vivo predictive model.

References

1. Bellucci M, Agostini F, Masin M, Tartaglia GG (2011) Predicting protein associations with long noncoding RNAs. Nat Methods 8:444–445. https://doi.org/10.1038/nmeth.1611
2. Zhang J, Ferré-D'Amaré AR (2014) Dramatic improvement of crystals of large RNAs by cation replacement and dehydration. Structure 22:1363–1371. https://doi.org/10.1016/j.str.2014.07.011
3. Kertesz M, Wan Y, Mazor E et al (2010) Genome-wide measurement of RNA secondary structure in yeast. Nature 467:103–107. https://doi.org/10.1038/nature09322
4. Wilkinson KA, Merino EJ, Weeks KM (2006) Selective 2′-hydroxyl acylation analyzed by primer extension (SHAPE): quantitative RNA structure analysis at single nucleotide resolution. Nat Protoc 1:1610–1616. https://doi.org/10.1038/nprot.2006.249
5. Delli Ponti R, Marti S, Armaos A, Tartaglia GG (2017) A high-throughput approach to profile RNA structure. Nucleic Acids Res 45:e35. https://doi.org/10.1093/nar/gkw1094
6. Zemora G, Waldsich C (2010) RNA folding in living cells. RNA Biol 7:634–641. https://doi.org/10.4161/rna.7.6.13554
7. Zhou H-X, Rivas G, Minton AP (2008) Macromolecular crowding and confinement: biochemical, biophysical, and potential physiological consequences. Annu Rev Biophys 37:375–397. https://doi.org/10.1146/annurev.biophys.37.032807.125817
8. Delli Ponti R, Armaos A, Vandelli A, Tartaglia GG (2020) CROSSalive: a web server for predicting the in vivo structure of RNA molecules. Bioinformatics 36:940–941. https://doi.org/10.1093/bioinformatics/btz666
9. Gruber AR, Lorenz R, Bernhart SH et al (2008) The Vienna RNA websuite. Nucleic Acids Res 36:W70–W74. https://doi.org/10.1093/nar/gkn188
10. Agostini F, Zanzoni A, Klus P et al (2013) catRAPID omics: a web server for large-scale prediction of protein-RNA interactions. Bioinformatics 29:2928–2930. https://doi.org/10.1093/bioinformatics/btt495
11. Wan Y, Qu K, Zhang QC et al (2014) Landscape and variation of RNA secondary structure across the human transcriptome. Nature 505:706–709. https://doi.org/10.1038/nature12946
12. Watts JM, Dang KK, Gorelick RJ et al (2009) Architecture and secondary structure of an entire HIV-1 RNA genome. Nature 460:711–716. https://doi.org/10.1038/nature08237
13. Spitale RC, Flynn RA, Zhang QC et al (2015) Structural imprints in vivo decode RNA regulatory mechanisms. Nature 519:486–490. https://doi.org/10.1038/nature14263
14. Zhang S-W, Wang Y, Zhang X-X, Wang J-Q (2019) Prediction of the RBP binding sites on lncRNAs using the high-order nucleotide encoding convolutional neural network. Anal Biochem 583:113364. https://doi.org/10.1016/j.ab.2019.113364
15. Smola MJ, Christy TW, Inoue K et al (2016) SHAPE reveals transcript-wide interactions, complex structural domains, and protein interactions across the Xist lncRNA in living cells. Proc Natl Acad Sci U S A 113:10322–10327. https://doi.org/10.1073/pnas.1600008113
16. Brion P, Westhof E (1997) Hierarchy and dynamics of RNA folding. Annu Rev Biophys Biomol Struct 26:113–137. https://doi.org/10.1146/annurev.biophys.26.1.113
17. Lowman HB, Draper DE (1986) On the recognition of helical RNA by cobra venom V1 nuclease. J Biol Chem 261:5396–5403
18. Liu J, Jia G (2014) Methylation modifications in eukaryotic messenger RNA. J Genet Genomics 41:21–33. https://doi.org/10.1016/j.jgg.2013.10.002

Chapter 3

RBPmap: A Tool for Mapping and Predicting the Binding Sites of RNA-Binding Proteins Considering the Motif Environment

Inbal Paz, Amir Argoetti, Noa Cohen, Niv Even, and Yael Mandel-Gutfreund

Abstract

RNA-binding proteins (RBPs) play a key role in post-transcriptional regulation via binding to coding and non-coding RNAs. Recent development in experimental technologies, aimed to identify the targets of RBPs, has significantly broadened our knowledge on protein-RNA interactions. However, for many RBPs in many organisms and cell types, experimental RNA-binding data is not available. In this chapter we describe a computational approach, named RBPmap, available as a web service via http://rbpmap.technion.ac.il/ and as a stand-alone version for download. RBPmap was designed for mapping and predicting the binding sites of any RBP within a nucleic acid sequence, given the availability of an experimentally defined binding motif of the RBP. The algorithm searches for a sub-sequence that significantly matches the RBP motif, considering the clustering propensity of other weak matches within the motif environment. Here, we present different applications of RBPmap for discovering the involvement of RBPs and their targets in a variety of cellular processes, in health and disease states. Finally, we demonstrate the performance of RBPmap in predicting the binding targets of RBPs in large-scale RNA-binding data, reinforcing the strength of the tool in distinguishing cognate binding sites from weak motifs.

Key words RNA-binding proteins (RBPs), Binding site prediction, RBP motifs, Protein-RNA, RBPmap

1 Introduction

In eukaryotes, post-transcriptional regulation is known to be coordinated by hundreds of RNA-binding proteins (RBPs), which bind in a specific manner to both coding and non-coding RNAs [1]. The interactions between RBPs and their targets form highly dense interconnected networks, orchestrating the different steps of the gene expression pathway [2]. Consistently, perturbations in post-transcription regulatory interactions have been shown to affect key cellular processes [3] and to be involved in many human diseases

Erik Dassi (ed.), *Post-Transcriptional Gene Regulation*, Methods in Molecular Biology, vol. 2404,
https://doi.org/10.1007/978-1-0716-1851-6_3,

[4]. In recent years, there has been great progress in the development of experimental technologies for identifying the targets of RBPs [5]. Current in-vivo technologies are based on cross-linking and immunoprecipitation (CLIP). The CLIP based technologies have provided extensive amounts of information on the binding sites of RBPs in the transcriptomes of human and other model organisms [6]. The enhanced CLIP (eCLIP) methodology [7] was recently employed to profile the targets of 150 RBPs in different human cell lines [8]. The data, generated as part of the Encyclopedia of DNA Elements (ENCODE) project, is a very useful resource for studying protein-RNA interactions [9]. In accordance with the advances in the experimental methodologies, different computational approaches have been developed and integrated to analyze large-scale protein-RNA binding data and to extract the functional elements associated with RBPs [8, 10]. However, for many RBPs and for the majority of organisms and cell types such data is not available. In the last decade, diverse algorithms and tools were designed for de-novo prediction of protein-RNA interactions (for review, *see* ref. 11). Several computational methods were custom-made for predicting the binding sites of RBPs with particular functions (such as splicing) (e.g., [12, 13]). In addition, more general computational models were introduced for accurate detection of RBP target sites and their binding preferences [14, 15]. These are incorporated in methods that rely on sequence attributes (such as RBPmap [16]) and in methodologies that combine both sequence and structural information (e.g., RNAcontext [17, 18], GraphPROT [19], RCK [20], and SMARTIV [21, 22]). A unique feature of RBPmap is that it takes the information from the motif environment into account in evaluating the probability of the sub-sequence to be a true binding site. Specifically, RBPmap employs a weighted rank approach that reflects the tendency of weak RBP motifs to cluster around a pivot sequence, which significantly matches the experimentally defined binding motif of the given RBP [23]. In addition, RBPmap enables the incorporation of conservation information in order to filter out non-probable regulatory sites. RBPmap algorithm was implemented in a web service http://rbpmap.technion.ac.il/ and is also available for download via the web site [16]. For efficient use of the algorithm, RBPmap web service and the download version provide a comprehensive database of RBP motifs, derived from high-throughput experiments conducted for human, mouse, and *Drosophila melanogaster* RBPs [16]. The majority of the motifs that are available in the database are derived from the RNA-binding compendium [24]. Recently we have expanded the RBPmap database, including motifs generated by the RNA bind-n-seq (RBNS) methodology [25, 26]. Here we present different applications of RBPmap for discovering the involvement of RBPs in key cellular processes as well as for predicting novel RBP-target interactions in disease states

and in viral infections. We demonstrate that RBPmap can accurately distinguish true RBP binding sites from non-specific targets and identify putative RBP binding sites in genomic sequences as well as in synthetic data.

2 The RBPmap Tool

The algorithm incorporated in RBPmap was originally developed for predicting the binding sites of splicing factors [23] and was implemented in the SFmap web server [13]. The algorithm was extended to support prediction and mapping of binding sites for any RBP within a given sequence (or sequences), relying on the availability of an experimental binding motif (or motifs) for the RBP of interest [16]. The RBPmap tool accepts as an input a nucleic acid (DNA or RNA) sequence or their genomic coordinates and outputs the location of the predicted binding sites within the sequence along with their calculated binding score and statistical significance.

2.1 RBPmap Algorithm

2.1.1 The Match Score

RBPmap computes a match score between a given RBP motif, provided as a Position Specific Scoring Matrix (PSSM) or a consensus, and all sub-sequences in the query sequence, calculated in overlapping windows. The match score (S_{match}) for a PSSM is defined as following:

$$S_{\text{match}} = \sum_{i=1}^{L} f_{N_i}$$

where L is the motif length, N_i is the specific nucleotide in position i, and f_{Ni} is the frequency of the nucleotide as defined in the PSSM. In cases where the motif is represented as a simplified consensus, the score is calculated as the Hamming distance between the motif and a sub-sequence of length L. The match scores calculated for each sub-sequence are negatively correlated with the distance, with values ranging between 0 (no match) and 1 (best match). For each match score RBPmap evaluates the significance of the score by comparing it to a mean match score, which is calculated for a randomly selected set of genomic regions defined as the background. The background set is composed of exonic and intronic sequences within protein coding and non-coding regions. The mean match scores for the background sequences are pre-calculated for all motifs that are available in the RBPmap dataset or calculated on the fly for new motifs provided by users. The significance of the match scores are defined by a *p-value* and are filtered according to two thresholds, a *significant threshold (sig)* and a *suboptimal threshold (sub)* (*see* Fig. 1). The significant threshold is used to define the pivot site and the suboptimal threshold is used to calculate the final RBPmap score, namely the Weighted Rank

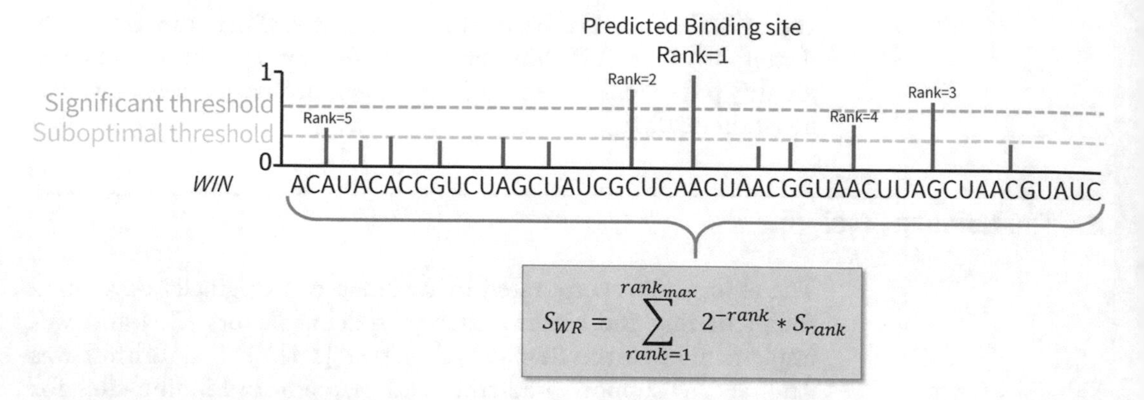

Fig. 1 A schematic representation of RBPmap algorithm for predicting an RBP binding site. For each significant match, defined as the predicted binding site, RBPmap calculates the Weighted Rank (WR) score, which is a weighted sum of the suboptimal matches within a window of 50 nts around the significant match. The significant threshold is used to define the significant match site and the suboptimal threshold, which is less stringent, filters the sites that are clustered around the putative binding site and are further ranked based on their match score

(WR) (*see* below). RBPmap allows to control the stringency level of the binding site predictions by selecting different combinations of the significant and suboptimal thresholds. The default stringency in RBPmap version 1.2 *is* *p-value*[sig] < 0.01 and *p-value* [sub] < 0.02.

2.1.2 The Weighted Rank (WR) Score

RBPmap calculates the WR score for all the sub-sequences in the input that have a significant match with the RNA-binding motif of interest. The WR score takes the propensity of similar motifs to cluster around the match site within a window of length 50 nts (25 nts up- and down-stream from the motif start position) into account. In order to calculate the WR score for a sub-sequence located less than 25 nts from the start and end of the sequence, RBPmap retrieves the genomic sequence from the reference genome. In case the input is a FASTA sequence, RBPmap uses the BLAST Like Alignment Tool (BLAT) [27] to determine the query sequence coordinates. Further, sub-sequences within the window, which pass the suboptimal threshold, are ranked according to their match scores and are included in the WR score calculation as defined in the following formula:

$$S_{\mathrm{WR}} = \sum_{\mathrm{rank}=1}^{\mathrm{rank}_{\max}} 2^{-\mathrm{rank}} * S_{\mathrm{rank}}$$

where $\mathrm{rank}_{\max}$ is the number of suboptimal sites within the window and S_{rank} is the match score of each suboptimal site.

To evaluate the significance of the final WR score, RBPmap calculates the WR score for a set of randomly selected 50 nt genomic regions, defined as the background. RBPmap uses two background sets (originating from human and *Drosophila melanogaster* genomes), each is composed of all genomic regions (exonic and intronic sequences, within protein coding and non-coding regions as well as untranslated regions (UTRs) of protein coding genes and intergenic regions) [16]. The WR score of each putative binding site is compared to the mean WR score calculated for the given RBP motif in the background model to evaluate the final statistical values for each predicted site. RBPmap reports only sites with a significant WR score ($p\text{-}value < 0.05$).

2.2 RBPmap Features

2.2.1 RBPmap Motif Database

RBPmap depends on the availability of an RNA-binding motif for predicting and mapping the binding site of an RBP of interest within a given sequence. To this end, the RBPmap tool provides an internal database of RNA-binding motifs for an extensive set of RBPs from human, mouse, and *Drosophila melanogaster*. A user can manually select any combination of motifs to be mapped to the input sequence (sequences) or choose to run RBPmap on the entire list of motifs in the dataset. In addition, the user can search the database for an RBP of interest via the RBPmap search engine by entering an RBP name, a symbol or common alias. The majority of the motifs available in the RBPmap database are from the compendium of RNA-binding motifs [24] generated by the RNAcompete methods [28]. Other motifs are extracted from high-throughput in-vivo binding assays and independent low-throughput assays. In the recent version of RBPmap (version 1.2), we have added 131 motifs that were generated by the RBNS methodology [25]. RBNS was originally developed to characterize the binding specificities of RBPs adapting the bind-n-seq approach used for determining DNA binding specificities [26]. Together RBPmap motif database includes 274 motifs for 182 different RBPs, all available as PSSMs. In addition to the motifs in the RBPmap database, users can provide their own motifs, as a PSSM in MEME format [29] or as a consensus motif, using the IUPAC symbols.

2.2.2 Output Features

RBPmap predicts binding sites, that are mapped to the given sequence or sequences, for the RBPs of interest. In addition, RBPmap provides detailed information on the motif predictions for each RBP that was chosen by the user. This information includes:

1. The location of the predicted site: The location includes the start and end positions within the input sequence as well as the exact genomic coordinates, when this information is available.
2. The matched motif: The motif that matches the given sequence, provided as a consensus sequence in IUPAC symbols.

3. The sequence of the predicted site: The exact sequence of the predicted site (within the input sequence) that matches the motif. In the web-based presentation of the summary table, the exact sequence is shown in the context of its flanking sequences (25 nts each side).
4. Statistical measures.

In addition to the summary table, RBPmap web service provides a visualized display of the binding sites using the UCSC Genome Browser. The Genome Browser visualization results for each RBP are presented as an independent custom track. The predicted binding sites custom tracks are shown in full mode, where each RBP track is colored differently. The black line represents the *p-value* threshold and the height of each predicted binding site is determined according to its *Z*-score. The predicted binding sites are displayed on the start position of the predicted match on the plus strand. In cases that the input sequence is on the minus strand, the start position is depicted on the plus strand but the motif should be read from right to left.

3 RBPmap Applications

The interactions of RBPs with specific functional elements on their RNA target sequences are required for the control of gene expression regulation at many different processing steps. RBPmap has been applied for de-novo prediction of regulatory elements of RBPs in many biological contexts, revealing the role of RBPs in novel post-transcriptional mechanism and their involvement in human diseases.

3.1 Predicting RBP-RNA Interactions Involved in Post-transcriptional Regulation

RBPmap is most commonly used for discovering novel RBP-target interactions that control mRNA processing events (such as splicing, editing, translation, etc.), which are involved in different cellular processes (e.g., [30]), developmental stages (e.g., [31]), and disease conditions (e.g., [32]). Several studies have applied RBPmap for mapping binding sites of known RBPs in long non-coding RNAs (lncRNAs), proposing novel functions for RBP-lncRNA interactions. For example, RBPmap was successfully used for predicting interactions of SR proteins with the *Paupar* lncRNA and directing the formation of a specific PAX6 isoform, which is essential for α cell development in mice [33]. RBPmap was also employed for discovering new unexpected roles for RBPs. Siam et al. have discovered a novel role for an epigenetic factor, acetyltransferase p300, in regulating alternative splicing. Employing RBPmap we found a significant enrichment of binding sites of two splicing factors (Sam68 and hnRNP M) within exons that were skipped upon knockdown of p300. Based on RBPmap results we predicted the involvement of these RBPs in p300 mediated splicing, which was further confirmed experimentally [34]. In another work, studying the role of alternative polyadenylation (APA) in cellular senescence, RBPmap was employed to identify the potential RBPs that regulate cell senescence via downregulating the Rras2 proteins. In the latter study, the authors ran RBPmap on a 34 nt region within the 3′UTR of the Rras2 gene, that was found to be essential for regulating its protein level. RBPmap results indicated four RBPs that may potentially interact with specific binding sites within that region. Further functional experiments confirmed that the predicted RBP, TRA2B, indeed binds to that region and that the binding of TRA2B to its predicted motif AGAA is essential for repressing the Rras2 protein levels [35].

3.2 Predicting RBP-RNA Interactions Associated with Viral Infections

As aforementioned, RBPmap has been mainly applied to study RBP-RNA interactions within human and other model organisms. Another commonly used application of RBPmap is predicting putative binding sites of human RBPs in viral transcripts. It is well-established that RNA viruses strongly depend on host post-transcriptional processes (capping, polyadenylation, localization, RNA stability, translation) during their infection cycle. In a previous study, we have employed RBPmap to investigate viral-host interaction in Kaposi's sarcoma-associated herpesvirus (KSHV). Specifically, we asked how a viral protein distinguishes between the viral and host transcripts. Using RBPmap we predicted an interaction between a family of human splicing factors and a viral mRNA ORF57. Based on RBPmap predictions it was further shown that the host RBPs are essential for enhancing the expression of the ORF57 viral mRNA [36]. In a recent study, RBPmap was employed to identify putative interactions between human RBPs and SARS-CoV-2, which are potentially involved in modulating the stability of the viral transcripts [37]. Interestingly, based on RBPmap predictions the authors concluded that mutations within the UTRs in 28 variants of SARS-CoV-2 are not likely to change the binding of the human RBPs to these regions. To examine the applicability of RBPmap to predict RBP binding sites within synthetic sequences, we applied it to the synthetic RNA, Pfizer-BioNtech BNT162b2 vaccine. Figure 2 demonstrates the predicted human RBPs within the 3′UTR region of the synthetically designed spike mRNA sequence of SARS-CoV-2, originated from the TLE5 gene [38]. Based on RBPmap predictions we conclude that the 3′UTR of the designed mRNA includes many binding sites of human RBPs involved in different RNA processes, such as splicing, stability, and mRNA transport.

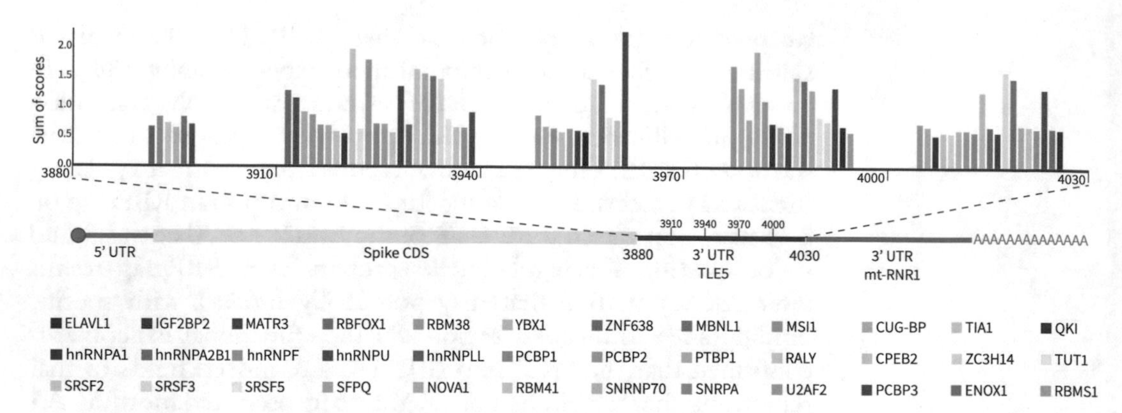

Fig. 2 RBPmap predictions on synthetic RNA sequences. RBPmap was applied on the 3′UTR of the synthetic sequences used for Pfizer-BioNtech BNT162b2 vaccine against SARS-CoV-2. Shown are the RBPs predicted by RBPmap to bind at three different regions within the 3′UTR of the synthetic sequence, originated from the TLE5 gene. The height of the bars demonstrates the sum of RBPmap scores for each RBP found to be bound within the region

3.3 Evaluating RBPmap Predictions Based on High-Throughput Binding Data

To evaluate RBPmap predictions we tested it on eCLIP data for 34 different RBPs available via ENCODE [9]. From the ENCODE database we chose 34 different RBPs for which eCLIP data was available from at least one cell line and their motifs were found in the RBPmap database. For 17 of the RBPs motif information was extracted from the RNA-binding site compendium [24], and 17 RBPs had available motifs generated by the RBNS methodology [26]. We extracted the processed peak file (available in bed narrow-peak format) and ranked the target sequences according to the "Signal value" column. Subsequently we selected the top 2000 sequences as strong binding sequences and the bottom 2000 sequences as weak binders (sequences shorter than 21 nts were excluded). Next, for each protein we ran RBPmap independently on the top (strong binders) and bottom (weak binders) sets, selecting for each dataset the corresponding RBP motif from the RBPmap database, using the default parameters (RBPmap version 1.2). For each protein we repeated the process for at least two independent data sets conducted on the same cell line (technical repeats). In Fig. 3 we present the average precision (specificity), i.e., the fraction of true positive predictions among all sites predicted by RBPmap, and recall (sensitivity), i.e., the fraction of true positives among the true binding sites, that were calculated for each of the 34 different proteins. The RBP names and motifs used to run RBPmap (extracted from RBPmap database) are depicted. Overall, our results reinforce that RBPmap identifies the binding targets of RBPs with high accuracy (0.79, 0.65, median precision and recall, respectively). These results reinforce that RBPmap can successfully identify true binding targets of RBPs and more so, to distinguish between strong and week binding sites. We recommend using RBPmap as an initial step for discovering potential RBP-target interactions that can be further confirmed experimentally.

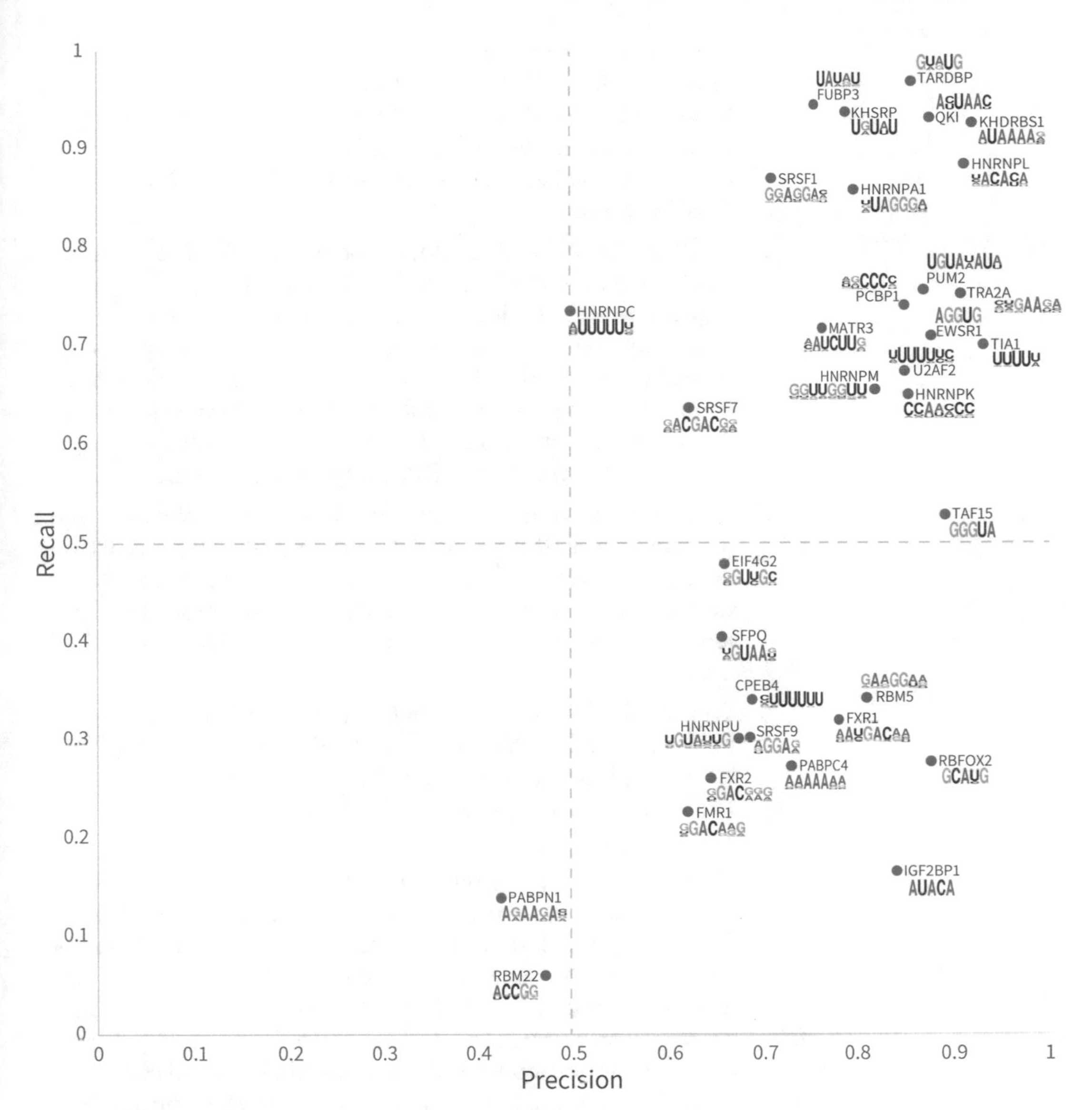

Fig. 3 RBPmap performance on high-throughput binding data. RBPmap was tested on eCLIP data extracted from the ENCODE database for 34 different RBPs. Shown are the average precision (specificity) and recall (sensitivity) values that were calculated based on RBPmap predictions in top 2000 sequences (strong binders) and bottom 2000 sequences (weak binders) for each RBP. The RBP official name and the motif used for running RBPmap are depicted on the graph. As shown, for the majority of RBPs RBPmap identified the true binding site with high specificity (precision >0.6)

4 Notes

1. RBPmap can be used to predict binding sites of RBPs within a nucleic acid sequence and can be applied for any organism or to synthetic data given the RBP binding motif and the sequence, which should be provided in FASTA format. RBPmap enables

full functionality for the following genomes assemblies: Human (GRCh38/hg38, GRCh37/hg19, NCBI36/hg18), Mouse (GRCm38/mm10, NCBI37/mm9) and *Drosophila melanogaster* (BDGP R5/dm3). For these genomes, sequences can be provided as genomic coordinates (in BED format) or in FASTA format.

2. RPPmap can also be employed for predicting binding sites of RBPs in synthetic (RNA or DNA) sequences, as long as the sequences are provided in FASTA format. As for all other sequences that are not fully supported by RBPmap, for synthetic sequences the significance of the score is calculated by comparing it to a theoretical background model. Expectedly, RBPmap cannot display the results in the UCSC genome browser when mapping RBPs on synthetic sequences.
3. The minimal and maximal length for an individual input sequence in RBPmap are 21 and 10,000 nts, respectively. Sequences that do not match the length restriction will be exempted. To overcome the technical restriction long input sequences can be divided to shorter segments up to 10,000 nts long.
4. To provide users with results in a timely and efficient manner, RBPmap restricts the total number of entries per run to 5000. However, if a user is interested in analyzing large amount of sequences it is possible to divide them into several RBPmap jobs.
5. Given that regulatory regions, and specifically binding sites of RBPs, tend to be evolutionary conserved, RBPmap allows to filter out predicted sites that are non-conserved. Non-conserved sites are defined when the mean conservation score that is calculated for their window is lower than the mean conservation score calculated for intronic regulatory regions. Filtering for non-conserved regions is optional and can only be applied to binding sites that are mapped to intronic and intergenic regions within human, mouse, or *Drosophila melanogaster* genomes. When this advanced option is selected, non-conserved sites are removed from the results.

Acknowledgments

The maintenance of RBPmap and the development of new features added to version 1.2 were supported by the ISF grant number 1182/16. We would like to thank Tamar Hashimshony for helpful comments on this chapter.

References

1. Hentze MW, Castello A, Schwarzl T, Preiss T (2018) A brave new world of RNA-binding proteins. Nat Rev Mol Cell Biol 19 (5):327–341. https://doi.org/10.1038/nrm.2017.130
2. Mittal N, Roy N, Babu MM, Janga SC (2009) Dissecting the expression dynamics of RNA-binding proteins in posttranscriptional regulatory networks. Proc Natl Acad Sci U S A 106(48):20300–20305. https://doi.org/10.1073/pnas.0906940106
3. Conlon EG, Manley JL (2017) RNA-binding proteins in neurodegeneration: mechanisms in aggregate. Genes Dev 31(15):1509–1528. https://doi.org/10.1101/gad.304055.117
4. Gebauer F, Schwarzl T, Valcárcel J, Hentze MW (2021) RNA-binding proteins in human genetic disease. Nat Rev Genet 22 (3):185–198. https://doi.org/10.1038/s41576-020-00302-y
5. Dasti A, Cid-Samper F, Bechara E, Tartaglia GG (2020) RNA-centric approaches to study RNA-protein interactions in vitro and in silico. Methods 178:11–18. https://doi.org/10.1016/j.ymeth.2019.09.011
6. Nechay M, Kleiner RE (2020) High-throughput approaches to profile RNA-protein interactions. Curr Opin Chem Biol 54:37–44. https://doi.org/10.1016/j.cbpa.2019.11.002
7. Van Nostrand EU, Pratt GU, Shishkin AA, Gelboin-Burkhart CU, Fang MU, Sundararaman BU, Blue SU, Nguyen TU, Surka C, Elkins KU, Stanton RU, Rigo F, Guttman M, Yeo GU (2016) Robust transcriptome wide discov ery of RNA-binding protein binding sites with enhanced CLIP (eCLIP). Nat Methods 13 (6):508–514. https://doi.org/10.1038/nmeth.3810
8. Van Nostrand EL, Pratt GA, Yee BA, Wheeler EC, Blue SM, Mueller J, Park SS, Garcia KE, Gelboin-Burkhart C, Nguyen TB, Rabano I, Stanton R, Sundararaman B, Wang R, Fu XD, Graveley BR, Yeo GW (2020) Principles of RNA processing from analysis of enhanced CLIP maps for 150 RNA binding proteins. Genome Biol 21(1):90. https://doi.org/10.1186/s13059-020-01982-9
9. Van Nostrand EL, Freese P, Pratt GA, Wang X, Wei X, Xiao R, Blue SM, Chen JY, Cody NAL, Dominguez D, Olson S, Sundararaman B, Zhan L, Bazile C, Bouvrette LPB, Bergalet J, Duff MO, Garcia KE, Gelboin-Burkhart C, Hochman M, Lambert NJ, Li H, McGurk MP, Nguyen TB, Palden T, Rabano I, Sathe S, Stanton R, Su A, Wang R, Yee BA, Zhou B, Louie AL, Aigner S, Fu XD, Lecuyer E, Burge CB, Graveley BR, Yeo GW (2020) A large-scale binding and functional map of human RNA-binding proteins. Nature 583(7818):711–719. https://doi.org/10.1038/s41586-020-2077-3
10. Uhl M, Houwaart T, Corrado G, Wright PR, Backofen R (2017) Computational analysis of CLIP-seq data. Methods 118-119:60–72. https://doi.org/10.1016/j.ymeth.2017.02.006
11. Dvir S, Argoetti A, Mandel-Gutfreund Y (2018) Ribonucleoprotein particles: advances and challenges in computational methods. Curr Opin Struct Biol 53:124–130. https://doi.org/10.1016/j.sbi.2018.08.002
12. Hiller MG, Pudimat R, Busch A, Backofen R (2006) Using RNA secondary structures to guide sequence motif finding towards single-stranded regions. Nucleic Acids Res 34(17): e117. https://doi.org/10.1093/nar/gkl544
13. Paz I, Akerman M, Dror I, Kosti I, Mandel-Gutfreund Y (2010) SFmap: a web server for motif analysis and prediction of splicing factor binding sites. Nucleic Acids Res 38(Web Server issue):W281–W285. https://doi.org/10.1093/nar/gkq444
14. Li XC, Kazan H, Lipshitz HD, Morris QD (2014) Finding the target sites of RNA-binding proteins. Wiley Interdis Rev RNA 5(1):111–130. https://doi.org/10.1002/wrna.1201
15. Sasse A, Laverty KU, Hughes TR, Morris QD (2018) Motif models for RNA-binding proteins. Curr Opin Struct Biol 53:115–123. https://doi.org/10.1016/j.sbi.2018.08.001
16. Paz I, Kosti I, Ares M Jr, Cline M, Mandel-Gutfreund Y (2014) RBPmap: a web server for mapping binding sites of RNA-binding proteins. Nucleic Acids Res 42(Web Server issue): W361–W367. https://doi.org/10.1093/nar/gku406
17. Kazan H, Morris Q (2013) RBPmotif: a web server for the discovery of sequence and structure preferences of RNA-binding proteins. Nucleic Acids Res 41(Web Server issue): W180–W186. https://doi.org/10.1093/nar/gkt463
18. Kazan H, Ray D, Chan ET, Hughes TR, Morris Q (2010) RNAcontext: a new method for learning the sequence and structure binding preferences of RNA-binding proteins. PLoS

Comput Biol 6:e1000832. https://doi.org/10.1371/journal.pcbi.1000832

19. Maticzka D, Lange SJ, Costa F, Backofen R (2014) GraphProt: modeling binding preferences of RNA-binding proteins. Genome Biol 15(1):R17. https://doi.org/10.1186/gb-2014-15-1-r17
20. Orenstein Y, Shamir R (2017) Modeling protein-DNA binding via high-throughput in vitro technologies. Brief Funct Genomics 16(3):171–180. https://doi.org/10.1093/bfgp/elw030
21. Polishchuk M, Paz I, Kohen R, Mesika R, Yakhini Z, Mandel-Gutfreund Y (2017) A combined sequence and structure based method for discovering enriched motifs in RNA from in vivo binding data. Methods 118-119:73–81. https://doi.org/10.1016/j.ymeth.2017.03.003
22. Polishchuk M, Paz I, Yakhini Z, Mandel-Gutfreund Y (2018) SMARTIV: combined sequence and structure de-novo motif discovery for in-vivo RNA binding data. Nucleic Acids Res 46(W1):W221–w228. https://doi.org/10.1093/nar/gky453
23. Akerman M, David-Eden H, Pinter RY, Mandel-Gutfreund Y (2009) A computational approach for genome-wide mapping of splicing factor binding sites. Genome Biol 10(3):R30. https://doi.org/10.1186/gb-2009-10-3-r30
24. Ray D, Kazan H, Cook KB, Weirauch MT, Najafabadi HS, Li X, Gueroussov S, Albu M, Zheng H, Yang A, Na H, Irimia M, Matzat LH, Dale RK, Smith SA, Yarosh CA, Kelly SM, Nabet B, Mecenas D, Li W, Laishram RS, Qiao M, Lipshitz HD, Piano F, Corbett AH, Carstens RP, Frey BJ, Anderson RA, Lynch KW, Penalva LO, Lei EP, Fraser AG, Blencowe BJ, Morris QD, Hughes TR (2013) A compendium of RNA-binding motifs for decoding gene regulation. Nature 499(7457):172–177. https://doi.org/10.1038/nature12311
25. Dominguez D, Freese P, Alexis MS, Su A, Hochman M, Palden T, Bazile C, Lambert NJ, Van Nostrand EL, Pratt GA, Yeo GW, Graveley BR, Burge CB (2018) Sequence, structure, and context preferences of human RNA binding proteins. Mol Cell 70 (5):854–867.e859. https://doi.org/10.1016/j.molcel.2018.05.001
26. Lambert N, Robertson A, Jangi M, McGeary S, Sharp PA, Burge CB (2014) RNA Bind-n-Seq: quantitative assessment of the sequence and structural binding specificity of RNA binding proteins. Mol Cell 54 (5):887–900. https://doi.org/10.1016/j.molcel.2014.04.016
27. Kent WJ (2002) BLAT—the BLAST-like alignment tool. Genome Res 12(4):656–664. https://doi.org/10.1101/gr.229202
28. Ray D, Kazan H, Chan ET, Pena Castillo L, Chaudhry S, Talukder S, Blencowe BJ, Morris Q, Hughes TR (2009) Rapid and systematic analysis of the RNA recognition specificities of RNA-binding proteins. Nat Biotechnol 27(7):667–670. https://doi.org/10.1038/nbt.1550
29. Bailey TL, Johnson J, Grant CE, Noble WS (2015) The MEME suite. Nucleic Acids Res 43(W1):W39–W49. https://doi.org/10.1093/nar/gkv416
30. Fiszbein A, Krick KS, Begg BE, Burge CB (2019) Exon-mediated activation of transcription starts. Cell 179(7):1551–1565.e1517. https://doi.org/10.1016/j.cell.2019.11.002
31. Takahashi K, Yamanaka S (2006) Induction of pluripotent stem cells from mouse embryonic and adult fibroblast cultures by defined factors. Cell 126(4):663–676. https://doi.org/10.1016/j.cell.2006.07.024
32. Tang SJ, Shen H, An O, Hong H, Li H, Song Y, Han J, Tay DJT, Ng VHE, Bellido Molias F, Leong KW, Pitcheshwar P, Yang H, Chen L (2020) Cis- and trans-regulations of pre-mRNA splicing by RNA editing enzymes influence cancer development. Nat Commun 11(1):799. https://doi.org/10.1038/s41467-020-14621-5
33. Singer RA, Arnes L, Cui Y, Wang J, Gao Y, Guney MA, Burnum-Johnson KE, Rabadan R, Ansong C, Orr G, Sussel L (2019) The long noncoding RNA Paupar modulates PAX6 regulatory activities to promote alpha cell development and function. Cell Metab 30(6):1091–1106.e1098. https://doi.org/10.1016/j.cmet.2019.09.013
34. Siam A, Baker M, Amit L, Regev G, Rabner A, Najar RA, Bentata M, Dahan S, Cohen K, Araten S, Nevo Y, Kay G, Mandel-Gutfreund Y, Salton M (2019) Regulation of alternative splicing by p300-mediated acetylation of splicing factors. RNA 25(7):813–824. https://doi.org/10.1261/rna.069856.118
35. Chen M, Lyu G, Han M, Nie H, Shen T, Chen W, Niu Y, Song Y, Li X, Li H, Chen X, Wang Z, Xia Z, Li W, Tian XL, Ding C, Gu J, Zheng Y, Liu X, Hu J, Wei G, Tao W, Ni T (2018) 3' UTR lengthening as a novel mechanism in regulating cellular senescence. Genome Res 28(3):285–294. https://doi.org/10.1101/gr.224451.117
36. Vogt C, Hackmann C, Rabner A, Koste L, Santag S, Kati S, Mandel-Gutfreund Y, Schulz TF, Bohne J (2015) ORF57 overcomes the

detrimental sequence bias of Kaposi's sarcoma-associated herpesvirus lytic genes. J Virol 89 (9):5097–5109. https://doi.org/10.1128/jvi.03264-14

37. Mukherjee M, Goswami S (2020) Global cataloguing of variations in untranslated regions of viral genome and prediction of key host RNA binding protein-microRNA interactions modulating genome stability in SARS-CoV-2. PLoS One 15(8):e0237559. https://doi.org/10.1371/journal.pone.0237559

38. Orlandini von Niessen AG, Poleganov MA, Rechner C, Plaschke A, Kranz LM, Fesser S, Diken M, Löwer M, Vallazza B, Beissert T, Bukur V, Kuhn AN, Türeci Ö, Sahin U (2019) Improving mRNA-based therapeutic gene delivery by expression-augmenting 3' UTRs identified by cellular library screening. Mol Ther 27(4):824–836. https://doi.org/10.1016/j.ymthe.2018.12.011

Part II

Expression Studies

Chapter 4

Analysis of mRNA Translation by Polysome Profiling

Anne Cammas, Pauline Herviou, Leïla Dumas, and Stefania Millevoi

Abstract

mRNA translation is a key step in gene expression that allows the cell to qualitatively and quantitatively modulate the cell's proteome according to intra- or extracellular signals. Polysome profiling is the most comprehensive technique to study both the translation state of mRNAs and the protein machinery associated with the mRNAs being translated. Here we describe the procedure commonly used in our laboratory to gain insights into the molecular mechanisms underlying translation regulation under patho-physiological conditions.

Key words mRNA translation, Polysome profiling, Translatome, Ribosome

1 Introduction

After synthesis and maturation in the nucleus, m7G-capped mRNAs are exported and translated in the cytoplasm by a sophisticated machinery that integrates intra- and extracellular signals to adjust protein synthesis levels to the specific needs of the cell and its environment. The translational machinery is composed of ribosomes (large ribonucleoprotein particles consisting of small (40S) and large (80S) ribosomal subunits), protein factors dedicated to protein synthesis (i.e., translation initiation, elongation, and termination factors), and RNA-binding proteins (RBPs).

Because each ribosome's binding site is limited to a few nucleotides [1], depending on the combination of various *cis/trans* mRNA factors, multiple ribosomes can bind and translate the same mRNA simultaneously, forming the so-called polysome. Polysomal assembly has an impact on translation efficiency, and is therefore analyzed in depth to study its functions and mechanisms acting on individual mRNAs, subgroups of mRNAs or more globally on the translatome (i.e., the set of mRNAs undergoing active translation). Among the various techniques for studying mRNA translation, polysome profiling is the most common and versatile because

Erik Dassi (ed.), *Post-Transcriptional Gene Regulation*, Methods in Molecular Biology, vol. 2404,
https://doi.org/10.1007/978-1-0716-1851-6_4,

it allows both target mRNAs and protein regulatory factors to be studied in the same experiment in a biased (i.e., using a candidate-gene approach) and/or unbiased (i.e., performing a large-scale analysis) manner.

Here we describe the procedure that we routinely use in the laboratory to study the regulation of translation by RNA structures [2, 3], translation factors [4], or RBPs [5]. As shown in Fig. 1, the protocol consists of three main steps: (a) cell lysis and isolation of the cytoplasmic fraction; (b) separation of the ribonucleoprotein complexes by sucrose gradient centrifugation; and (c) fractionation, which consists of recovering the different complexes monitored by following their absorbance at 254 nm using a UV detector (Fig. 1). These complexes are: mRNPs, 40S and 60S ribosomal subunits, monosomes (80S), and polysomes that can be further sorted into light (LP) and heavy polysomes (HP), depending on the number of ribosomes associated with the mRNAs. Then, RNAs and/or proteins from each fraction (which may be pooled together in some cases for convenience) can be studied in a biased manner to track the distribution of single mRNA targets or protein regulators along the sucrose gradient, or holistically using large-scale approaches (including high-throughput RNA-sequencing, DNA microarrays, mass spectrometry).

The polysomal profiling approach can be used to infer the translational status in healthy (e.g., neurons [6] or adipocytes [7]) or pathological situations (e.g., in cancer cells [8] or after virus infection, as in [9]) and in different organisms (e.g., *Drosophila* [10] or sea urchin [11]). It is also used to understand the contribution of translational regulation to gene expression in response to different stresses (e.g., UVB irradiation [12] or cold shock [13]) or to drugs (including inhibitors of cancer cell oncogenic signaling pathways [14–16]).

2 Materials

Prepare all solutions using DNase/RNase-free water and filter-sterilize the solutions through a 0.22μm filter.

2.1 Cell Lysis

1. Tissue culture dishes, regular media, and cell culture equipment.
2. Cell scrapers.
3. Pre-chilled 1.5 mL safe-lock microcentrifuge tubes.
4. 1.5 mL microcentrifuge tubes.
5. Filter pipette tips.
6. Refrigerated microcentrifuge.
7. 100 mg/mL cycloheximide (CHX) in DMSO. Store aliquots at −20 °C.

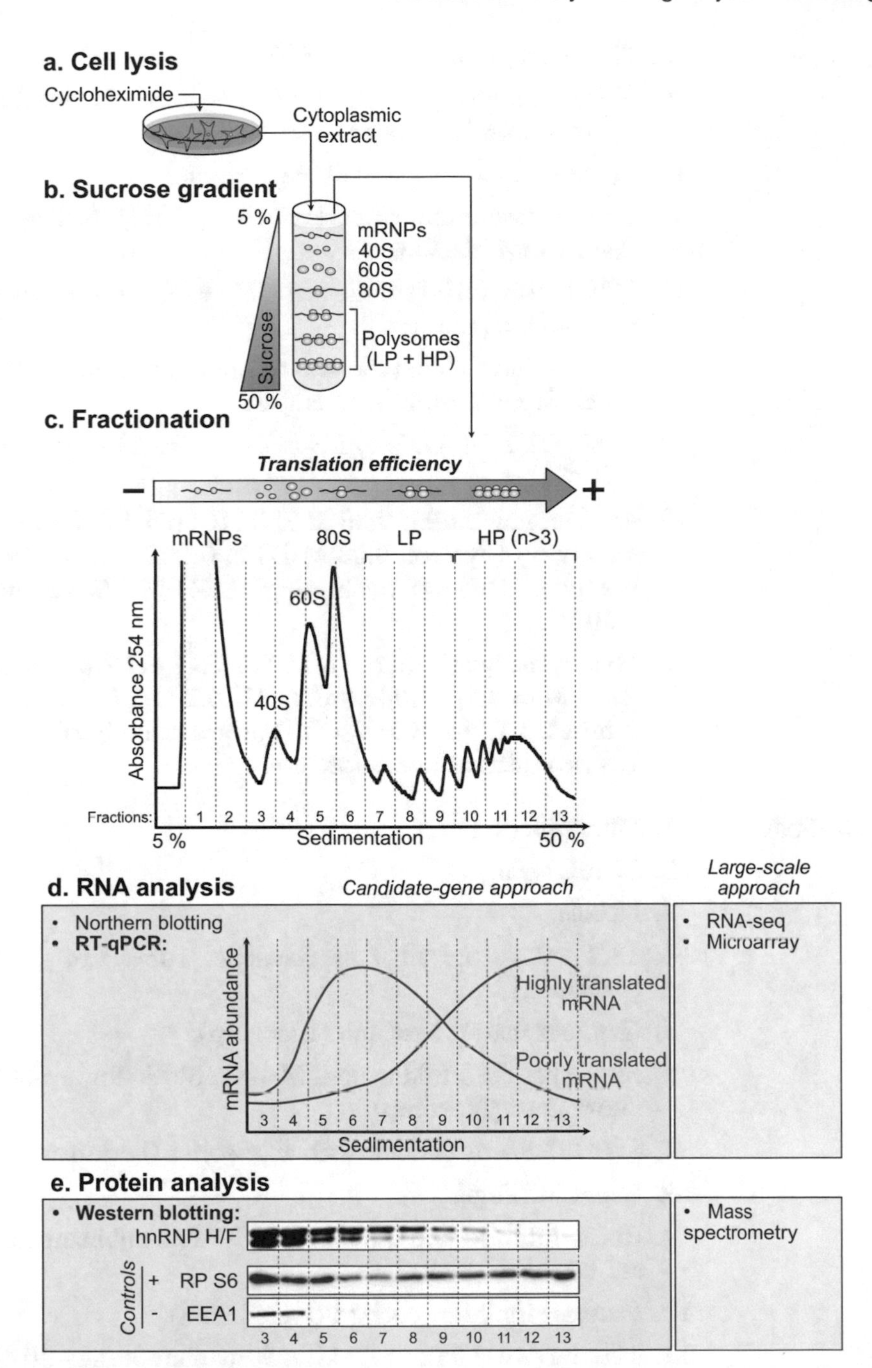

Fig. 1 Polysome fractionation workflow: After cycloheximide pre-treatment, cells are lysed (**a**) and cytoplasmic extracts are loaded onto sucrose gradients where polysomes, monosomes (80S), ribosomal subunits (40S and 60S), and messenger ribonucleoprotein particles (mRNPs) are separated by centrifugation (**b**). Fractions are collected and the polysome profile is recorded by measuring the UV absorbance at 254 nm (**c**). RNA and protein distribution across the sucrose gradient could be analyzed in candidate-gene or large-scale approaches. Northern blotting or RT-qPCR (reverse transcription-quantitative PCR) are used to determine changes in the distribution of specific mRNAs extracted from each fraction (**d**). Protein distribution throughout the gradient is visualized by western blotting. Proteins known to be associated with polysomes (e.g., the ribosomal protein RPS6) or not (e.g., the lysosomal protein EEA1) could be used as positive or negative controls, respectively (**e**)

8. Phosphate buffered saline (PBS).
9. RNase inhibitor (e.g., RNaseOUT™ Recombinant Ribonuclease Inhibitor, Invitrogen).
10. Protease inhibitor cocktail (e.g., Sigma).
11. Phosphatase inhibitor cocktail (e.g., 2 mM NaPyroP, 1 mM NaF, 2 mM Na_3VO_4).
12. 10% Triton X-100 solution in DNase/RNase-free water. Store at room temperature.
13. 10% Sodium deoxycholate solution in DNase/RNase-free water. Store at room temperature.
14. 1 M DTT in DNase/RNase-free water. Store aliquots at −20 °C.
15. 4× Laemmli buffer: 0.25 M Tris HCl pH 6.8, 8% (w/v) SDS, 40% (v/v) Glycerol, 0.08 M DTT, 0.002% (w/v) Bromophenol blue. Dissolve by heating at 37 °C. Store aliquots at −20 °C.
16. Hypotonic lysis buffer: 5 mM Tris HCl pH 7.6, 1.5 mM KCl, 2.5 mM $MgCl_2$, 100μg/mL CHX, 0.2 U/μL RNase inhibitor, 2 mM DTT. Add CHX, DTT, protease inhibitor cocktail and RNase inhibitor just before use.

2.2 Sucrose Gradient

1. Filter pipette tips.
2. 50 mL syringe.
3. Needle.
4. 13.2 mL Thinwall Polypropylene Tubes, 14 × 89 mm (Beckman).
5. Gradient master base unit (Biocomp).
6. Magnabase™ Holder and Marker Block for Sw41 Rotors, Short Caps (Biocomp).
7. SW41Ti Rotor Package with accessories (Beckman).
8. Ultracentrifuge.
9. RNase inhibitor (e.g., RNaseOUT™ Recombinant Ribonuclease Inhibitor, Invitrogen).
10. Protease inhibitor cocktail (e.g., Sigma).
11. 100 mg/mL CHX in DMSO. Store aliquots at −20 °C.
12. 60% (w/v) sucrose solution in DNase/RNase-free water.
13. 10× gradient buffer: 200 mM HEPES pH 7.6, 1 M KCl, 50 mM $MgCl_2$.
14. 50% sucrose solution: 124.5 mL 60% sucrose solution, 15 mL 10× gradient buffer. Adjust the volume to 150 mL with DNase/RNase-free water. Filter with 0.22μm filter. Store at 4 °C.

15. 5% sucrose solution: 12 mL 60% sucrose solution, 15 mL 10× gradient buffer. Adjust the volume to 150 mL with DNase/RNase-free water. Filter with 0.22μm filter. Store at 4 °C.

2.3 Fractionation

1. 2 mL microcentrifuge tubes.
2. Filter pipette tips.
3. Forceps.
4. Automated syringe pump (e.g., KDS Legato™ 100, single syringe infusion pump, Sigma).
5. Density gradient fractionation system with a UA-6 UV detector and a fraction collector (recommended: Teledyne Isco).
6. Computer with a data acquisition software (recommended: TracerDAQ pro).
7. DNase/RNase-free water.
8. Liquid nitrogen.
9. Chasing solution: Add 0.02% (w/v) bromophenol blue to 150 mL 60% (w/v) sucrose solution in water. Store at 4 °C.

2.4 RNA Analysis

1. 2 mL microcentrifuge tubes.
2. 1.5 mL safe-lock microcentrifuge tubes.
3. 1.5 mL microcentrifuge tubes.
4. Filter pipette tips.
5. Refrigerated centrifuge.
6. −20 °C freezer.
7. −80 °C freezer.
8. Vortex.
9. Thermoblock.
10. TRIzol LS (Invitrogen).
11. Chloroform.
12. Glycogen.
13. Isopropanol. Keep at 4 °C.
14. 70% Ethanol. Keep at −20 °C.
15. DNase/RNase-free water.

2.5 Protein Analysis

1. 2 mL microcentrifuge tubes.
2. 1.5 mL safe-lock microcentrifuge tubes.
3. Vortex.
4. Refrigerated centrifuge.
5. −20 °C freezer.
6. −80 °C freezer.

7. Thermoblock.
8. DNase/RNase-free water.
9. Isopropanol.
10. 100% ethanol.
11. Laemmli buffer.

3 Methods

To protect RNA from nuclease and heat degradation: Clean the workbench with nuclease-inhibitor solution, always wear gloves, use ice-cold solutions and pre-chilled tubes, perform all the lysis steps on ice and keep samples at 4 °C as much as possible.

3.1 Cell Lysis

1. Grow adherent cells in culture dishes to ~80% confluence (*see* **Note 1**).
2. Prior to lysis, treat cells with 100μg/mL CHX. Incubate for 5 min at 37 °C and 5% CO_2 (*see* **Notes 2** and **3**) (Fig. 1a).
3. Remove the growth medium from the dish and wash cells twice with 5 mL ice-cold PBS supplemented with 100μg/mL CHX.
4. Scrape cells in ice-cold PBS-CHX solution and collect them in pre-chilled tubes.
5. Pellet cells at 200 × *g* at 4 °C for 5 min.
6. Remove supernatant.
7. Add CHX, DTT, protease inhibitor cocktail, and RNase inhibitor to the hypotonic lysis buffer (*see* **Note 4**). Resuspend pellet in 450μL of ice-cold hypotonic lysis buffer.
8. Vortex for 5 s and incubate for 5 min on ice.
9. Add 25μL of 10% Triton X-100 and 25μL of 10% sodium deoxycholate.
10. Vortex for 5 s and incubate for 5 min on ice.
11. Centrifuge at 16,000 × *g* for 7 min at 4 °C to pellet the nuclei and cellular debris.
12. Transfer supernatant into a new pre-chilled tube.
13. Save 10% of the lysate for RNA and/or protein input. For RNA input, complete up to 400μL with hypotonic lysis buffer and proceed with RNA extraction as for fractions. For protein input, add Laemmli and load onto SDS-PAGE gel at constant $OD_{260\ nm}$ (*see* **step 15** and **Note 5**).
14. Flash-freeze the lysate in liquid nitrogen and store at −80 °C or immediately load lysates on the sucrose density gradients.

15. Dilute (1:5) 2μL of each sample in DNase/RNase-free water and measure the $OD_{260\ nm}$ (*see* **Note 6**).
16. For each sample, transfer the same $OD_{260\ nm}$ (ideally, 10–20 $OD_{260\ nm}$) into a new pre-chilled tube and adjust volume to 450μL with hypotonic lysis buffer (*see* **Note 7**).

3.2 Sucrose Gradient Formation, Loading, and Ultracentrifugation (Fig. 1b)

1. Pre-chill SW41Ti rotor, buckets, and the ultracentrifuge at 4 °C.
2. Mark the half-full point of each polypropylene tube by using the SW41 marker block provided with the gradient maker.
3. Transfer marked tubes into the tube holder provided with the gradient maker.
4. Add protease inhibitor, 10 U/mL RNase inhibitor, and 100μg/mL CHX to 5 and 50% sucrose solutions. Prepare 6 mL of each solution per gradient.
5. Pour 5% sucrose until it reaches the mark at the half-full point (*see* **Note 8**).
6. Use a 50 mL syringe with its needle to add the 50% sucrose solution from the bottom of the polypropylene tube upwards until it reaches the mark at the half-full point (*see* **Note 9**).
7. Check that no air bubbles are trapped in the sucrose and close the tubes with the caps supplied with the gradient maker. Remove the traces of the sucrose solution in the cap with a 1000μL pipette (*see* **Note 10**).
8. Switch on the gradient maker and adjust the magnetic support in a horizontal position using "up" and "down" switches by means of a level indicator. Gently place the tube holder with the sealed tubes on the magnetic support. Select the manufacturer program in the gradient list "SW41 5–50%" and run the program. *If your lysate is not ready to use, you may keep your gradients at 4 °C.*
9. Carefully remove 450μL at the surface of each gradient and slowly add the same volume of cell lysate on top of the gradient (*see* **Note 11**).
10. Place the gradients into the pre-chilled buckets.
11. Weight and balance each gradient before the ultracentrifugation by gently adding hypotonic lysis buffer at the surface (*see* **Note 12**).
12. Ultracentrifuge the gradients at 222,228 × *g* for 2 h at 4 °C using SW41Ti rotor (*see* **Note 13**).

3.3 Fractionation

1. About 15 min before the end of the centrifugation, prepare the fraction collector by (1) cleaning it with hot DNase/RNase-free water, (2) switching on the (a) UV detector (the green

light indicates when the UV lamp is warmed up), (b) computer, (c) syringe pump, and (d) fraction collector. Use the 50 mL syringe to load the chasing solution until it reaches the needle. Make sure to see the chasing solution coming out from the needle so that no bubbles are introduced into the gradient.

2. Set the pump at 1.3 mL/min and the fraction collector on 35 s. Launch the acquisition program TracerDAQ on the computer. Place 2 mL tubes in the fraction collector.
3. Carefully remove the buckets from the rotor and then the polypropylene tubes from the buckets (*see* **Note 14**).
4. Screw the first polypropylene tube into the UV detector and pierce it with the needle.
5. Start the pump and the fraction collector at the same time. Acquire data using TracerDAQ on the computer and the graph paper of the UA-6 UV detector. The UV absorbance at 254 nm of the gradient is detected and ~ 800μL fractions are collected (Fig. 1c).
6. Place fractions on ice before flash-freezing them in liquid nitrogen and store at −80 °C (*see* **Note 15**).
7. Stop gradient fractionation as soon as the chasing solution falls in a 2 mL collecting tube (usually 13 fractions were collected).
8. Save the acquisition on TracerDAQ in .csv format.
9. If you have another gradient to fractionate, use the syringe to remove the chasing solution from the fraction collector. Throw away the pierced polypropylene tube. Load the chasing solution as described in **step 1** and fractionate the next gradient following the same protocol. After the last gradient fractionation, remove all the chasing solution and keep the polypropylene tube in place.
10. Wash the fractionator tubing with 6 × 50 mL of hot, DNase/RNase-free water to remove the sucrose (*see* **Note 16**).
11. Soak all components of the fractionation system (tubing, needle, joints, screws, etc.) in hot, DNase/RNase-free water to wash them.

3.4 RNA Analysis

You can either isolate RNA from individual fractions or from pooled fractions (e.g., non-polysomes (NP), light polysomes (LP) and heavy polysomes (HP)) (*see* **step 9** and **Note 17**).

1. Pre-chill the centrifuge at 4 °C.
2. Add 1 mL TRIzol LS to 400μL of the half fractions or to the inputs (*see* **Note 18**).
3. Homogenize by flipping the tubes.
4. Incubate for 5 min at room temperature.

5. Add 265μL of chloroform.
6. Mix thoroughly by vortexing.
7. Incubate for 10 min at room temperature.
8. Centrifuge at 12,000 × *g* for 15 min at 4 °C.
9. Carefully recover the upper aqueous phase in a fresh safe-lock 1.5 mL tube. Fractions can be pooled at this step in 15 mL falcon tubes. In this case, scale the following volumes accordingly.
10. Add 1μL of glycogen and 665μL of cold isopropanol (*see* **Note 19**).
11. Homogenize by flipping the tubes.
12. Incubate at −20 °C overnight.
13. Centrifuge at 12,000 × *g* for 30 min at 4 °C.
14. Remove the supernatant.
15. Wash twice with 70% ethanol. Centrifuge at 12,000 × *g* for 5 min at 4 °C.
16. Air-dry the pellet for 5–15 min at room temperature.
17. Resuspend the pellet in DNase/RNase-free water (*see* **Note 20**).
18. Transfer resuspended RNAs in fresh 1.5 mL tubes. RNAs can be frozen or used directly in reverse transcription reactions (Fig. 1d).
19. Quantify the amount of RNA in each fraction by measuring the OD at 260 nm. Also measure the OD at 230 nm and 280 nm to assess the purity of your RNAs with respect to contaminants, i.e., phenol and proteins, respectively (*see* **Note 21**).

3.5 Protein Analysis

1. Add 1 mL of 100% isopropanol to each 400μL half fraction. For the fractions with highest sucrose density, sucrose must be diluted 1:2 and divided in two separate tubes (200μL fraction + 200μL H_2O + 1 mL 100% isopropanol per tube) (*see* **Note 22**).
2. Mix thoroughly by vortexing for 30 s and incubate at −80 °C overnight.
3. Vortex.
4. Centrifuge at 16,000 × *g* for 20 min at 4 °C.
5. Remove the supernatant. Be careful to not aspirate the white RNA-containing pellets.
6. Wash with 1 mL of 100% ethanol.
7. Vortex.
8. Centrifuge at 16,000 × *g* for 10 min at 4 °C.

9. Remove the supernatant.
10. Centrifuge at 16,000 × *g* for 1 min at 4 °C.
11. Carefully remove the supernatant completely.
12. Air-dry the pellet for 10–30 min at room temperature.
13. For each fraction, resuspend the pellet in 40μL 1× Laemmli buffer.
14. Vortex.
15. If the pellet is difficult to resuspend, heat 10 min at 95 °C.
16. Proteins can be directly loaded on a SDS-PAGE gel (Fig. 1e) or stored at −20 °C.

4 Notes

1. Since translation efficiency is directly correlated to cell proliferation and density, it is recommended to (1) seed the cells 2 days before lysis and recover them in an exponential growth phase, and (2) not exceed 80% confluency the day of the recovery. Cell density and culture conditions are cell line/treatment dependent and need to be optimized (e.g., for HEK293T cells two 15-cm Petri dishes are seeded with 3×10^6 cells; for HCT116 cells two 15-cm Petri dishes are seeded with 4×10^6 cells and for U251 and U87 cells six 15-cm Petri dishes are seeded with 1.25×10^6 cells).
2. CHX is a translation elongation inhibitor that binds the E-site of the ribosome and impedes tRNA translocation. CHX treatment stalls translating ribosomes and prevents ribosome runoff resulting in a "freezing picture" of the translational status. Since the effect of CHX is reversible, it is advised to include this inhibitor during all the fractionation procedure. Unlike CHX, emetine irreversibly inhibits protein synthesis, making this elongation inhibitor an interesting alternative to CHX. It should be used for 15 min at 37 °C in 20μg/mL concentration.
3. In order to demonstrate that the transcript/protein repartition toward the sucrose gradient mirrors their association with translating polysomes, their responsiveness to treatments affecting polysomal integrity should be tested. This control is often carried out with puromycin, a structural analog of aminoacyl tRNAs that is incorporated into the nascent polypeptide chain, leading to the termination of mRNA translation and dissociation of elongating ribosomes. To this end, before lysate preparation, cells should be incubated with 100μg/mL puromycin (Sigma, #P8833) for 1 h at 37 °C.
4. Phosphatase inhibitors should be added to hypotonic lysis buffer when studying the polysomal repartition of phosphorylated proteins.

5. Extraction of RNA/protein from the input gives information on their steady state levels in the cytosol.
6. Since RNAs and proteins absorb UV light, the $OD_{260\ nm}$ measurement gives an estimation of the amount of material in the cell lysate that could be used to standardize the quantities to be loaded on the gradient between each condition. Hypotonic buffer containing protease inhibitor cocktail and RNase inhibitor should be used as blank.
7. It is important to load on the gradient a minimum of 10 $OD_{260\ nm}$ of cell lysate, otherwise the recovery yields might not be sufficient.
8. Pouring the 5% sucrose solution prior to the 50% sucrose solution minimizes interface disruption.
9. Between two gradients, the needle used for pouring the 50% sucrose solution should be cleaned with paper towel to avoid sucrose gradient solution mixing.
10. Since floating bubbles will disrupt gradient linearity, be careful not to create bubbles against the tube wall when loading each concentrated gradient solutions (these bubbles could be removed using a needle or a pipet tip) or not to trap bubbles when sealing the tubes with caps (a hole in the caps impedes the formation of these bubbles). Generating gradients that are as similar as possible is determinant for the quality and reproducibility of the fractionation.
11. Be cautious when manipulating the sucrose gradients (e.g., when removing 450μL from the top or loading the cell lysate) not to disrupt them. Placing the pipette tips close to the surface and against the tube wall is recommended.
12. Before running ultracentrifugation, ensure that the buckets containing gradient tubes are correctly equilibrated with a maximal difference of 0.01 g in weight. Hypotonic lysis buffer should be added with caution (*see* **Note 11**) on top of the gradient to adjust tube weights.
13. The acceleration and braking parameters of the ultracentrifuge can be set to "high" to save time as they do not disturb the gradients.
14. After ultracentrifugation and while waiting for fractionation, rotor buckets are placed in a pre-formed hole in the ice very carefully to avoid gradients from being altered. Forceps could be used to carefully remove the polypropylene tubes.
15. If 10–20 $OD_{260\ nm}$ of cell lysate were loaded on top of the gradient, fractions could be split so that RNAs are extracted from one half of the fraction and the proteins from the other half.

16. To save time, the pump flow could be set at 30 mL/min during washing steps.
17. For genome-wide studies, the RNA Clean & Concentrator™ Kit from Zymo can be used to achieve the required high quality of extracted RNA (RNA Integrity Number (RIN) >9). The manufacturer's instruction should be followed with a scale-up of the volumes and an accurate homogenization of the sucrose containing fractions in the RNA binding buffer during the first steps of the procedure.
18. TRIzol LS designed for processing liquid samples and containing guanidium isothiocyanate for RNase/DNase inhibition should be preferred over Phenol or Trizol reagents. It is recommended to add TRIzol to frozen samples or as soon they are thawed.
19. Addition of the RNA carrier glycogen is optional but recommended since it helps to better visualize RNA pellets after precipitation.
20. For an efficient RNA resuspension, incubation at 55–60 °C for 10–15 min is recommended.
21. Ensure that the quality of the extracted RNA fulfills the requirements for RT-qPCR or genome-wide analysis by assessing (1) RNA integrity using agarose gel electrophoresis and (2) RNA purity by estimating the 260/230 and 260/280 ratios (ideally, they should be >1.8) as well as the RIN (ideally, it should be >8). When extracting RNA from each fraction using Trizol LS, low RNA recovery yields might impact the RNA quality, giving a 260/230 ratio below 1.7. A second round of overnight precipitation, using sodium acetate 3 M, pH 4.0 and isopropanol, considerably improves RNA quality.
22. Since high concentrations of sucrose impedes protein precipitation, it is recommended to split the last three fractions in half and dilute them 1:2 in DNase/RNase-free water before adding isopropanol. For the same reason, ensure to accurately homogenize isopropanol with the sucrose containing fractions with vigorous vortexing.

References

1. Ingolia NT (2016) Ribosome footprint profiling of translation throughout the genome. Cell 165(1):22–33. https://doi.org/10.1016/j.cell.2016.02.066
2. Cammas A, Dubrac A, Morel B, Lamaa A, Touriol C, Teulade-Fichou MP, Prats H, Millevoi S (2015) Stabilization of the G-quadruplex at the VEGF IRES represses cap-independent translation. RNA Biol 12 (3):320–329. https://doi.org/10.1080/15476286.2015.1017236
3. Cammas A, Lacroix-Triki M, Pierredon S, Le Bras M, Iacovoni JS, Teulade-Fichou MP, Favre G, Roche H, Filleron T, Millevoi S, Vagner S (2016) hnRNP A1-mediated translational regulation of the G quadruplex-containing RON receptor tyrosine kinase mRNA linked to tumor progression.

Oncotarget 7(13):16793–16805. https://doi.org/10.18632/oncotarget.7589

4. Bertorello J, Sesen J, Gilhodes J, Evrard S, Courtade-Saïdi M, Augustus M, Uro-Coste E, Toulas C, Cohen-Jonathan Moyal E, Seva C, Dassi E, Cammas A, Skuli N, Millevoi S (2020) Translation reprogramming by eIF3 linked to glioblastoma. Nucleic Acids Res Cancer 2:zcaa020
5. Herviou P, Le Bras M, Dumas L, Hieblot C, Gilhodes J, Cioci G, Hugnot JP, Ameadan A, Guillonneau F, Dassi E, Cammas A, Millevoi S (2020) hnRNP H/F drive RNA G-quadruplex-mediated translation linked to genomic instability and therapy resistance in glioblastoma. Nat Commun 11(1):2661. https://doi.org/10.1038/s41467-020-16168-x
6. Blair JD, Hockemeyer D, Doudna JA, Bateup HS, Floor SN (2017) Widespread translational remodeling during human neuronal differentiation. Cell Rep 21(7):2005–2016. https://doi.org/10.1016/j.celrep.2017.10.095
7. Spangenberg L, Shigunov P, Abud AP, Cofre AR, Stimamiglio MA, Kuligovski C, Zych J, Schittini AV, Costa AD, Rebelatto CK, Brofman PR, Goldenberg S, Correa A, Naya H, Dallagiovanna B (2013) Polysome profiling shows extensive posttranscriptional regulation during human adipocyte stem cell differentiation into adipocytes. Stem Cell Res 11(2):902–912. https://doi.org/10.1016/j.scr.2013.06.002
8. Cerezo M, Guemiri R, Druillennec S, Girault I, Malka-Mahieu H, Shen S, Allard D, Martineau S, Welsch C, Agoussi S, Estrada C, Adam J, Libenciuc C, Routier E, Roy S, Desaubry L, Eggermont AM, Sonenberg N, Scoazec JY, Eychene A, Vagner S, Robert C (2018) Translational control of tumor immune escape via the eIF4F-STAT1-PD-L1 axis in melanoma. Nat Med 24(12):1877–1886. https://doi.org/10.1038/s41591-018-0217-1
9. Johannes G, Carter MS, Eisen MB, Brown PO, Sarnow P (1999) Identification of eukaryotic mRNAs that are translated at reduced cap binding complex eIF4F concentrations using a cDNA microarray. Proc Natl Acad Sci U S A 96(23):13118–13123. https://doi.org/10.1073/pnas.96.23.13118
10. Qin X, Ahn S, Speed TP, Rubin GM (2007) Global analyses of mRNA translational control during early Drosophila embryogenesis. Genome Biol 8(4):R63. https://doi.org/10.1186/gb-2007-8-4-r63
11. Chasse H, Boulben S, Costache V, Cormier P, Morales J (2017) Analysis of translation using polysome profiling. Nucleic Acids Res 45(3):e15. https://doi.org/10.1093/nar/gkw907
12. Powley IR, Kondrashov A, Young LA, Dobbyn HC, Hill K, Cannell IG, Stoneley M, Kong YW, Cotes JA, Smith GC, Wek R, Hayes C, Gant TW, Spriggs KA, Bushell M, Willis AE (2009) Translational reprogramming following UVB irradiation is mediated by DNA-PKcs and allows selective recruitment to the polysomes of mRNAs encoding DNA repair enzymes. Genes Dev 23(10):1207–1220. https://doi.org/10.1101/gad.516509
13. Bastide A, Peretti D, Knight JR, Grosso S, Spriggs RV, Pichon X, Sbarrato T, Roobol A, Roobol J, Vito D, Bushell M, von der Haar T, Smales CM, Mallucci GR, Willis AE (2017) RTN3 is a novel cold-induced protein and mediates neuroprotective effects of RBM3. Curr Biol 27(5):638–650. https://doi.org/10.1016/j.cub.2017.01.047
14. Hulea L, Gravel SP, Morita M, Cargnello M, Uchenunu O, Im YK, Lehuede C, Ma EH, Leibovitch M, McLaughlan S, Blouin MJ, Parisotto M, Papavasiliou V, Lavoie C, Larsson O, Ohh M, Ferreira T, Greenwood C, Bridon G, Avizonis D, Ferbeyre G, Siegel P, Jones RG, Muller W, Ursini-Siegel J, St-Pierre J, Pollak M, Topisirovic I (2018) Translational and HIF-1alpha-dependent metabolic reprogramming underpin metabolic plasticity and responses to kinase inhibitors and biguanides. Cell Metab 28(6):817–832.e8. https://doi.org/10.1016/j.cmet.2018.09.001
15. Larsson O, Morita M, Topisirovic I, Alain T, Blouin MJ, Pollak M, Sonenberg N (2012) Distinct perturbation of the translatome by the antidiabetic drug metformin. Proc Natl Acad Sci U S A 109(23):8977–8982. https://doi.org/10.1073/pnas.1201689109
16. Muller D, Shin S, Goullet de Rugy T, Samain R, Baer R, Strehaiano M, Masvidal-Sanz L, Guillermet-Guibert J, Jean C, Tsukumo Y, Sonenberg N, Marion F, Guilbaud N, Hoffmann JS, Larsson O, Bousquet C, Pyronnet S, Martineau Y (2019) eIF4A inhibition circumvents uncontrolled DNA replication mediated by 4E-BP1 loss in pancreatic cancer. JCI Insight 4(21):e121951. https://doi.org/10.1172/jci.insight.121951

Chapter 5

Exploring Ribosome-Positioning on Translating Transcripts with Ribosome Profiling

Alexander L. Cope, Sangeevan Vellappan, John S. Favate, Kyle S. Skalenko, Srujana S. Yadavalli, and Premal Shah

Abstract

The emergence of ribosome profiling as a tool for measuring the translatome has provided researchers with valuable insights into the post-transcriptional regulation of gene expression. Despite the biological insights and technical improvements made since the technique was initially described by Ingolia et al. (Science 324 (5924):218–223, 2009), ribosome profiling measurements and subsequent data analysis remain challenging. Here, we describe our lab's protocol for performing ribosome profiling in bacteria, yeast, and mammalian cells. This protocol has integrated elements from three published ribosome profiling methods. In addition, we describe a tool called RiboViz (Carja et al., BMC Bioinformatics 18:461, 2017) (https://github.com/riboviz/riboviz) for the analysis and visualization of ribosome profiling data. Given raw sequencing reads and transcriptome information (e.g., FASTA, GFF) for a species, RiboViz performs the necessary pre-processing and mapping of the raw sequencing reads. RiboViz also provides the user with various quality control visualizations.

Key words Ribosome profiling, Ribo-seq, RNA, Translation initiation, Antibiotic inhibitors, Footprinting

1 Introduction

The development of ribosome profiling (or Ribo-Seq) has allowed researchers to measure what is often termed as the translatome [1]. Briefly, ribosome profiling combines high-throughput sequencing with ribosome (or nuclease) footprinting, an older technique that assigns ribosomes to specific positions within an mRNA [2–4]. This allows researchers to identify and quantify the number of ribosomes actively translating an mRNA, reflecting the protein production level of a gene. Ribosome profiling has proven to be a powerful complement to other high-throughput omics

Alexander L. Cope and Sangeevan Vellappan contributed equally.

Erik Dassi (ed.), *Post-Transcriptional Gene Regulation*, Methods in Molecular Biology, vol. 2404,
https://doi.org/10.1007/978-1-0716-1851-6_5,

measurements. Ribosome profiling is often used in conjunction with RNA-seq (transcriptome) to measure translation efficiency, allowing researchers to estimate how changes in conditions can impact post-transcriptional regulation. Ribosome profiling has also been used to identify novel open-reading frames (ORFs), particularly small ORFs (<100 nucleotides), that are often missed by ORF predictors when annotating genomes [5]. Ribosome profiling has been used in conjunction with mass spectrometry-based proteomics measurements to improve protein identification efficiency [6].

Besides providing gene-specific estimates of expression and translation efficiency, ribosome profiling has the advantage that it provides the location of actively translating ribosomes along an mRNA transcript. Although ribosome pauses are often deleterious to protein production, empirical work has consistently demonstrated that regions of slow translation and ribosomal pausing can be functionally important in some instances [7–9]. Ribosome profiling allows for the identification of such translational pauses. Combining translational pause estimates with other sequence features (e.g., codon usage, downstream mRNA secondary structure) through mathematical models can be used to determine the biological causes of translational pausing.

A standard ribosome profiling measurement is essentially four key steps: (1) translation inhibition, (2) digestion and purification of mRNA transcripts, (3) creation of sequencing libraries for ribosome-protected fragments (RPFs), and (4) sequencing and data analysis (Fig. 1). The translation inhibition step has proven crucial for appropriate interpretation during data analysis. Early ribosome profiling experiments relied upon the antibiotic cycloheximide (CHX) to stall translation. CHX does not stall translation initiation, resulting in the buildup of ribosomes early in the transcript, resulting in a 5′-ramp [10–12]. Given that CHX binding to the ribosome is both non-instantaneous and reversible, the kinetics of CHX binding and dissociation presumably allow newly initiated ribosomes to translocate beyond the start codon. Another possible effect of CHX treatment is that ribosomes might preferentially arrest at specific codons that do not necessarily correspond to codons that are more abundantly occupied by ribosomes in untreated cells. This may explain why ribosome profiling experiments using cycloheximide often failed to find a correlation between codon-specific tRNA abundances and codon-specific ribosome densities. Such a correlation was observed in ribosome profiling measurements using liquid nitrogen to freeze cells, halting translation rapidly. The use of CHX appears only to be a concern if codon-level information is relevant to the investigation. Gene-level analysis (e.g., differential analysis of expression, translation efficiency estimation, etc.) should not be affected by the use of CHX.

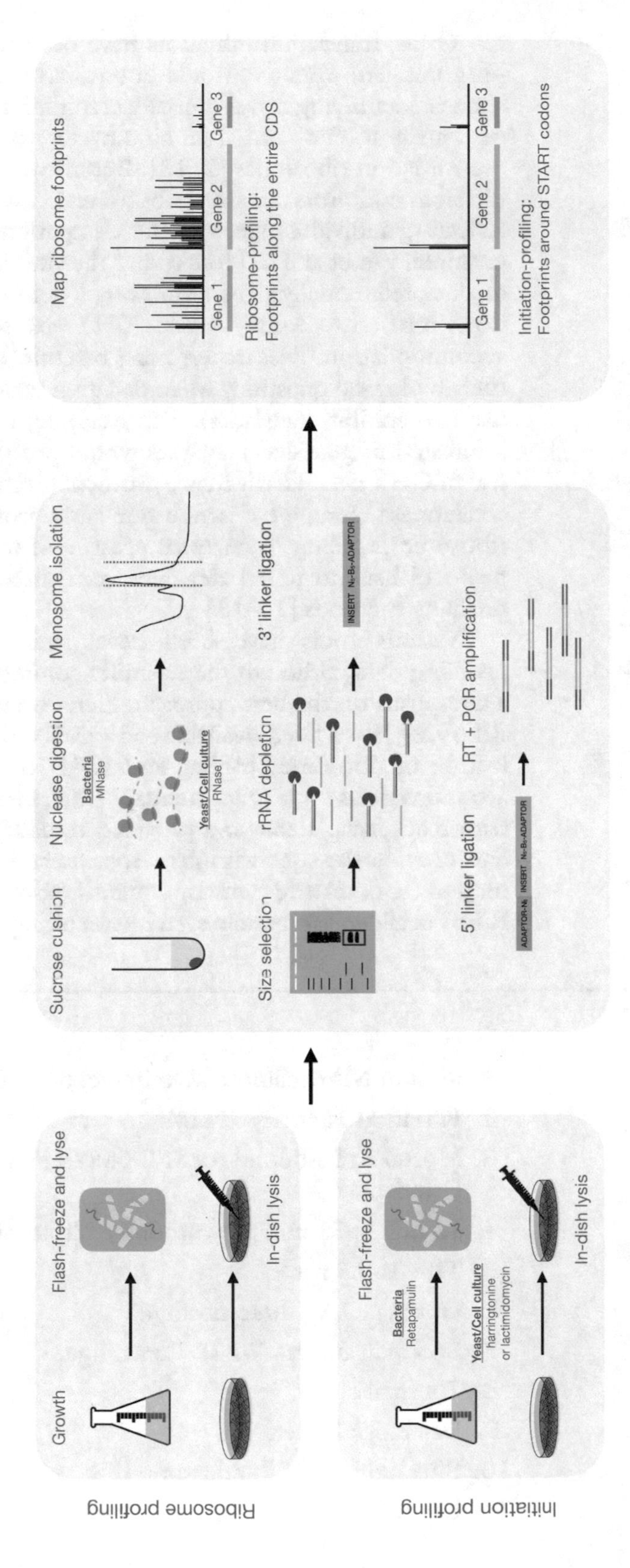

Fig. 1 Schematic of ribosome- and initiation profiling to study ribosome-positioning on mRNA and dynamics of protein synthesis

Other translation inhibitors have been used for more specific purposes. For example, studies interested in translation initiation, such as identifying novel translation initiation start sites, make use of harringtonine and lactimidomycin to preferentially pause pre-initiation ribosomes [2, 13]. Recent work has also shown that different inhibitors can arrest ribosomes in different conformations, reflecting individual steps in the elongation cycle [14, 15]. For example, Wu et al. [15] found that the translation inhibitor anisomycin preferentially pauses ribosomes before tRNA accommodation into the A-site, while CHX pauses ribosomes post-accommodation. Researchers must carefully consider the needs of their biological questions when deciding which method should be used to inhibit translation. For example, a researcher primarily interested in gene-level analyses would probably benefit from the use of CHX over a flash-freeze protocol given the latter's technical challenges. Here, we describe our lab's protocol for performing ribosome profiling in bacteria, yeast, and mammalian cells. This protocol has integrated elements from three published ribosome profiling methods [16–18]

Various tools have been developed to analyze ribosome profiling data, although the scientific community has yet to reach a consensus on the best approach. Here, we will outline the use of RiboViz [19], a freely available software for the analysis and visualization of ribosome profiling and RNA-seq data. RiboViz wraps around various tools (e.g., cutadapt [20], hisat2 [21]) to filter and trim reads, map them to a provided transcriptome (with UTRs), and provide the user with gene-specific and codon-specific information (e.g., relative transcript quantifications). Recent updates to RiboViz allow it to handle even hierarchically multiplexed samples.

2 Materials

2.1 Library Preparation Components

1. 90 mm Nitrocellulose Membrane, pore: 0.45 μm.
2. Filtration assembly 90MM.
3. Mortar & Pestle and/or SPEX 6875 Freezer/Mill® Cryogenic Grinder.
4. 13 mm × 51 mm polycarbonate ultracentrifuge tube.
5. TLA 100.3 rotor.
6. Optima TLX Ultracentrifuge.
7. Beckman Allegra X-22R Centrifuge.
8. Thermal cycler.
9. Dry block heater.
10. Blue light Transilluminator.
11. LB Broth.

12. RNase-free water.
13. 1 M Tris*Cl pH 8, RNase-free.
14. 5 M NaCl, RNase-free.
15. 1 M MgCl2, RNase-free.
16. Chloramphenicol.
17. Triton X-100, molecular biology grade.
18. SUPERase*In, 20 U/μl.
19. Turbo DNase, 2 U/μl.
20. Sucrose, molecular biology grade.
21. Micrococcal Nuclease.
22. Direct-zol RNA MiniPrep.
23. TRIzol.
24. Ethanol, molecular biology grade.
25. Isopropanol, molecular biology grade.
26. GlycoBlue, 15 mg/ml.
27. 3 M NaOAc pH 5.5, RNase-free.
28. 0.5 M EDTA, RNase-free.
29. EGTA.
30. Bromophenol blue.
31. 5 M NaCl, RNase-free.
32. Ficoll 400, BioXtra for molecular biology.
33. Denaturing 15% polyacrylamide TBE-urea gel, 12 wells.
34. Novex 10% TBE-Urea Gels, 10 well.
35. O'RangeRuler 10 bp Ladder.
36. 10× TBE, RNase-free.
37. 10,000× SYBR Gold.
38. RNase-Free Disposable Pellet Pestles.
39. Spin-X Centrifuge Tube Filters.
40. T4 polynucleotide kinase.
41. T4 RNA Ligase 2, truncated K227Q supplied with PEG 8000 50% w/v and 10× T4 RNA ligase buffer.
42. 5′ DNA Adenylation Kit.
43. Yeast 5′-deadenylase.
44. RecJ exonuclease.
45. Oligo Clean & Concentrator.
46. Ribo-zero Plus rRNA Depletion.
47. 6-Tube Magnetic Separation Rack.

48. T4 RNA Ligase 1.
49. Mth RNA Ligase in 5′ Adenylation Kit supplied with 1 mM ATP and 10× Buffer.
50. ProtoScript II.
51. SR RT Primer.
52. Phusion High-Fidelity DNA Polymerase.
53. dNTPS.
54. Non-denaturing 8% polyacrylamide TBE gel, 12 wells.
55. Concentrated HCl.
56. Qubit RNA BR Assay Kit.
57. Qubit RNA HS Assay Kit.
58. 15 ml and 50 ml conical tubes.
59. PCR tubes.
60. 1.5 ml Non-stick RNase, DNase-free microfuge tubes.
61. 10 μl, 20 μl, 200 μl, 1000 μl RNase, DNase-free filter tips.
62. Gel Loading tips.
63. Corning® Costar® Spin-X® centrifuge tube filters.
64. Custom oligos (*see* Table 1).
65. *E coli* lysis buffer (*see* Table 2): 20 mM TrisHCl pH 8.0, 150 mM $MgCl_2$, 100 mM NH_4Cl, 5 mM $CaCl_2$, 0.4% v/v Triton X-100, 0.1% NP-40 (mild), MilliQ H_2O. *See* **Note 1**.
66. Yeast and mammalian cells polysome buffer (*see* Table 3): 20 mM Tris pH 7.4, 150 mM NaCl, 150 mM $MgCl_2$, MilliQ H_2O.
67. Yeast and mammalian cells lysis buffer (*see* Table 4): Yeast and mammalian cells polysome buffer, 1% v/v Triton X-100.
68. Sucrose cushion recipe (*see* Table 5): 20 mM TrisHCl pH 8.0, 10 mM $MgCl_2$, 500 mM NH_4Cl, 5 mM $CaCl_2$, 1 mM DTT, 1 M sucrose, 20 U/ml SUPERase In.
69. Resuspension buffer for *E. coli* (*see* Table 6): 20 mM TrisHCl pH 8.0, 10 mM $MgCl_2$, 100 mM NH_4Cl, 5 mM $CaCl_2$.
70. Resuspension buffer for yeast and mammalian cells (*see* Table 7): Yeast and mammalian cells polysome buffer, 20 mM Tris pH 7.4, 150 mM NaCl, 50 mM $MgCl_2$, MilliQ H_2O.
71. 2× denaturing sample loading buffer (*see* Table 8): 95% v/v deionized formamide, 0.5 M EDTA, 20% SDS, bromophenol blue.
72. RNA gel extraction buffer (*see* Table 9): 300 mM NaOAc pH 5.5, 1 mM EDTA, 0.25% w/v SDS (0.00866 M).
73. DNA gel extraction buffer (*see* Table 10): 300 mM NaCl, 1 mM EDTA, 10 mM Tris pH 8.

Table 1
Custom oligos (*See* Note 2)

Oligo description	Sequence	Type	Barcode
Size control 15 nt oligoribonucleotide	5′-AUGUACACGGAGUCG	RNA	N/A
Size control 45 nt oligoribon ucleotide	5′-AUGUACACGGAGUCGA CCCGCAACGCGAUGUACA CGGAGUCGACC	RNA	N/A
5′ Adaptor	5′-GUUCAGAGUUCUACAGU CCGACGAUCNNNN	RNA	N/A
3′ Linker	5′-Phos-NNNNNJJJJJAGATCGG AAGAGCACACGTCTGAA-3′ddC	DNA	"JJJJJ" represents barcodes. The following barcodes can be used—ATCGT, AGCTA, CGTAA, CTAGA, GATCA, GCATA, TAGAC, TCTAG, ACTGA, CAGTA, GTACT, TGCAT
SR RT Primer for Illumina	5′-AGACGTGTGCTCTTCC GATCT-3′	DNA	N/A
SR Primer for Illumina	5′-AATGATACGGCGACC ACCGAGATCTACACG TTCAGAGTTCTACAGT CCG*A-3′	DNA	N/A
Index Primers for Illumina	5′-CAAGCAGAAGACGGCAT ACGAGATJJJJJJGTGACTG GAGTTCAGACGTGTGC TCTTCCGATC-s-T-3′	DNA	"JJJJJJ" represents barcodes. The following barcodes can be used—ATCACG, CGATGT, TTAGGC, TGACCA, ACAGTG, GCCAAT, CAGATC, ACTTGA, GATCAG, TAGCTT, GGCTAC, CTTGTA

Table 2
***E. coli* lysis buffer (*See* Note 1)**

Stock	Amount (1 ml)	Final concentration
1 M Tris–HCl, pH 8.0	20 μl	20 mM
1 M $MgCl_2$	150 μl	150 mM
4 M NH_4Cl	25 μl	100 mM
1 M $CaCl_2$	5 μl	5 mM
Triton X-100	4 μl	0.4% v/v
NP-40	1 μl	0.1% w/v
MilliQ H_2O	Fill to 1 ml	N/A

Table 3
Yeast and mammalian cells polysome buffer (*See* Note 3)

Stock	Amount (1 ml)	Final concentration
1 M Tris–HCl, pH 7.4	20 μl	20 mM
5 M NaCl	30 μl	150 mM
1 M $MgCl_2$	150 μl	150 mM
MilliQ H_2O	Fill to 1 ml	N/A

Table 4
Yeast and mammalian cells lysis buffer

Stock	Amount (1 ml)	Final concentration
1 M Tris–HCl, pH 7.4	20 μl	20 mM
5 M NaCl	30 μl	150 mM
1 M $MgCl_2$	150 μl	150 mM
Cycloheximide (100 mg/ml)	1 μl	100 μg/ml
Triton X-100	10 μl	1% v/v
MilliQ H_2O	Fill to 1 ml	N/A

Table 5
Recipe for sucrose cushion

Stock	Amount (1 ml)	Final concentration
1 M Tris–HCl, pH 8.0	20 μl	20 mM
1 M $MgCl_2$	10 μl	10 mM
4 M NH_4Cl	125 μl	500 mM
1 M $CaCl_2$	5 μl	5 mM
1 M DTT	1 μl	1 mM
Sucrose	0.34 g	1 M
SUPERase In (20 U/μl)	1 μl	20 U/ml
MilliQ H_2O	Fill to 1 ml	N/A

74. 6× non-denaturing sample loading buffer (*see* Table 11): 10 mM Tris pH 8, 1 mM EDTA, 15% w/v Ficoll 400, 0.25% bromophenol blue.

Table 6
Resuspension buffer for *E. coli*

Stock	Amount (1 ml)	Final concentration
1 M Tris–HCl, pH 8.0	20 μl	20 mM
1 M $MgCl_2$	10 μl	10 mM
4 M NH_4Cl	25 μl	100 mM
1 M $CaCl_2$	5 μl	5 mM
MilliQ H_2O	Fill to final volume	N/A

Table 7
Resuspension buffer for Yeast and mammalian cells

Stock	Amount (1 ml)	Final concentration
1 M Tris–HCl, pH 7.4	20 μl	20 mM
5 M NaCl	30 μl	150 mM
1 M $MgCl_2$	150 μl	150 mM
MilliQ H_2O	Fill to final volume	N/A

Table 8
2× Denaturing sample loading buffer

Stock	Amount (10 ml)	Final concentration
95% deionized formamide	9.5 ml	95% v/v
0.5 M EDTA	0.5 ml	25 mM
20% SDS	12.5 μl	0.025%
Bromophenol blue	N/A	~300 μg/ml

Table 9
RNA gel extraction buffer

Stock	Amount (1 ml)	Final concentration
3 M NaOAc, pH 5.5	100 μl	300 mM
0.5 M EDTA, pH 8.0	2 μl	1 mM
10% SDS	25 μl	0.25% w/v
MilliQ H_2O	Fill to 1 ml	N/A

Table 10
DNA gel extraction buffer

Stock	Amount (1 ml)	Final concentration
5 M NaCl	60 μl	300 mM
0.5 M EDTA, pH 8.0	2 μl	1 mM
1 M Tris–HCl, pH 8.0	10 μl	10 mM
MilliQ H_2O	Fill to 1 ml	N/A

Table 11
6× Non-denaturing sample loading buffer

Stock	Amount (1 ml)	Final concentration
1 M Tris–HCl, pH 8.0	10 μl	10 mM
0.5 M EDTA, pH 8.0	2 μl	1 mM
Ficoll 400	0.15 g	15% w/v
Bromophenol blue	N/A	0.25%
MilliQ H_2O	Fill to 1 ml	N/A

3 Methods

3.1 Culture and Pre-treatments

3.1.1 Culture: E. coli

1. Start the overnight bacterial culture at 37 °C, 250 rpm.
2. Transfer 200 μl (1:500) of O/N Culture into 100 ml of the medium, 37 °C, 225 rpm, until OD600 is about 0.4.
3. For initiation profiling experiments—add retapamulin at the final of 0.025 mg/ml for 5 min @37 °C, 225 rpm. Then, harvest the cells by filtering 50 ml cultures at a time, as described in Subheading 3.2.1.
4. For initiation profiling experiment using retapamulin, *see* **Note 4**.

3.1.2 Culture: S. cerevisiae

1. Start the overnight yeast culture at 30 °C, 250 rpm.
2. Transfer 200 μl (1:500) of O/N Culture into 100 ml of the medium, 30 °C, 225 rpm, until OD600 is about 0.4.
3. For initiation profiling experiments, *see* **Note 5**.

3.1.3 Culture: Mammalian Cells

1. Use one 10 cm dish of subconfluent adherent cells.
2. Wash the cells gently with 5 ml PBS and aspirate the PBS thoroughly. Make sure to place the dish on ice while doing this and use ice-cold PBS.
3. For initiation profiling experiments, *see* **Note 5**.

3.2 Cell Harvest by Filtration

3.2.1 *E. coli* and *S. cerevisiae*

1. Prepare a liquid nitrogen bath while cells are growing. Use a needle to punch several holes in the cap of a 50 ml conical tube and place it directly into liquid nitrogen and fill the tube with liquid nitrogen.
2. We use the Kontes 90 mm filtration apparatus with 0.45 μm Whatman membrane to harvest the cells. Pre-wet the filter with distilled water.
3. With vacuum on, pour in 50 ml culture and allow the media to drain.
4. *[Critical]* Use a new and sterilized metal spatula for each sample. Be quick and gentle while scraping the cells along the filter's surface *while* the media is filtering. Plunge the spatula into the conical tube, and using a second spatula, dislodge the pellets into the tube (*see* **Note 6**).
5. Keep pellets from each 50 ml culture separately.
6. *[Pause]* Using a pierced cap, cap the tube and place it upright at −80 °C to allow the liquid nitrogen to evaporate. Make sure to avoid the thawing of the pellets.

3.2.2 Mammalian Cells

See Subheading 3.3.3.

3.3 Cell Lysis

Always prepare the lysis buffer (*see* Table 2 for E. coli/Table 4 for yeast and mammalian cells) immediately prior to use and store on ice. Due to the viscous nature of Triton X-100 and NP-40, avoid any pipetting error while preparing the buffer. Always prepare additional lysis buffer as you may need more for balancing during sucrose cushion, *see* Subheading 3.4, **step 1**.

3.3.1 Cell Lysis Using Freezer Mill: *E. coli* and *S. cerevisiae* (See **Note 7**)

1. Pre-chill mixer mill grinding cylinders and mortar on liquid nitrogen.
2. Drip 1.5 ml of lysis buffer to a mortar filled with liquid nitrogen and allow it to freeze. Transfer the frozen pellets to the grinding cylinder.
3. Transfer frozen culture pellets into the grinding cylinder containing the frozen lysis buffer. Make sure to avoid the thawing of the pellets.
4. Grind using the following cycles: 10 cycles, 10 Hz, 5 min pre-cool, 1 min run, 1 min cool.
5. Transfer the pulverized cells into pre-chilled 50 ml conical tubes.
6. Clarify lysate as per Subheading 3.3.2, **steps 5–8**.

3.3.2 Cell Lysis Using Mortar and Pestle: *E. coli* and *S. cerevisiae*

1. Pre-chill the mortar, pestle, and sterilized metal spatula with liquid nitrogen.

2. Transfer frozen cell paste to mortar and make sure to avoid thawing of the pellets. Drip 1.5 ml lysis buffer to each sample (placed in a mortar filled with liquid nitrogen). Add the lysis buffer slowly to prevent large chunks of frozen droplets. Smaller droplets are easier to grind.
3. Grind frozen cell pellet with combined lysis buffer to a very fine powder. Make sure to avoid the thawing of the pulverized cells.
4. Transfer the pulverized cells into a 50 ml pre-chilled conical tube. If not, proceed with the next steps, store at −80 °C.
5. Thaw the pulverized cells gently at room temperature (it takes approximately 30–40 min to thaw). Mix the lysate gently and place on ice for 10–15 min. We observed that placing on ice for 10–15 min before clarifying the lysate increases lysis efficiency.
6. Centrifuge the tube at 3000 × *g*, at 4 °C for 5 min, and recover the supernatant into a pre-chilled non-stick RNase-free 2 ml microfuge tube.
7. Spin down the lysate at 20,000 × *g* for 10 min at 4 °C.
8. Transfer supernatant to pre-chilled non-stick RNase-free 2 ml Eppendorf tube.
9. *[Pause]* Lysate can be stored by immersion in liquid nitrogen, and stored at −80 °C.

3.3.3 Cell Lysis: Mammalian Cells

1. Aspirate media from the dish of adherent cells and wash the cells gently with PBS.

   [Critical] Make sure to place the dish on ice while doing this and use ice-cold PBS for the wash.
2. Aspirate excess PBS.
3. Drip 400 μl of lysis buffer slowly onto the cells, covering the entire surface of the dish.
4. Swirl the dish and scrape cells to collect the lysates on one side of the dish. Pipette lysates from this pool back towards the top and scrape again to one side of the dish.
5. Pipette the lysate gently to mix the cells, disperse cell clumps and collect them in a pre-chilled non-stick RNase-free 2 ml microfuge tube. Place the microfuge tube on ice for 10 min.
6. Triturate cells ten times through a 26-gauge needle for efficient cell lysis.
7. Follow the steps in Subheading 3.3.2, **steps 7–9**.

3.4 Pellet Ribosome-Associated mRNAs

1. For each sample, create a sucrose cushion of 900 μl. Once the sucrose cushions are cooled, and the samples are thawed, load the samples onto the top of the cushion. Do so using the same apparatus that you use to balance the cushion. Add the sample to the side of the tube so that it gently flows onto the top of the

cushion. You should see a distinct interface. Load up to 1.5 ml of the lysate. After loading, check to make sure everything is still balanced. Use the lysis buffer to balance the tubes.

2. Use a centrifuge with a fixed-angle rotor (Beckman Coulter) at 71,000 rpm and 4 °C for 2.5 h.
3. Remove supernatant from the ribosome pellet and mark the ultracentrifuge tube's outside edge, where the ribosome pellet is glassy and translucent. The pellet may not be visible until after the supernatant is removed.
4. *[Optional]* Rinse each pellet with 150 μl of pre-chilled resuspension buffer without disturbing the pellet. Squirt the resuspension buffer on the tube walls and *not* on the pellet. *[Critical]* If the pellet dislodges even partially, proceed immediately to Subheading 3.4. **step 6**.
5. Resuspend the ribosomal pellet in 150 μl of resuspension buffer.
6. Vortex until the pellet is fully dissolved. It can take 3–5 min to dislodge and dissolve the pellet completely.
7. *[Pause]* The resuspension can be flash-frozen in liquid N_2 and stored at −80 °C.

3.5 Quantification for Nuclease Digestion

1. Measure A260 of a 1:100 dilution in water (1 μl sample + 99 μl water) using a Nanodrop. Use 1 μl Lysis Buffer +99 μl water for the blank. Alternatively, a Denovix or Qubit fluorometer can be used to measure A260 of a 1:10 dilution of the sample using the Qubit RNA BR Assay kit.
2. Roughly 500 μg/12.5 AU of RNA will be used for nuclease digestion (*see* **Note 18**).
3. Once the amount needed for 500 μg/12.5 AU of RNA is calculated, add the resuspension buffer to bring the final volume to 200 μl.
4. *[Pause]* Freeze at −80 °C to store or continue.

3.5.1 Nuclease Digestion: E. coli (See Note 9)

1. Use 500 μg/12.5 AU of the sample in 200 μl that was pelleted for nuclease digestion. Scale the MNase accordingly if the sample is less than 500 μg/12.5 AU.
2. Add 6 μl of SUPERase In to stop other RNAse activity.
3. Add 2 μl of 375 U/μl MNase (total of 750 U).
4. Incubate the samples in a thermomixer at 1400 rpm, 25 °C for 1 h.
5. Add 2 μl of 0.5 M EGTA pH 8 to quench the reaction. This is critical when MNase is used, as MNase is a calcium-dependent enzyme, and EGTA chelates calcium to stop the further reaction.

6. Place the samples on ice while preparing for sucrose gradient. Do not pause or freeze the sample at this point.

3.5.2 Nuclease Digestion: S. cerevisiae and Mammalian Cells

1. Use 30–100 μg of the sample in 200 μl that was pelleted for nuclease digestion.
2. Add 1.5 μl RNase I (10 U/μl by the Epicentre definition) for 30 μg of RNA. Scale the volume of RNAse I accordingly if the sample has more than 30 μg of RNA.
3. Incubate the samples at room temperature for 45 min with gentle agitation.
4. Add 10 μl SUPERase*In RNase inhibitor to stop nuclease digestion and transfer the digestion to ice.

3.6 Sucrose Gradient and Monosome Isolation

While the sucrose cushion step is meant to isolate translating 70S/80S ribosomes, other large particles, including small and large ribosomal subunits, also sediment during this step. This increases the amount of rRNA in the sample; hence it is essential to remove them. Additionally, if the enzymatic digestion is incomplete, the sample will have many disomes and polysomes. To ensure that we select only monosome footprints, we perform a second ultracentrifugation with a linear 10–50% w/v sucrose gradient and recover monosome peaks. We use the Biocomp Instruments Gradient Station to create sucrose gradients and fractionate polysomes, and a Beckman Optima L-90K ultracentrifuge with an SW-41Ti rotor for ultracentrifugation.

3.6.1 Monosome Isolation

1. Layer the sample atop a sucrose gradient. Be careful not to let the sample enter into the gradient.
2. Centrifuge the sample for 2–3 h at 41k RPM (207,000 x g) at 4 °C. For bacterial samples, we have observed better results with a spin time of 3 h. *See* **Note 10**.
3. Remove samples carefully from the buckets. Keep the assembly of gradients and buckets in the ice/4 °C fridge.
4. Monosome peaks are isolated from the gradient by operating the fractionator in manual mode. The spectral profile is monitored in real time, and peaks are split into separate tubes.

3.6.2 Extract RNA from Monosome Fraction

1. Add 600 μl of TRIzol reagent to the monosome fraction (or 1:1 volume—whichever is smaller). Mix well by vortexing and incubate the mixture for 5 min at room temperature. Purify RNA by using the Direct-zol kit (Zymo R2050) (*see* below).
2. Add an equal volume of ethanol (95–100%) to a sample in TRIzol and mix thoroughly.
3. Transfer the mixture into a Zymo-Spin™ IICR Column in a collection tube and centrifuge at 15,000 × *g* for 30 s at room temperature. Transfer the column into a new collection tube and discard the flow-through.

4. Add 400 μl Direct-zol™ RNA PreWash to the column and centrifuge. Discard the flow-through and *repeat this step.*
5. Add 700 μl RNA Wash Buffer to the column and centrifuge at 15,000 × *g* at room temperature for 1 min to ensure complete removal of the wash buffer. Transfer the column carefully into an RNase-free tube.
6. To elute RNA, add 50 μl of room temperature DNase/RNase-free water directly to the column matrix and centrifuge at 15,000 × *g* at room temperature for 1 min (*see* **Note 11**).
7. Precipitate RNA from the elution by adding 38.5 μl water, 1.5 μl GlycoBlue, and 10 μl 3 M NaOAc pH 5.5, followed by ice-cold 150 μl isopropanol.
8. Chill precipitation for 1 h on ice or at −20 °C overnight.
9. Pellet RNA by centrifugation for 30 min at 20,000 × *g*, 4 °C. Carefully pipette all liquid from the tube. Wash three times with 1 ml of 80% cold EtOH and spin down at 20,000 × *g* for 5 min at 4C. Allow it to air-dry for 5–10 min. *See* **Note 12**.
10. Resuspend the RNA in 20 μl 10 mM TrisHCl pH 8.0 and place it on ice.
11. Quantify RNA amount using Denovix fluorometer at 1:10 dilution using Qubit RNA BR Assay kit. You will run approximately 20 μg on the size-selection gel next. Increase the amount of RNA accordingly if needed. Save the remaining RNA at −80 °C indefinitely and can be used if there is a need to repeat the experiment.

3.7 Footprint Fragment Purification

3.7.1 Gel Size Selection

1. Use a 15% polyacrylamide TBE-Urea gel to size select 15–45 nt footprint fragments. Pre-run the gel at 200 V for at least 10 min in 1× TBE.
2. *See* Table 8 for the recipe of 2× denaturing sample loading buffer. Store it at 4 °C or make aliquots and store at −20 °C.
3. Prepare a control oligo sample for two lanes with 3 μl lower marker oligo 10 μM, 3 μl upper marker oligo 10 μM, and 6 μl 2× denaturing sample loading buffer. 6 μl will be loaded per lane.
4. Add 20 μg sample RNA and an equal volume of 2× denaturing sample loading buffer. For example, for 10 μl of sample RNA, add 10 μl of 2× denaturing sample loading buffer.
5. Denature samples for 90 s at 80 °C.
6. Flush the wells of the gel with the TBE buffer to remove any debris. Load 2 wells per sample on the polyacrylamide gel with the control oligo sample (mixed upper and lower markers) on either side of the RNA samples. If possible, leave an empty lane between different samples to avoid contamination.

7. Run the gel at 200 V until bromophenol blue reaches 3/4 down the gel (takes approximately 40–50 min). Do not risk overrunning the gel as the 15 nt region runs at the bottom of the gel, and you may end up losing the right size footprint fragments.
8. Stain the gel for 2–3 min with SYBR Gold staining solution (1:10,000 SYBR Gold in 1× TBE). *See* **Note 13**.
9. Visualize and take an image using the blue light illuminator using the fluorescence setting for SYBR Gold.
10. On the illuminator, excise the 15–45 nt region demarcated by the lower and upper size marker oligos from each footprinting sample.
11. Excise the 15 nt and 45 nt markers and treat as a control sample. They will be used as a positive control in ligation and reverse transcription later on.
12. Re-image gel to confirm proper cutting and take an image.
13. While waiting for the gel to run, prepare a 0.5 ml tube (per lane) with three to four holes poked through the bottom using an 18 G needle. The gel cutout will be placed inside this tube. The 0.5 ml tube will then be placed inside a 1.7 ml tube.
14. Spin for 5 min at maximum speed to crush and extrude the gel into the 1.7 ml tube.
15. After spinning, most of the gel would have been extruded into the 1.7 ml tube. However, there may be small debris that sticks to the 0.5 ml tube. Ensure that any remaining gel from the 0.5 ml tube is collected into the 1.7 ml tube using a pipette tip.

3.7.2 Gel Extraction

1. Add 400 μl RNA gel extraction buffer (*see* Table 9) and freeze samples for at least 30 min on dry ice or −80 °C.
2. Thaw the samples on ice and add 2.5 μl of SUPERase-Inhibitor to each sample.
3. Elute RNAs from the gel debris by incubating the samples at 4 °C, 300 rpm, overnight.
4. Briefly, quick spin gel extractions to collect the eluate at the bottom of the tube.
5. Transfer approximately 400 μl of the eluate into a 1.7 ml tube.
6. Transfer remaining gel slurry to the top of the Spin-X Cellulose acetate filter (0.45 μm, *Costar*). Using wide orifice pipette tips makes the transfer easier. Make sure to transfer any remaining gel debris into the Spin-X filter.
7. Spin for 3 min at 20,000 × *g* at room temperature.
8. Transfer the eluate into the 1.7 ml tube. Repeat the spin to collect any remaining eluate from the gel debris.
9. Precipitate RNA by adding 1.5 μl GlycoBlue.

10. Add 500 μl cold isopropanol to each sample and mix gently.
11. Chill precipitation for 1 h on ice or at −20 °C overnight.
12. Recover RNA as described in Subheading 3.6.2, **steps 9** and **10**, except elute RNA in 8 μl RNAse-free water. Use 1 μl for quantification using Denovix.

3.8 Ribosomal RNA Depletion Using RiboZero Plus (Illumina, 20036696) [Optional: See Note 14]

[Critical] Skip this step for 15 nt and 45 nt control oligos.

1. Dilute 100 ng total RNA in 11 μl using nuclease-free ultrapure water in a PCR tube.
2. We use the RiboZero Plus assay according to the manufacturer's instruction to hybridize the probes, deplete rRNA, and remove the probes.
3. We perform the clean-up RNA by using Zymo Oligo clean and concentrator, *see* Subheading 3.8.1.

3.8.1 Clean-up RNA (Using Zymo Oligo Clean and Concentrator)

1. Before starting, check if the DNA Wash Buffer was previously diluted. If not, add 96 ml 100% ethanol (104 ml 95% ethanol) to the 24 ml DNA Wash Buffer concentrate (D4060) to create 1× DNA Wash Buffer.
2. Add MilliQ water to each sample to bring the sample volume up to 50 μl.
3. Add 100 μl or 2× total sample volume of Oligo Binding Buffer to 50 μl sample.
4. Add 400 μl or 8× total sample volume of ethanol (100%), mix gently by pipetting, and transfer the mixture to a provided Zymo-Spin™ Column in a Collection Tube. Make sure to collect any remaining sample into the column.
5. Centrifuge at ≥10,000 × *g* for 30 s.
6. Discard the flow-through.
7. Add 750 μl DNA Wash Buffer to the column. Centrifuge at 15,000 × *g* for 30 s and discard the flow-through.
8. Then centrifuge at maximum speed for 1 min. Transfer the column to a microcentrifuge tube.
9. Add 8 μl water directly to the column matrix and let it sit for 10 min. Then, centrifuge at 12,000 × *g* for 30 s to elute the oligonucleotide.
10. Use 1 μl from each tube to quantify RNA concentration using Denovix. Transfer the columns to a fresh microcentrifuge tube.
11. Add 10 μl water directly to the column matrix, and centrifuge at 12,000 × *g* for 30 s to elute the remaining oligonucleotide. Use 1 μl from each tube to quantify RNA concentration using Denovix. Combine with Tube 1 ONLY if the RNA concentration of Tube 1 is low. Scale subsequent reaction volumes accordingly. Otherwise, keep this as a backup.

3.9 3′ Linker Ligation

3.9.1 Preparation of 20 μM Preadenylated 3′ Linker

1. Combine the following components.

Linker oligonucleotide at 100 μM (*see* Table 1)	1.2 μl
10× 5′ DNA adenylation reaction buffer	2.0 μl
1 mM ATP	2.0 μl
Mth RNA Ligase	2.0 μl
Water	12.8 μl
Total	20 μl

2. Incubate for 1 h at 65 °C, then heat-inactivate the enzyme at 85 °C for 5 min.
3. Purify using Zymo Oligo Clean and Concentrator (Subheading 3.8.1, **steps 1–9**). Elute in 6.0 μl nuclease-free water.

3.9.2 3′ Dephosphorylation

1. Prepare 3′ dephosphorylation reaction using only Tube 1. Create another reaction using the 15 nt + 45 nt control oligos purified from gel size selection. This control will be referred to as insert-length control. *[Critical] do not* add ATP to the reaction, as T4 PNK stoichiometrically transfers 3′-phosphates to itself.

	Volume (μl)
RNA sample	7
T4 PNK Buffer (10×)	1
T4 PNK (10 U/μl)	1
SUPERase In	1

2. Incubate at 37 °C for 1 h.

3.9.3 3′ Ligation with Pre-adenylated Linker

1. In the same tube, perform linker ligation by adding the following components.

[Critical] Be careful when pipetting 50% PEG 8000, it is viscous, and we recommend using wide orifice pipette tips.

	Volume (μl)
50% (w/v) PEG-8000	7
10× T4 RNA ligase buffer	1
Pre-adenylated linker (20 μM)	1
T4 RnI2(tr) K227Q (200 U/μl)	1

2. At this point, in addition to the insert-length control, create a no-insert control by adding the following components.

	Volume (μl)	Remark
5′ adaptor (100 μM, denatured)	2 μl	Incubate at 70 °C for 2 min and place on ice immediately. Use within 30 min
Pre-adenylated linker (20 μM)	2 μl	
50% (w/v) PEG-8000	7 μl	
10× T4 RNA ligase buffer	2 μl	
SUPERase In	2 μl	
T4 RnI2(tr) K227Q (200 U/μl)	2 μl	
Nuclease-free water	3 μl	

3. Incubate at 28 °C for 4 h.

3.9.4 Selective Degradation of Pre-adenylated Linkers

1. Add the following components in the same tube for selective degradation of pre-adenylated linkers.

	Volume (μl)
5′ deadenylase (50 U/μl)	1
Rec J exonuclease (10 U/μl)	1
Total	22

2. Incubate at 30 °C for 45 min.

3.9.5 Pool, Clean, and Concentrate Samples

1. Pool all the samples together, except the controls. If the samples vary significantly in their RNA concentrations at Subheading 3.8.1, **step 10** or at Subheading 3.7.2, **step 12** if rRNA depletion was omitted, adjust the volumes of each sample in the pooled sample to a more uniform distribution. If unsure about pooling, proceed with each sample separately.
2. If the total pooled (or individual) sample volume is less than 50 μl, add additional MilliQ water for a final volume of 50 μl. If the total sample volume is greater than 50 μl, no extra water is needed.
3. Purify using the Oligo Clean & Concentrator (Subheading 3.8.4, **steps 1–9**), except elute in 9 μl nuclease-free water. Adjust volume to 9 μl after elution, if below 9 μl.

3.9.6 Evaluate Samples and Controls

At this point, you should have at least three tubes: one with all the pooled samples (or separate tubes for each sample if skipping the pooling step), one with the insert-length control, and one with the no-insert control.

3.10 5′ Linker Ligation

3.10.1 5′ Phosphorylation

1. Mix the following components, incubate at 37 °C for 30 min.
   [Critical] Skip this step for no-insert control.

	Volume
RNA	10.5 μl
10× PNK buffer	1.5 μl
PNK	1 μl
ATP	2 μl
Total	15 μl

2. Purify using the Oligo Clean and Concentrator (Subheading 3.8.4, **steps 1–9**). Elute in 10 μl nuclease-free water. *See* **Note 15**.

3.10.2 5′ Ligation

[Critical] Skip this step for no-insert control.

1. Incubate 5′ adapter at 70 °C for 2 min and place on ice immediately. Use it within 30 min in the following reaction.
2. Mix the following, incubate at 16 °C for 16 h.

	Volume	Remark
RNA	10 μl	
5′ adaptor (100 μM, denatured)	3 μl	Incubate at 70 °C for 2 min and place on ice immediately. Use within 30 min
10× T4 ligation buffer	3 μl	
ATP (10 mM)	3 μl	
T4 RNA ligase 1	2 μl	
50% (w/v) PEG-8000	7 μl	
SUPERase In	2 μl	
Total	30 μl	

3. Purify using the Oligo Clean & Concentrator (Subheading 3.8.4, **steps 1–9**), except elute in 10 μl nuclease-free water. *See* **Note 16**.

3.11 Hybridization of RT Primer

1. Mix the following:

	Volume
RNA	10 μl
SR RT Primer (10 μM)	2.0 μl
Total	12 μl

2. Hybridize the probe at 65 °C for 10 min. Keep hybridized RNAs on ice immediately until ready to use.

3.12 Reverse Transcription

3.12.1 Reaction Mix

1. Mix the following and incubate at 50 °C for 30 min.

	Volume
RNA (hybridized with RT primer)	12 μl
5× First-Strand Buffer	4 μl
dNTPs (10 mM each)	1 μl
0.1 M DTT	1 μl
SUPERase In (20 U/μl)	1 μl
Protoscript II (200 U/μl)	1 μl
Total	20 μl

3.12.2 RNA Hydrolyzation and Neutralization

1. Hydrolyze the RNA template by adding 2.2 μl 1 M NaOH to each reaction and incubating at 70 °C for 20 min and then cool to room temperature.
2. Add 1.8 μl of 1.2 M HCl to the reaction above to neutralize the reaction.

3.12.3 RT Size-Selection Gel

1. The insert-length control and no-insert control will be used to indicate proper ligation and reverse transcription bands. They will be used as size markers for gel excision.
2. Pre-run a 10% TBE-Urea 1.0 mm gel at 200 V for at least 10 min in 1× TBE.
3. Prepare a ladder: 3.5 μl 10 bp ladder 0.1 μg/μl.
4. Add 24 μl 2× denaturing sample loading buffer to each RT sample (pooled sample, no-insert control, and insert-length control).
5. Follow the steps in Subheading 3.7.1, **steps 5–9**.
6. On the illuminator, excise the 15–45 nt region demarcated by the lower and upper size marker oligos from each footprinting sample.

7. Cut from the top of the upper marker (45 nt) to the bottom of the lower marker (15 nt). Cut a separate band for the no-insert control, which will be below the 15 nt band but above the excess RT-primer band.
8. Re-image gel to confirm proper cutting.
9. Follow **step 15** from Subheading 3.7.1.

3.12.4 RT Gel Extraction

1. Add 500 μl DNA gel extraction buffer (*see* Table 10) and freeze samples for at least 30 min on dry ice or −80 °C.
2. Incubate samples at room temperature (25 °C) on a thermo-mixer, 300 rpm, overnight.
3. Follow the steps in Subheading 3.7.2, **steps 4–12**, except precipitate cDNA by adding 2 μl Glycoblue and then 650 μl cold isopropanol. After precipitation, resuspend cDNA in 20 μl MilliQ water.

3.13 PCR

See **Note 17**. *[Critical]* Perform this for the pooled sample and controls.

3.13.1 Pilot PCR

1. Prepare the reaction mix (4 PCR cycle conditions for 1 sample and 2 controls).

	Volume
5× Phusion HF buffer	26.4 μl
dNTPs (10 mM each)	3 μl
SR Primer (10 μM)	4.8 μl
Index Primer (10 μM)	4.8 μl
Phusion Polymerase	1.8 μl
MilliQ Water	79.2 μl
Total	120

2. In a 0.5 ml tube, combine 40 μl Reaction mix with 4 μl Sample.
3. Divide this into 4 × 10 μl aliquots into 0.5 ml tubes.
4. Pilot PCR protocol is as follows:

Cycle step	Temperature	Time	Cycles
Initial denaturation	98 °C	30 s	1
Denaturation Annealing Extension	98 °C 65 °C 72 °C	10 s 10 s 5 s	15
Hold	4 °C		

[Critical] Each sample will be aliquoted into 4 PCR tubes. Take a tube out at the end of 6, 9, 12, and 15 PCR amplification cycles to determine the right cycle required for proper amplification.

5. *For removing PCR tubes at the end of each cycle condition*, pause the machine at 72 °C, wait for 5 s and quickly remove tubes placing them directly on ice. Then, quickly close the lid and continue the program. Each sample will now have 4 PCR products made from 4 different PCR cycle numbers.

3.13.2 Pilot PCR Gel

1. Pre-run an 8% TBE-Native 1.0 mm gel at 200 V for at least 10 min in 1× TBE.
2. Prepare a ladder sample: 3.5 μl 10 bp ladder 0.1 μg/μl O'RangeRuler.
3. Add 2 μl 6× non-denaturing sample loading buffer (*see* Table 11) to each PCR sample.
4. Follow Subheading 3.7.2., **steps 6–9**.
5. Determine the right number of cycles needed to obtain a good band without any larger molecular weight products. *See* **Note 18**.

3.13.3 Preparative PCR Amplification

1. Based on the pilot PCR, identify the ideal cycle count for each sample.
2. Follow Subheading 3.13.1, **steps 1–4**, but prepare only half of the amount of reaction mix and do not divide the reaction. For the main sample, in a 0.5 ml tube, combine 40 μl Reaction mix with 4 μl sample.
3. For the controls, combine 9 μl reaction mix with 1 μl control samples.
4. Run PCR using the cycle count determined from the pilot PCR.

3.13.4 Preparative PCR Gel

1. Follow the steps from Subheading 3.13.2, **steps 1–4**. Adjust the volume of the 6× non-denaturing sample loading buffer according to the sample volume.
2. On the illuminator, excise the 152–182 bp region demarcated by the lower and upper size marker oligos from each footprinting sample. *See* **Note 19**.
3. Re-image gel to confirm proper cutting.
4. For gel crushing and elution, follow Subheading 3.7.1, **steps 13–15**.

3.13.5 Preparative Gel Extraction

1. Add 500 μl DNA gel extraction buffer (*see* Table 10) and freeze samples for at least 30 min on dry ice or −80 °C.
2. Incubate samples at room temperature (25 °C) on a thermomixer, 300 rpm, overnight.
3. Follow the steps in Subheading 3.7.2, **steps 4–12**, except precipitate cDNA by adding 2 μl Glycoblue and then 650 μl cold isopropanol. After precipitation, resuspend pellets in 21 μl of 10 mM Tris pH 8.
4. Use 1 μl for quantification using Denovix or Qubit.
5. 20 μl of the final library is ready for sequencing.

3.14 Data Analysis

High-throughput sequencing of ribosomal profiling libraries allows for the in-depth interrogation of the translatome. Modern ribosomal profiling libraries can contain tens to hundreds of millions of reads. Often, these libraries are contaminated with high levels of rRNA and tRNA reads. Both in vitro and in silico methods are used to reduce contaminant reads prior to aligning to protein-coding sequences. Ribosome profiling libraries can have complex structures that allow for the multiplexing of samples using barcodes and the deduplication of PCR duplicates using unique molecular identifiers. Ribosome profiling reads contain information on the location of ribosomes, which can be used to investigate biological problems beyond gene expression. Reads can be assigned to specific codons on transcripts, and read abundances can be used to analyze ribosome stalls or other characteristics of translation. Finally, with the standardization of ribosome profiling library preparation protocols, ribosome profiling is applied to increasingly diverse organisms, including yeast [1], humans [22], mice [2], bacteria [16], and Archaea [23]. These diverse organisms often require special considerations during the data analysis step. For example, the method of assigning A-sites to a read varies between eukaryotes and bacteria due to differences in the nucleases used during mRNA digestion [16].

To analyze ribosome profiling data while keeping these details in mind, we utilize the RiboViz pipeline. RiboViz is a scalable, flexible, and automatic command-line tool that is a wrapper of other tools commonly used in RNA-seq analysis such as cutadapt [20], hisat2 [21], and UMI-tools [24], while also utilizing the R and Python programming languages to perform analyses and visualizations. A full description of the tool is available on Github (https://github.com/riboviz/riboviz). Briefly, RiboViz takes raw sequencing data (single samples or multiplexed libraries), removes the sequencing adapters, demultiplexes the reads (if necessary), performs PCR deduplication and removal of contaminant sequences (if necessary), maps reads to a user-provided transcriptome, and performs A-site mapping of reads based on a user-provided file that details A-site displacement. This information, as well as additional files and parameters, are specified in a central YAML file that is referenced in the command-line call to RiboViz.

For each sample, RiboViz outputs files that can be used as the starting points for other downstream analyses. Aligned reads are returned in the BAM format along with their index, allowing them to be directly viewed in genome browsers or used in other applications. Quality control metrics such as three-nucleotide periodicity, read length distributions, and analysis of reading frames are provided in both graphical and flat-file formats. The results of the mapping are also conveniently stored in an H5 file. For each sample, the H5 file summarizes the number of reads of a given length that map to a gene (including UTRs). These H5 files are easily parsable, with genes indexed by the provided gene name, allowing for more specific and specialized analyses and visualizations. Other flat-files and visualizations detail the relationship between TPM values of mRNA abundance and translation rates of genes with their sequence features such as gene length, GC content, UTR length, and RNA folding energies. We also quantify codon-specific ribosome densities and quantify position-specific biases in sequencing datasets. Expression metrics may be used for the basis of differential expression analysis and can be combined with RNA-seq measurements to estimate translation efficiency.

4 Notes

1. If the RNA yield after lysis remains low, a harsher detergent can be used. Replace NP-40 with 0.1% sodium deoxycholate. This can also be useful when working with other bacteria with sturdier cell walls. In *E. coli* MG1655 strain, we did not see a significant difference in RNA yield when either NP-40 or 0.1% sodium deoxycholate was used.
2. When possible, work with gel-purified oligos. In some cases, size selection of oligos might be needed to see clearer bands during library preparation.
3. In recent protocols, the use of cycloheximide has been replaced with a higher salt concentration in the lysis buffer [16]. However, if you still observe ribosomal runoff (lower ribosome densities near the start codon), we recommend adding 100 μg/ml cycloheximide in the yeast polysome and lysis buffer.
4. In wild-type *E. coli* cells, multidrug efflux pump TolC can pump out retapamulin, and the activity of retapamulin is limited to stalling ribosomes at initiation sites. However, the use of retapamulin in *tolC* mutants has been successful in increasing ribosome distribution peaks at the start codon and mapping the translation initiation sites in *E. coli* [25].

5. For initiation profiling experiments in eukaryotic cells, ribosomes can be arrested at the start codons using three different methods.
   (a) Pre-treatment with 2 μg/ml harringtonine in DMSO at 37 °C for 2 min [2].
   (b) Pre-treatment with 50 μM lactimidomycin in DMSO at 37 °C for 30 min [13].
   (c) Cell lysis in the presence of 5 μM lactimidomycin, followed by a puromycin treatment to remove non-initiating ribosomes [26].
6. Be careful not to tear the filter. This may lead to sample loss. Do not let the sample dry while harvesting as it may stress the cells. For 50–100 ml *E. coli* culture, you should be able to do this at least once in about 60 s.
7. Using the freezer mill dramatically increases RNA yield compared to harvesting using mortar and pestle.
8. We have had success with using as little as 30 μg of RNA. Using additional material at this stage enables you to save partial material at several downstream steps, providing redundancy in case any particular step does not work.
9. Micrococcal nuclease (MNase) is used to generate footprints in *E. coli*, and its activity can be regulated by quenching calcium ions. However, MNase has sequence specificity biased towards A or T and may not be suitable for GC-rich bacteria.
10. Although a 2-h spin is sufficient to separate monosomes from bacterial samples, a 3-h spin is recommended as it would allow for better separation between the monosome and bacterial ribosome subunits.
11. From this point onwards, please use rigorous techniques to avoid RNase contamination. This includes the use of gloves, RNase-free reagents and consumables, filtered pipette tips, and regularly ethanol wiped benchtops.
12. Ethanol wash is critical to get good quality RNA without any salt contamination arising from isopropanol precipitation. We recommend at least one ethanol wash after isopropanol precipitation.
13. Staining with SYBR for too long and/or vigorous shaking with SYBR staining solution increases the risk of eluting the RNA and/or primers from the gel. It may cause contamination and/or loss of samples.
14. The RiboZero+ rRNA depletion kit works by RNase H digestion of ribosomal RNA targeted by DNA probes. It has worked well for RNA sequencing studies but has been reported to cause bias in ribosome profiling footprint libraries [27]. As

such, we recommend skipping the rRNA depletion and proceeding directly to linker ligation. If skipping rRNA depletion, increase the sequencing depth per sample by two- to threefold.

15. Removing PNK after 5′ phosphorylation is critical to ensure that PNK does not continue its phosphorylation activity on the 5′ adapter in the following ligation step. Heat inactivation of the enzyme is not recommended as magnesium ions from PNK buffer will degrade RNA at elevated temperature.
16. The purification step is critical after 5′ ligation to ensure complete removal of magnesium ions, which may end up degrading RNA under high temperature during RT primer hybridization.
17. Pilot PCR helps determine the right number of PCR cycles required for proper amplification and identifying the cycle at which over-amplification occurs.
18. Higher molecular weight products appear with an increasing number of PCR cycles. You will likely get a thick band without smears between cycles 6 and 12. If there are any smears at the appropriate number of cycles, run the preparative PCR at one fewer cycle to avoid high heavy bands.
19. Indexing primers are unstable even when placed at 4 °C. As such, make aliquots of the primers and store at −20 °C. While analyzing the gel image, expect to see primer bands near 50 bp (SR primer) and 60–70 bp (Index primer). The absence of either of the primers indicates non-specific amplification of random product, which may sometimes appear at the right size range of the expected library. This is where the insert-length and no-insert controls would help in extracting the right size library.

References

1. Ingolia NT, Ghaemmaghami S, Newman JRS, Weissman JS (2009) Genome-wide analysis in vivo of translation with nucleotide resolution using ribosome profiling. Science 324 (5924):218–223. https://doi.org/10.1126/science.1168978
2. Ingolia NT, Lareau LF, Weissman JS (2011) Ribosome profiling of mouse embryonic stem cells reveals the complexity and dynamics of mammalian proteomes. Cell 147(4):789–802. https://doi.org/10.1016/j.cell.2011.10.002
3. Steitz JA (1969) Polypeptide chain initiation: nucleotide sequences of the three ribosomal binding sites in bacteriophage R17 RNA. Nature 224(5223):957–964. https://doi.org/10.1038/224957a0
4. Wolin SL, Walter P (1988) Ribosome pausing and stacking during translation of a eukaryotic mRNA. EMBO J 7(11):3559–3569. https://doi.org/10.1002/j.1460-2075.1988.tb03233.x
5. Weaver J, Mohammad F, Buskirk AR, Storz G (2019) Identifying small proteins by ribosome profiling with stalled initiation complexes. MBio 10(2):e02819-18. https://doi.org/10.1128/mBio.02819-18
6. Koch A, Gawron D, Steyaert S, Ndah E, Crappé J, De Keulenaer S et al (2014) A proteogenomics approach integrating proteomics and ribosome profiling increases the efficiency of protein identification and enables the discovery of alternative translation start sites. Proteomics 14(23–24):2688–2698. https://doi.org/10.1002/pmic.201400180
7. Fluman N, Navon S, Bibi E, Pilpel Y (2014) mRNA-programmed translation pauses in the

targeting of E. coli membrane proteins. elife 3: e03440. https://doi.org/10.7554/eLife.03440

8. Walsh IM, Bowman MA, Soto Santarriaga IF, Rodriguez A, Clark PL (2020) Synonymous codon substitutions perturb cotranslational protein folding in vivo and impair cell fitness. Proc Natl Acad Sci U S A 117(7):3528–3534. https://doi.org/10.1073/pnas.1907126117
9. Xu Y, Ma P, Shah P, Rokas A, Liu Y, Johnson CH (2013) Non-optimal codon usage is a mechanism to achieve circadian clock conditionality. Nature 495(7439):116–120. https://doi.org/10.1038/nature11942
10. Gerashchenko MV, Gladyshev VN (2014) Translation inhibitors cause abnormalities in ribosome profiling experiments. Nucleic Acids Res 42(17):e134. https://doi.org/10.1093/nar/gku671
11. Hussmann JA, Patchett S, Johnson A, Sawyer S, Press WH (2015) Understanding biases in ribosome profiling experiments reveals signatures of translation dynamics in yeast. PLoS Genet 11(12):e1005732. https://doi.org/10.1371/journal.pgen.1005732
12. Weinberg DE, Shah P, Eichhorn SW, Hussmann JA, Plotkin JB, Bartel DP (2016) Improved ribosome-footprint and mRNA measurements provide insights into dynamics and regulation of yeast translation. Cell Rep 14(7):1787–1799. https://doi.org/10.1016/j.celrep.2016.01.043
13. Lee S, Liu B, Lee S, Huang S-X, Shen B, Qian S-B (2012) Global mapping of translation initiation sites in mammalian cells at single-nucleotide resolution. Proc Natl Acad Sci U S A 109(37):E2424–E2432. https://doi.org/10.1073/pnas.1207846109
14. Lareau LF, Hite DH, Hogan GJ, Brown PO (2014) Distinct stages of the translation elongation cycle revealed by sequencing ribosome-protected mRNA fragments. elife 3:e01257. https://doi.org/10.7554/eLife.01257
15. Wu CC-C, Zinshteyn B, Wehner KA, Green R (2019) High-resolution ribosome profiling defines discrete ribosome elongation states and translational regulation during cellular stress. Mol Cell 73(5):959–970.e5. https://doi.org/10.1016/j.molcel.2018.12.009
16. Mohammad F, Green R, Buskirk AR (2019) A systematically-revised ribosome profiling method for bacteria reveals pauses at single-codon resolution. elife:8. https://doi.org/10.7554/eLife.42591
17. McGlincy NJ, Ingolia NT (2017) Transcriptome-wide measurement of translation by ribosome profiling. Methods 126:112–129. https://doi.org/10.1016/j.ymeth.2017.05.028
18. Subtelny AO, Eichhorn SW, Chen GR, Sive H, Bartel DP (2014) Poly(A)-tail profiling reveals an embryonic switch in translational control. Nature 508(7494):66–71. https://doi.org/10.1038/nature13007
19. Carja O, Xing T, Wallace EWJ, Plotkin JB, Shah P (2017) riboviz: analysis and visualization of ribosome profiling datasets. BMC Bioinformatics 18(1):461. https://doi.org/10.1186/s12859-017-1873-8
20. Martin M (2011) Cutadapt removes adapter sequences from high-throughput sequencing reads. EMBnet J 17(1):10–12. http://journal.embnet.org/index.php/embnetjournal/article/view/200/479
21. Kim D, Paggi JM, Park C, Bennett C, Salzberg SL (2019) Graph-based genome alignment and genotyping with HISAT2 and HISAT-genotype. Nat Biotechnol 37(8):907–915. https://doi.org/10.1038/s41587-019-0201-4
22. Chen J, Brunner A-D, Zachery Cogan J, Nuñez JK, Fields AP, Adamson B et al (2020) Pervasive functional translation of noncanonical human open reading frames. Science 367(6482):1140–1146. https://doi.org/10.1126/science.aay0262
23. Gelsinger DR, Dallon E, Reddy R, Mohammad F, Buskirk AR, DiRuggiero J (2020) Ribosome profiling in archaea reveals leaderless translation, novel translational initiation sites, and ribosome pausing at single codon resolution. Nucleic Acids Res 48(10):5201–5216. https://doi.org/10.1093/nar/gkaa304
24. Smith T, Heger A, Sudbery I (2017) UMI-tools: modeling sequencing errors in Unique Molecular Identifiers to improve quantification accuracy. Genome Res 27(3):491–499. https://doi.org/10.1101/gr.209601.116
25. Meydan S, Marks J, Klepacki D, Sharma V, Baranov PV, Firth AE et al (2019) Retapamulin-assisted ribosome profiling reveals the alternative bacterial proteome. Mol Cell 74(3):481–493.e6. https://doi.org/10.1016/j.molcel.2019.02.017
26. Gao X, Wan J, Liu B, Ma M, Shen B, Qian S-B (2015) Quantitative profiling of initiating ribosomes in vivo. Nat Methods 12(2):147–153. https://doi.org/10.1038/nmeth.3208
27. Zinshteyn B, et al (2020) Nuclease-mediated depletion biases in ribosome footprint profiling libraries. RNA 26(10):1481–1488

Chapter 6

Identification of RNA Binding Partners of CRISPR-Cas Proteins in Prokaryotes Using RIP-Seq

Sahil Sharma and Cynthia M. Sharma

Abstract

CRISPR-Cas systems consist of a complex ribonucleoprotein (RNP) machinery encoded in prokaryotic genomes to confer adaptive immunity against foreign mobile genetic elements. Of these, especially the class 2, Type II CRISPR-Cas9 RNA-guided systems with single protein effector modules have recently received much attention for their application as programmable DNA scissors that can be used for genome editing in eukaryotes. While many studies have concentrated their efforts on improving RNA-mediated DNA targeting with these Type II systems, little is known about the factors that modulate processing or binding of the CRISPR RNA (crRNA) guides and the *trans*-activating tracrRNA to the nuclease protein Cas9, and whether Cas9 can also potentially interact with other endogenous RNAs encoded within the host genome. Here, we describe RIP-seq as a method to globally identify the direct RNA binding partners of CRISPR-Cas RNPs using the Cas9 nuclease as an example. RIP-seq combines co-immunoprecipitation (coIP) of an epitope-tagged Cas9 followed by isolation and deep sequencing analysis of its co-purified bound RNAs. This method can not only be used to study interactions of Cas9 with its known interaction partners, crRNAs and tracrRNA in native systems, but also to reveal potential additional RNA substrates of Cas9. For example, in RIP-seq analysis of Cas9 from the foodborne pathogen *Campylobacter jejuni* (CjeCas9), we recently identified several endogenous RNAs bound to CjeCas9 RNP in a crRNA-dependent manner, leading to the discovery of PAM-independent RNA cleavage activity of CjeCas9 as well as non-canonical crRNAs. RIP-seq can be easily adapted to any other effector RNP of choice from other CRISPR-Cas systems, allowing for the identification of target RNAs. Deciphering novel RNA-protein interactions for CRISPR-Cas proteins within host bacterial genomes will lead to a better understanding of the molecular mechanisms and functions of these systems and enable us to use the in vivo identified interaction rules as design principles for nucleic acid targeting applications, fitted to each nuclease of interest.

Key words CRISPR-Cas, Prokaryotes, *Campylobacter jejuni*, Cas9, RNA-immunoprecipitation, Next-generation sequencing

1 Introduction

Post-transcriptional regulation represents an important layer of gene expression control in uni- and multicellular organisms. Hereby, RNA molecules (rRNA, mRNA, tRNA, and non-coding

Erik Dassi (ed.), *Post-Transcriptional Gene Regulation*, Methods in Molecular Biology, vol. 2404,
https://doi.org/10.1007/978-1-0716-1851-6_6,

RNAs) as well as RNA-binding proteins (RBPs) and ribonucleases (RNases) are central players in mediating and regulating gene expression. Most RNA molecules associate with proteins for exerting this control and co-exist as ribonucleoprotein (RNP) complexes. RNP complexes can execute numerous cellular functions including gene regulation, RNA modifications, mRNA translation, and RNA stability control [1, 2]. Studying these RNPs enables us to understand the role of RBPs and RNases in cellular physiology. Moreover, with the advent of numerous RNA-sequencing (RNA-seq) approaches using deep sequencing technologies it has become possible to study and analyze these RNPs on a transcriptome-wide scale in diverse organisms, including bacteria, and dissect their complex biological tasks [3, 4].

One diverse and widespread group of RNP complexes used by prokaryotes are the CRISPR-Cas (Clustered Regularly Interspaced Short Palindromic Repeats and CRISPR-associated) systems. These RNA-based prokaryotic immune systems are present in approximately half of bacterial and 85% archeal genomes, which have evolved over the course of many years to create an immunological memory, guided by the incorporated nucleic acids, forming the basis of an organism's adaptive immunity [5–7]. Such CRISPR-Cas adaptive immune systems have been extensively studied over the last decade for their role as defense systems in prokaryotes to protect against foreign mobile genetic elements such as phages and plasmids and more recently as tools for enabling precise genome editing in various organisms [8–10]. CRISPR-Cas systems primarily function in the form of active RNPs with Cas nuclease effector complexes, which are guided by CRISPR RNA (crRNA) guides encoded in CRISPR repeat-spacer arrays to target and destroy specific DNA/RNA sequences. CRISPR-Cas systems are divided into two broad classes, with class I systems being composed of multiple Cas proteins, which form the core effector module, whereas class II systems comprise a single, multidomain effector nuclease [11]. Each class of CRISPR-Cas is further divided into three types each, i.e., types I, III, and IV forming class 1, and types II, V, and VI in class 2. Each type further contains subtypes based on variations either in structure or function of CRISPR-Cas genes. CRISPR-Cas immune systems work by incorporating foreign DNA fragments, known as spacers into the host genome between specific repeat regions in the genome—the CRISPR locus. Once the spacers are acquired into these repeats, they are transcribed and processed into individual crRNAs, which interact with Cas proteins and mediate RNA-guided genome defense [12, 13]. Hereby, the sequence of the crRNAs defines the specificity of the adaptive immune response, both for the host and its progeny. This adaptive immune response can be achieved by targeting either DNA or RNA. While RNA targeting is specific to Type III and VI systems, DNA targeting is mediated by Type I, II, and V systems. Type III

CRISPR-Cas systems are the only exception that can target invading nucleic acids at the level of both DNA and RNA.

Of the six types of CRISPR-Cas systems, Cas9 nucleases from class 2, Type II are currently the most widely used for RNA-guided genome editing and various technological applications in bacterial and eukaryotic genomes [8, 14, 15]. This system uses a single Cas9 nuclease for targeting double-stranded DNA, using a crRNA guide and endonuclease activity of Cas9 via its RuvC and HNH nuclease domains [16, 17]. Type II-A, II-B, and II-C systems differ based on the size of their Cas9 protein and the presence of associated additional *cas* genes [18]. In addition to crRNA guides, Type II systems also require the presence of a *trans*-activating crRNA (tracrRNA) [19]. The tracrRNA has base-pairing complementarity to the repeat region of the crRNA guide and mediates processing and maturation of crRNAs via processing by the host factor RNase III. Moreover, tracrRNA is also required for Cas9 function as the mature tracrRNA-crRNA duplex is bound by Cas9, thereby guiding the Cas9:crRNA:tracrRNA RNP to a protospacer in the target DNA flanked by a protospacer-adjacent motif (PAM) [20]. For the purpose of genome editing, a chimeric RNA formed by the fusion of crRNA and tracrRNA called the single-guide RNA (sgRNA) is used, thus simplifying the system further [16].

Type II systems are widespread in pathogenic and commensal bacteria and there is emerging evidence that they could have additional functions beyond genome defense [21]. These unconventional functions are not restricted to Type II systems alone and involve regulation of bacterial virulence, group behavior dynamics, genome remodeling, DNA repair, antisense RNAs, and self-targeting mechanisms in numerous CRISPR-Cas types [22–29]. It is well known that Cas9:crRNA:tracrRNA RNP surveils the cytoplasm and targets DNA from invading genomes, but whether Cas9 has additional RNA partners involved in regulating host gene expression, remains unclear. Recently, the Cas9 homolog in *Francisella novicida* (FnoCas9) has been identified to target its own DNA with the help of a small CRISPR/Cas associated RNA termed scaRNA, leading to RNA-directed transcriptional repression [30]. Moreover, recent studies also indicate that certain members of the Cas9 family are capable of targeting RNA [14]. For example, using RNA co-immunoprecipitation combined with RNA-sequencing (RIP-seq) of epitope-tagged Cas9 from *Campylobacter jejuni* (CjeCas9), we recently uncovered that the CjeCas9 nuclease is capable of binding and cleaving endogenous RNAs in vivo, in addition to binding its canonical crRNA:tracrRNA pairs [31]. This crRNA-dependent targeting of endogenous RNAs is PAM-independent. Also, other Type II Cas9 family members from *Neisseria meningitidis* (NmeCas9) and *Staphylococcus aureus* (SauCas9) were shown to target single-stranded RNAs independently of a PAM in vitro [32, 33]. Moreover, Cas9 from

Streptococcus pyogenes (SpyCas9) can target RNA after addition of single-stranded DNA oligonucleotides with a PAM sequence in vitro and has been applied in human cells for RNA tracking and localization by replacing the active enzyme with catalytically dead SpyCas9 in vivo [34, 35]. These studies showed that even within the Type II system, Cas9 homologs from different species can have a variety of molecular activities. Moreover, while a lot of studies have concentrated on RNA-mediated DNA targeting with Type II Cas9 systems and its tremendous technological applications, very little is known about how these systems regulate activity of the RNA-guided effector nucleases and about their potential interactions with other RNAs encoded within the host genomes. With the exception of CjeCas9 and FnoCas9, most studies have described CRISPR-Cas Type II proteins in vitro or expressed exogenously in eukaryotic cells or other bacteria, which while providing valuable insights also limits the mechanistic understanding of these effector molecules in bacteria natively harboring these systems.

Here, we describe the so-called RIP-seq approach to identify the direct RNA substrates of CRISPR-Cas RNPs in prokaryotes (Fig. 1). RIP-seq combines co-immunoprecipitation (coIP) of an epitope-tagged RBP followed by isolation and deep sequencing analysis of its co-purified bound RNAs. Prior to the development of high-throughput sequencing technologies, co-purified transcripts from coIPs of RBPs were typically analyzed with either high-density microarrays or conventional low-throughput RNomics using direct RNA-sequencing or Sanger sequencing to identify, e.g., the targetome of the RNA chaperone Hfq in *Escherichia coli*, *Listeria monocytogenes*, and *Pseudomonas aeruginosa*, respectively [36–38]. The combination of coIP with RNA-seq for RIP-seq analyses of Hfq then allowed for identifying its small RNA (sRNA) and messenger RNA binding partners in *Salmonella* on a transcriptome-wide scale and with single-nucleotide resolution [39]. RIP-seq has also been successfully adapted and applied to study the regulons of a number of other RNA-binding proteins in bacterial species such as *E. coli*, *C. jejuni*, *Helicobacter pylori*, and *N. meningitidis* [40–47]. Using a similar coIP approach combined with deep sequencing in organisms harboring Type III CRISPR-Cas systems identified mature crRNAs bound to the effector proteins, which shed light into the mechanism governing targeting of both host and foreign RNAs in those systems [23, 48].

Based on RIP-seq to examine the direct RNA binding partners of the Cas9 nuclease of the foodborne pathogen C. *jejuni*, we had identified the abovementioned RNA-targeting activity of CjeCas9 and more recently also discovered non-canonical crRNAs [31, 67]. Hereby, we built on a RIP-seq protocol previously adapted for the use in Epsilonproteobacteria [40–47]. Using a genetically modified strain of *C. jejuni* expressing a 3xFLAG epitope at the C-terminus of CjeCas9 (CjeCas9-3xFLAG) at the

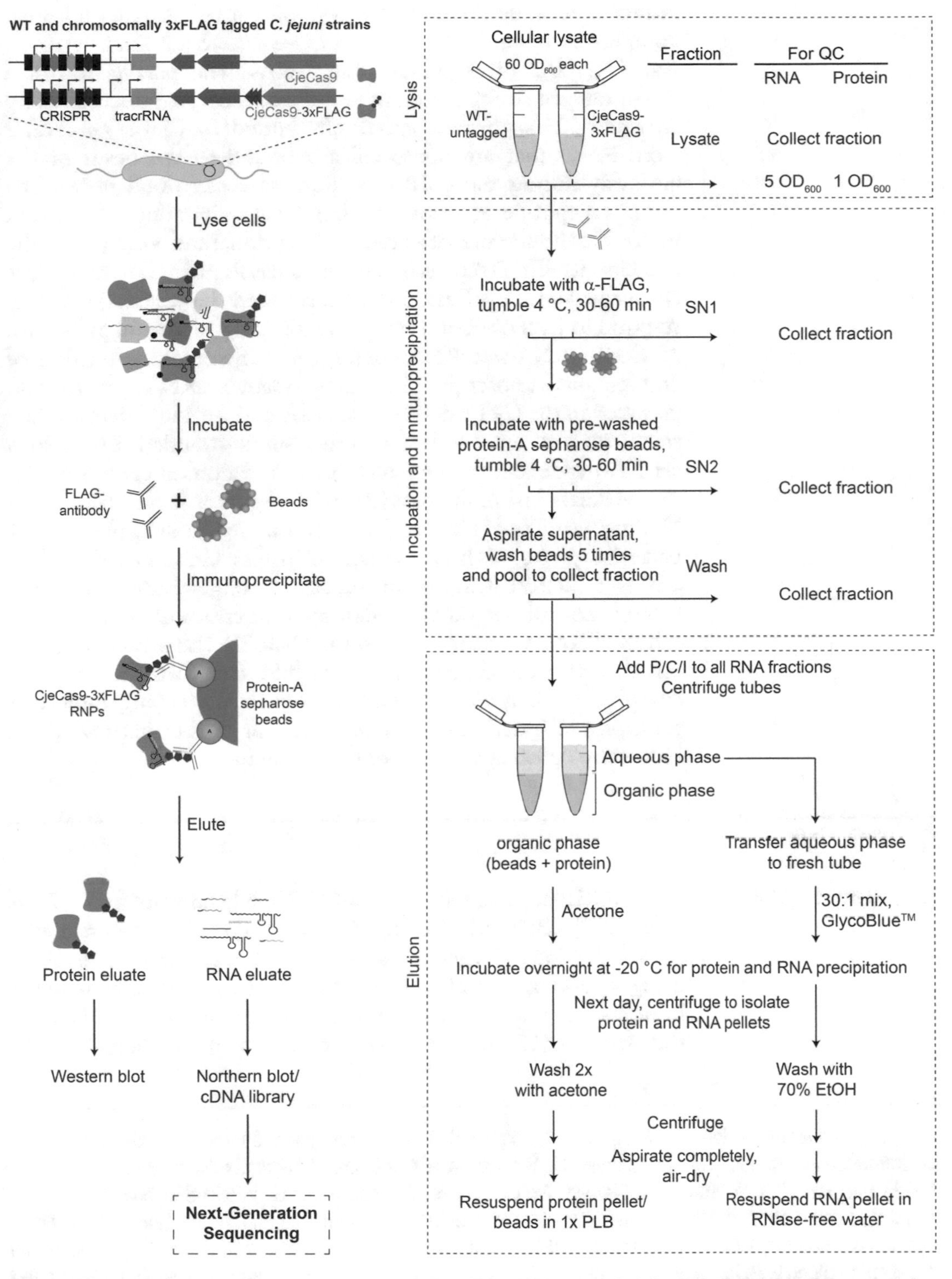

Fig. 1 RIP-seq analysis to identify the direct RNA substrates of Cas9 in *C. jejuni*. (*Left*) Schematic representation of co-immunoprecipitation of 3xFLAG-tagged CjeCas9. (*Right*) Detailed overview of the RIP-seq protocol used to identify CjeCas9-bound RNAs in *C. jejuni* strain NCTC11168. Bacterial lysates prepared from cell pellets corresponding to a cell number of 60 OD_{600} pellets using WT-untagged and CjeCas9-3xFLAG tagged

native locus and an antibody against the FLAG epitope allowed for immunoprecipitation and isolation of native ribonucleoprotein complexes comprised of CjeCas9-3xFLAG bound crRNA: tracrRNA and other endogenous RNAs. The parallel use of a WT-untagged strain as a negative control allows for discriminating enriched RNAs that are specifically bound to CjeCas9-3xFLAG from RNAs that are nonspecifically bound to the beads or the antibody during the coIP procedure as background noise. The use of an epitope tag allows for immunoprecipitating a protein of interest in the absence of a mono−/polyclonal antibody against the protein itself. Deep sequencing analysis of the transcripts co-purified with CjeCas9-3XFLAG revealed that the majority correspond to tracrRNA and crRNAs, indicating successful pull-down of Cas9-crRNA-tracrRNA complexes. Surprisingly, a fraction of endogenous transcripts, including many mRNAs, were also enriched in the CjeCas9-3xFLAG coIP, and we could demonstrate that CjeCas9 can bind and cleave single-stranded RNAs in a crRNA-dependent manner and that non-canonical crRNA guides can be derived from mRNAs [31, 67]. Similar as for studying RNA binding partners of CjeCas9, RIP-seq can be readily applied to any bacterial species of choice (Gram-positive or Gram-negative), harboring a CRISPR-Cas system and capable of genetic manipulation (chromosomally or via plasmids), for transcriptome-wide identification of RNAs interacting with the CRISPR-Cas protein of interest. A better understanding of the different CRISPR-Cas RNP complexes will facilitate deciphering their underlying molecular principles of function and to use them for further improving the biotechnological application of these systems.

2 Materials

2.1 Bacterial Strains

For the RIP-seq analysis, *C. jejuni* NCTC11168 wildtype (WT) as well as a CjeCas9-3xFLAG strain are used. The *C. jejuni* strain with a 3xFLAG-tagged CjeCas9 at the native locus was constructed using a double-stranded DNA construct and double-crossover homologous recombination, as described previously [31]. Hereby, the *cas9* (Cj1523c) gene was chromosomally tagged at its

Fig. 1 (continued) bacterial cells grown to exponential phase are used as the starting material for co-immunoprecipitation. This is followed by incubation with an anti-FLAG antibody, protein A-sepharose beads and washing to enrich for CjeCas9-3xFLAG RNPs. CjeCas9-3XFLAG protein and bound RNAs are separated using phenol/chloroform/isoamyl alcohol, precipitated, and eluted into protein and RNA fractions. Lysate, supernatant 1 (SN1), supernatant 2 (SN2), and wash fractions are collected during the course of the experiment along with the final eluate. All samples are used for quality control before proceeding with DNase I digestion and next-generation sequencing of eluted RNAs

C-terminus by introducing overlap PCR products containing 500 bp homologous ends encompassing a 3xFLAG sequence (DY KDHDGDYKDHDIDYKDDDDK) and a kanamycin resistance marker using electroporation into the *C. jejuni* NCTC11168 WT strain.

2.2 Bacterial Culture and Collection

1. Müller–Hinton agar plates and Brucella broth (BB), both supplemented with 10 μg/mL vancomycin. The agar was further supplemented with marker-selective antibiotics (50 μg/mL kanamycin) when required.
2. Resuspension buffer (Buffer A): 20 mM Tris–HCl, pH 8.0, 150 mM KCl, 1 mM $MgCl_2$ with 1 mM Dithiothreitol (DTT) (added freshly each time before use).
3. 1× Protein loading buffer (1× PLB): 62.5 mM Tris–HCl, pH 6.8, 100 mM DTT, 10% (v/v) Glycerol, 2% (w/v) SDS, 0.01% (w/v) Bromophenol blue.

2.3 RNA Co-immunoprecipitation

1. Buffer A.
2. Lysis buffer (per mL): 967 μL Buffer A (without DTT), 1 μL 1 mM DTT, 10 μL 0.1 M Phenylmethylsulfonylflouride (PMSF), 2 μL Triton X-100, 20 μL DNase I, and 10 μL Superase-In RNase Inhibitor.
3. 5× Protein loading buffer.
4. Eppendorf tubes (2 mL and 1.5 mL).
5. Fast-prep tubes with lysing Matrix B.
6. FastPrep-24^{TM} benchtop Homogenizer (MP Bio).
7. Monoclonal anti-FLAG M2 antibody (1 mg/mL).
8. Protein A-sepharose beads.
9. GlycoBlue™.
10. Roti-Aqua-P/C/I (phenol/chloroform/isoamyl alcohol).
11. RNA precipitation mix (30:1): Ethanol:3M NaOAc, pH 6.5.
12. 70% Ethanol (stored at −20 °C).
13. Acetone (stored at −20 °C).
14. RNase-free water.

2.4 Quality Control and DNase I Treatment

1. Antibodies: Monoclonal anti-FLAG antibody (1:1000 #F1804 Sigma-Aldrich) and anti-GroEL (1:10,000 #G6532 Sigma-Aldrich). Horseradish peroxidase-coupled anti-mouse IgG secondary antibody (1:10,000) and anti-rabbit IgG secondary antibody (1:10,000).
2. SDS-Polyacrylamide gels and electrophoresis equipment and buffers for western blots.

3. Polyacrylamide/Urea gels and electrophoresis equipment and buffers for northern blots.
4. RNA loading buffer II: 95% Formamide, 18 mM Ethylenediaminetetraacetic acid (EDTA), 0.025% w/v each of SDS, Xylene Cyanol and Bromophenol Blue.
5. DNase I.
6. 10× DNase I buffer containing $MgCl_2$.
7. Superase-In RNase Inhibitor.
8. Phase-lock tubes.

3 Methods

3.1 Bacterial Culture and Pellet Collection

1. Inoculate bacterial cultures of *C. jejuni* NCTC11168 WT control and the corresponding CjeCas9-3xFLAG strain in 100 mL (50 mL × 2 flasks) Brucella broth containing 10 mg/mL vancomycin to the selected growth phase, e.g., mid-exponential growth (OD_{600} = 0.5–0.6) at 37 °C under microaerobic (10% CO_2, 5% O_2) conditions as previously described [31] (*see* **Notes 1–4**).
2. Harvest cells by centrifugation at 6000 × *g* at 4 °C for 15 min using a refrigerated benchtop centrifuge. Resuspend cell pellets in 1 mL Buffer A (ice-cold), transfer to 1.5 mL eppendorf tube, and centrifuge at 11,000 × *g*, 4 °C for 3 min using a refrigerated microcentrifuge. Aspirate the supernatant as cleanly as possible and immediately snap-freeze the pellets in liquid nitrogen. Store the pellets at −80 °C (*see* **Note 5**).

3.2 Cell Lysis and Incubation with Antibody/Protein A-Sepharose Beads

1. Thaw frozen pellets (*see* Subheading 3.1, **step 2**) on ice and resuspend in 1 mL Lysis buffer by gentle pipetting. Transfer the 1 mL lysate from each sample onto one fast-prep tube for lysis (*see* **Note 6**). Place tubes on ice and avoid RNase contamination.
2. Perform lysis with FastPrep-24™ benchtop homogenizer (MP Bio) at settings of *4 M/s, 15 s* and clear the lysate in the fast-prep tube by centrifugation at 15,000 × *g*, 4 °C for 10 min to remove cellular debris generated during mechanical lysis (*see* **Note 7**).
3. Transfer the cleared lysate into a pre-chilled 2 mL eppendorf tube without disturbing the debris at the bottom and place the tube on ice. Measure the volume of the lysate collected and aliquot volumes equivalent to 1 and 5 OD_{600} for protein and RNA, respectively. As an example, for a lysate volume of approximately 700 μL collected from 60 OD_{600} culture, 11.5 μL (1/60 OD_{600}) for protein and 58 μL (5/60 OD_{600})

volume for RNA are transferred to chilled 2 mL eppendorf tubes. For the protein aliquot, add 38.5 μL buffer A and 50 μL of 5× Protein loading buffer to adjust to a final volume of 100 μL or 0.01 OD_{600}/μL. For the RNA aliquot, adjust the volume to 500 μL with Buffer A and store the tubes on ice (Fig. 1). These aliquots represent the amount of protein and RNA present in the sample after lysis and will be needed for quality control of the immunoprecipitation experiment (*see* **Note 8**).

4. Take the remaining lysate and add 35 μL anti-FLAG monoclonal antibody per tube. Incubate for 30–60 min at 4 °C, gently tumbling the tube (Fig. 1). While the antibody is incubated with the lysate, pre-wash 75 μL Protein A-Sepharose beads per sample with 500 μL of Buffer A by gently inverting tubes and centrifugation at 15,000 × *g* for 1 min at 4 °C. Wash at least three times to saturate the beads with Buffer A (*see* **Note 9**).
5. After 30–60 min incubation with the anti-FLAG antibody, aliquot volumes equivalent to 1 and 5 OD_{600} for protein and RNA, respectively (*see* Subheading 3.2, **step 3**). These aliquots are labeled as supernatant 1 (SN1) and stored at −20 °C for later use (Fig. 1).
6. After aliquoting SN1, transfer the rest of the supernatant to a fresh 2 mL tube containing 75 μL of pre-washed Protein A-sepharose beads per sample and incubate with gentle tumbling for 30–60 min at 4 °C (Fig. 1) (*see* **Note 10**).
7. After 30–60 min incubation, centrifuge at 15,000 × *g*, 4 °C for 1 min to spin down the beads and transfer the supernatant, now referred to as supernatant 2 (SN2) into fresh 2 mL tubes, without losing any beads (Fig. 1). From SN2, take an aliquot equivalent to 1 and 5 OD_{600} for protein and RNA quality control, respectively (*see* **Note 11**).
8. Wash the 2 mL eppendorf tube containing the beads with 500 μL Buffer A. Mix gently by inverting the tube (3–5 times) and centrifuge at 15,000 × *g*, 4 °C for 1 min. Discard the first wash fraction and repeat this step four times (Fig. 1). Collect fractions after each washing step into one tube (approx. 2 mL collected from four washes) for aliquoting protein and RNA samples equivalent to 1 and 5 OD_{600}, respectively (*see* **Note 12**).
9. After the last washing step, add 500 μL Buffer A to the 2 mL eppendorf tube containing the beads. Subsequently, add 500 μL phenol/chloroform/isoamyl alcohol (P/C/I) to separate protein and RNA fractions from the beads into aqueous and organic phase (Fig. 1). This tube is labeled as the elution fraction (*see* **Note 13**).

10. Similarly, add 500 μL P/C/I to all 2 mL eppendorf tubes on ice containing RNA aliquots collected during the course of the experiment, i.e., Lysate, SN1, SN2, and Wash aliquots (Fig. 1). The protein fractions collected during the course of the experiment in the 5× Protein loading buffer can be transferred to −20 °C for storage and later use.
11. Shake all tubes containing P/C/I vigorously for 15 s and incubate for 5 min at room temperature (Fig. 1).
12. Centrifuge at 15,000 × *g*, 4 °C for 30 min.
13. After separation of the aqueous and organic phase, transfer the aqueous phase (on top) from all aliquots into fresh 2 mL eppendorf tubes containing 1 mL of 30:1 mix of Ethanol:3 M sodium acetate and 1 μL GlycoBlue™. Store the tubes overnight at −20 °C for RNA precipitation (Fig. 1).
14. To the tube containing the Protein A-sepharose beads in organic phase (at bottom), add 1.4 mL acetone (ice-cold) and incubate overnight at −20 °C for protein precipitation.

RNA extraction:

15. On the next day, centrifuge the tubes containing the aqueous phases with 30:1 mix of Ethanol:3M sodium acetate at 15,000 × *g*, 4 °C for 30 min.
16. Carefully remove supernatant with a pipette without disturbing the pellet and wash once with 500 μL of ice-cold 70% EtOH. Centrifuge at 15,000 × *g*, 4 °C for 10 min. Carefully aspirate EtOH without touching the pellet (*see* **Note 14**).
17. Air-dry the pellets at room temperature in a laminar hood and resuspend them in 30 μL RNase-free water. Dissolve the pellet by heating the samples at 65 °C for 5 min on a heat block with constant shaking at 800 rpm and then store at −20 °C (Fig. 1). Use 5 μL for verification of successful coIP via northern blot analysis.

Protein extraction:

18. On the next day, centrifuge the tubes containing the beads and acetone at maximum speed ~21,000 × *g*, 4 °C for 60 min.
19. Carefully remove the supernatant and wash twice with 1 mL ice-cold acetone at maximum speed at 4 °C for 30 min (*see* **Note 15**).
20. Remove supernatant and air-dry the pellets at room temperature. Afterwards, resuspend in 120 μL of 1× Protein loading buffer. Boil the elution samples for 8 min at 95 °C on a heat block with constant shaking at 800 rpm and store at −20 °C (Fig. 1).

3.3 Quality Control for Verification of Successful CjeCas9-3xFLAG Immunoprecipitation

Prior to DNase I digestion, cDNA synthesis, and RNA-sequencing, the amount of immunoprecipitated CjeCas9-3xFLAG protein as well as co-purified crRNA and tracrRNA can be visualized by western and northern blot analysis, respectively (Fig. 2). This allows for quality control before proceeding with cDNA library preparation and next-generation sequencing (*see* **Note 16**).

3.3.1 Western Blot

1. Boil protein samples stored at −20 °C from all collected aliquots at 95 °C using a heat block with constant shaking at 800 rpm for 5–8 min.
2. 20 μL volume each from lysate, SN1, SN2, wash, and eluate fractions (corresponding to 0.2 OD_{600} for L, SN1, SN2, W, and 10 OD_{600} for eluate, i.e., 1/6 of a final of 120 μL lysate derived from 60 OD_{600} starting material) is used for western blot analysis after resolving on 10% SDS-PAGE. The blots are probed using monoclonal anti-FLAG and anti-GroEL antibodies (Fig. 2a). For western blot, a protocol can be found here [49].

3.3.2 Northern Blot

1. Transfer 5 μL volume of RNA from a total of 30 μL collected per fraction into a fresh 1.5 mL eppendorf tube, i.e., one each for Lysate, SN1, SN2, wash, and eluate.
2. Add 5 μL of RNA loading buffer II to all the tubes containing 5 μL RNA, for both WT control and CjeCas9-3xFLAG tagged samples.
3. Denature the mix at 95 °C on a heat block for 2 min and separate RNA samples by gel electrophoresis using a 6% PAA/7 M urea gel.
4. Transfer RNA from the PAA gel to a Hybond-XL membrane (GE-Healthcare) by electroblotting. After blotting, cross-link RNA to the membrane using UV irradiation and proceed with northern blotting to detect tracrRNA and crRNA3 in the eluted fractions as described previously (Fig. 2b) [31]. A protocol for northern blotting can be found here [50, 51].

3.4 DNase I Treatment

Prior to cDNA library preparations and after quality control, the remaining RNA samples must be treated with DNase I to remove any residual genomic DNA.

1. Denature 25 μL of the remaining RNA eluate samples at 65 °C on a heat block for 5 min. Afterwards, transfer the tube to ice for another 5 min.
2. Add 5 μL of 10× DNase I buffer containing $MgCl_2$, 0.5 μL Superase-In RNase Inhibitor, and 5 μL DNase I. Fill the total reaction volume to 50 μL by adding 14.5 μL of RNase-free water.

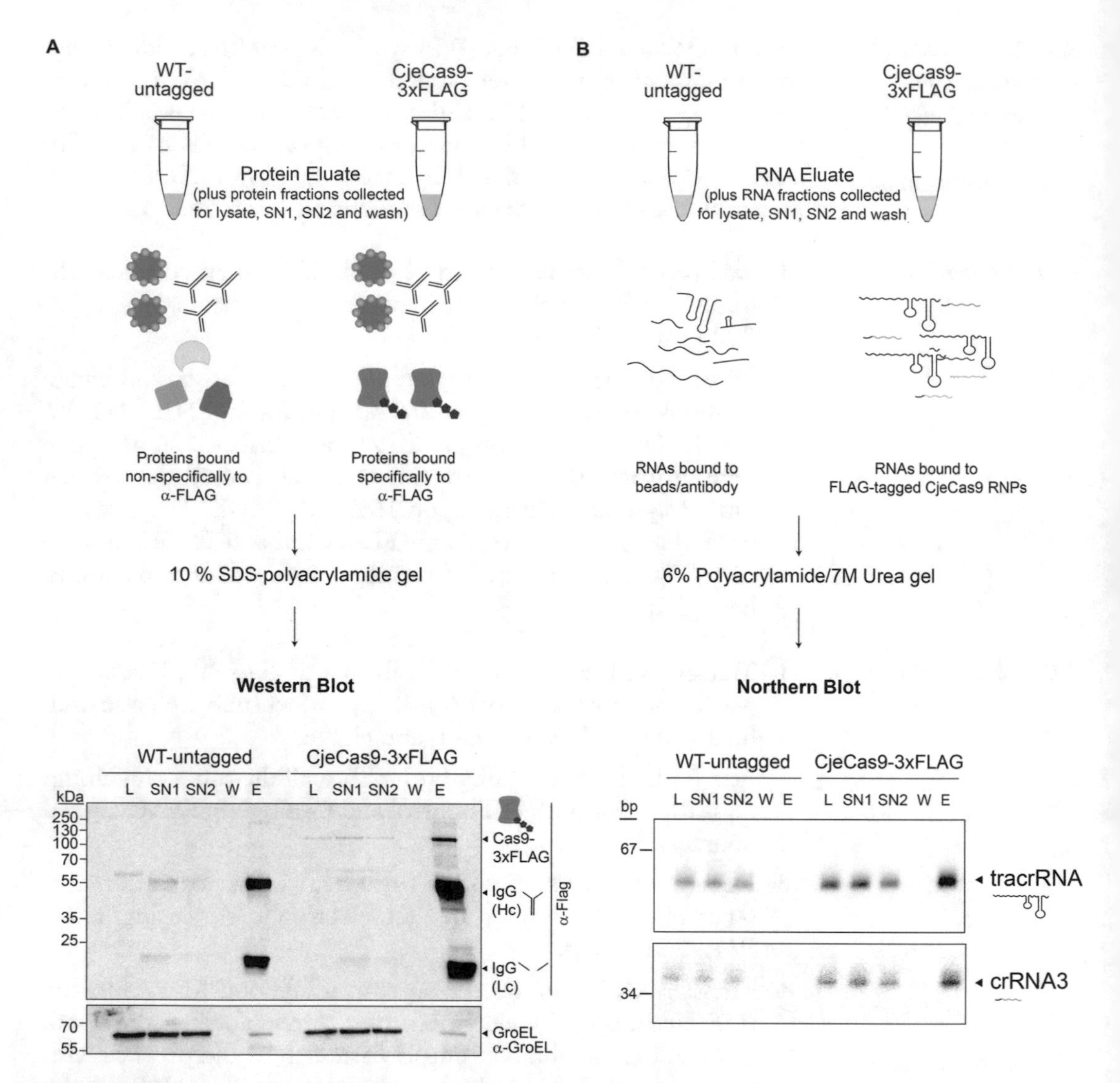

Fig. 2 Quality control for coIP of CjeCas9-3XFLAG and co-purified RNAs using western and northern blotting. (**a**) Protein eluates collected post IP are subjected to 10% SDS-polyacrylamide gel electrophoresis. 20μL sample from lysate, SN1, SN2, wash, and eluate fractions from the untagged WT control strain and the CjeCas9-3xFLAG tagged strain samples were probed with anti-FLAG and anti-GroEL (control) antibodies using the same blot. CjeCas9 is specifically enriched in the CjeCas9-3xFLAG samples and absent in untagged WT samples, confirming successful immunoprecipitation. The presence of IgG heavy (Hc) and light (Lc) chains in both WT and CjeCas9-3xFLAG elution samples indicates successful capture of the FLAG antibody by protein-A beads. (**b**) RNA eluates collected post immunoprecipitation are subjected to 6% PAA/7M urea gel electrophoresis followed by northern blotting. 5μL RNA sample each from lysate, SN1, SN2, wash, and eluate fractions of WT and CjeCas9-3xFLAG were used for northern blotting and blots were probed with ATP(γ-32P) labeled DNA oligonucleotides complementary to tracrRNA and crRNA3 to confirm their enrichment in CjeCas9-3xFLAG tagged samples. Representative western and northern blots are shown

3. Incubate the reaction mix for 45 min at 37 °C. Then add 1 μL DNase I to the tube, mix and spin down the reaction mix. Incubate at 37 °C for further 15 min.
4. After 15 min, add 50 μL RNase-free water to adjust the reaction volume to a total of 100 μL.
5. Add 100 μL P/C/I to stop the reaction for all samples and transfer the total volume to PLG tubes. Make sure to spin down the PLG tubes before use. Mix by rigorous shaking and afterwards centrifuge at 15,000 × *g*, 15 °C for 12 min.
6. Transfer the aqueous phase to a fresh 2 mL Eppendorf tube and add 2.5 volumes (~ 200 μL) of 30:1 mix of Ethanol:3M sodium acetate. Incubate overnight at −20 °C for RNA precipitation.
7. On the next day, centrifuge at 15,000 × *g*, 4 °C for 30 min. Carefully remove supernatant and wash the pellet with 350 μL of 70% Ethanol (stored at −20 °C). Centrifuge at 15,000 × *g*, 4 °C for 10 min.
8. Remove the supernatant and air-dry the RNA pellet in a laminar hood.
9. Resuspend the pellet in 30 μL RNase-free water, incubate at 65 °C on a heat block with constant shaking at 800 rpm for 5 min, and store at −20 °C for use later. Confirm successful depletion of genomic DNA using 1 μL of DNase I treated RNA as a template for a control PCR to amplify a gene of choice, with genomic DNA template serving as a positive control. If the product is still amplified in DNase I treated RNA sample, repeat DNase treatment for the sample(s).

3.5 cDNA Library Preparation, Sequencing, and Analysis

All RNA samples post DNase I treatment can be directly used for cDNA synthesis for Illumina sequencing without prior size-selection, fragmentation, or rRNA depletion. Methodological details of cDNA sequencing and analysis are beyond the scope of this chapter. An example protocol for cDNA library preparation has been previously described [31]. Other cDNA library preparation protocols can also be used based on instructions provided by the manufacturer. Here we describe the downstream data analysis workflow using our recently published data from our CjeCas9 RIP-seq analysis [31].

Briefly, RIP-seq analysis of CjeCas9 aims to identify enriched RNAs specifically bound to the protein. Sequencing of cDNA fragments generated from CjeCas9 co-purified RNAs leads to identification of individual reads, each of which can be computationally mapped to the reference genome. After sequencing and quality control steps, coverage plots are generated representing the number of mapped reads per nucleotide which can be used for visualization in a genome browser such as the Integrated Genome

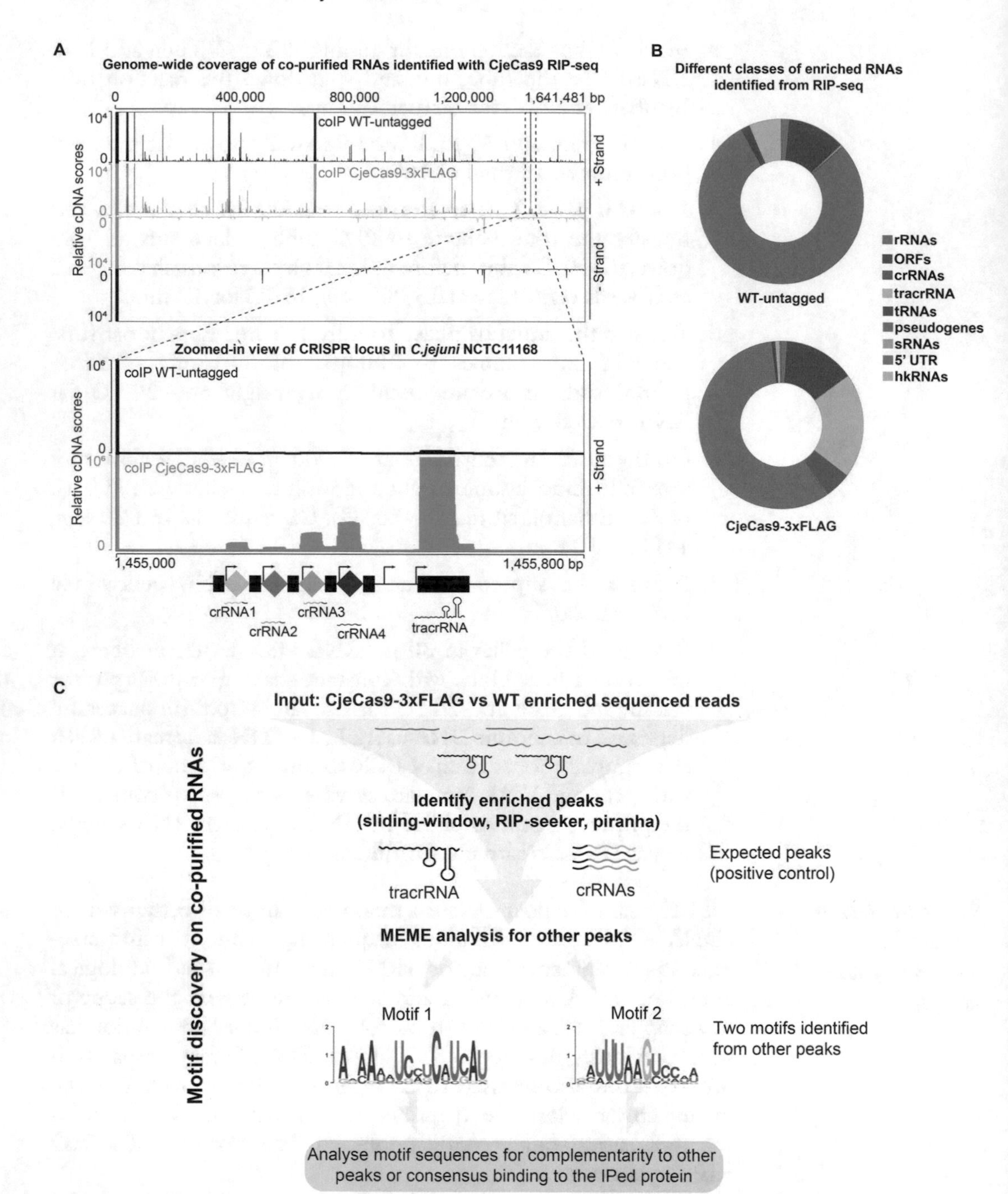

Fig. 3 Visualizing RIP-seq results and downstream analysis. (**a**) RIP-seq results for Cas9 from *C.jejuni* strain NCTC11168 [31]. (*Upper panel*) Screenshot of cDNA coverage plots of normalized RNA-seq reads from untagged WT control (black) and CjeCas9-3xFLAG coIP (green) libraries mapped to the *C. jejuni* NCTC11168 genome (both strands) and visualized using the Integrated Genome Browser. (*Lower panel*) A zoomed-in screenshot of the normalized RNA-seq reads from WT and CjeCas9-3xFLAG coIP libraries mapped to the CRISPR locus is shown to highlight the enrichment of crRNAs and tracrRNA in the CjeCas9-3xFLAG tagged

Browser (IGB) or Integrative Genomics Viewer (IGV) [52, 53]. This enables a direct visual comparison of relative cDNA reads for WT-untagged and epitope-tagged libraries mapped to a particular locus of choice, as shown for CRISPR locus, allowing for better interpretation of the data (Fig. 3a). Transcript quantification can then be performed between the CjeCas9-3xFLAG and WT coIP libraries and the resulting count data can be used for differential expression analysis via DESeq2 [55] to generate enrichment scores, thereby identifying RNA targets specifically bound to CjeCas9. Furthermore, the co-purified transcripts can be grouped according to functional classes based on the annotations. In the CjeCas9 RIP-seq example, the most abundantly co-purified RNA classes in the WT and CjeCas9-3xFLAG coIP library reads belong to ribosomal RNA (rRNA), transfer RNA (tRNA), and the housekeeping RNAs (hkRNAs). This is not surprising as rRNAs and tRNAs constitute more than 95% of the total cellular RNA pool and can be nonspecifically bound during the coIP approach. However, they are typically not enriched in the library of the tagged RBP vs the WT control and thus can be excluded as potential binding partners. On the other hand, a strong enrichment of the tracrRNA and all four crRNAs only in CjeCas9-3xFLAG coIP library indicates specific binding to CjeCas9. Moreover, a striking observation during the RIP-seq analysis of CjeCas9 was the enrichment of RNAs aside from tracrRNA and crRNAs in the CjeCas9 coIP library (Fig. 3b). Most of the additionally enriched RNAs corresponded to regions from open reading frames (ORFs), indicating co-purification of messenger RNAs with CjeCas9. Some of them

Fig. 3 (continued) versus WT coIP sample. Genomic coordinates and relative number of cDNA reads are indicated next to each screenshot. (**b**) Read distribution of different RNA classes depicted as doughnut charts showing the relative proportions of mapped cDNA reads in the CjeCas9-3xFLAG and WT coIP libraries. The RNA classes include ribosomal RNAs (rRNAs), messenger RNAs (ORFs), 5′ untranslated regions (5′ UTRs), pseudogenes, house-keeping RNAs (hkRNAs), transfer RNAs (tRNAs), small RNAs (sRNAs), the tracrRNA, and the four crRNAs of *C. jejuni* strain NCTC11168. ORFs (red), crRNAs (violet), and tracrRNA (green) were highly enriched while other RNA classes were depleted following immunoprecipitation and sequencing of co-purified RNAs from CjeCas9-3xFLAG coIP library versus the untagged WT coIP library. (**c**) Workflow for peak detection and motif discovery used for CjeCas9-3xFLAG bound co-purified RNAs. Normalized coverage files generated post cDNA sequencing of the CjeCas9 coIP vs. WT control coIP libraries were subjected to a peak-detection algorithm based on a sliding-window approach [31]. The resulting peaks were used as input for MEME analysis [54] and revealed the presence of two distinct motifs in several of the co-purified transcripts. These motifs were further analyzed regarding sequence and base-pairing probability to identify potential RNA:RNA interactions within the CjeCas9-bound interactome. Motif 1 and 2 displayed complementarity to the guide portion of crRNA3 and crRNA4 encoded in *C. jejuni* NCTC11168, indicating that these transcripts are bound by CjCas9 in a crRNA-dependent manner [31]

showed defined enriched peaks in the Cas9-3xFLAG coIP compared to the WT coIP sample.

Besides a manual annotation of such enriched peaks across the genome, numerous peak-detection tools are available for analyzing RIP-seq data such as RIP-seeker [56], Piranha [57], and our in-house peak-detection algorithm based on a sliding-window approach [31]. The resulting peaks enriched in the CjeCas9-3XFLAG coIP library can then be used to identify the presence of distinct motifs using MEME analysis [54]. In our study, MEME led to the identification of two motifs in most of the co-purified mRNAs. These motifs showed base-pairing complementarity to the sequences of the crRNA3 and crRNA4 guides encoded naturally in the host system, hinting at direct RNA:RNA interactions and association with the CjeCas9:crRNA:tracrRNA RNP (Fig. 3c). Depending on the RNA-binding protein used for coIP, the motifs could also indicate the binding site specificity of the RBP [40]. Bioinformatic programs such as IntaRNA or NUPACK can be used to examine base-pairing probabilities between the crRNAs and additional co-purified transcripts [58, 59]. Once computationally identified, it is important to confirm and validate the identified RNA: RNA interactions via other methods such as genetic manipulation of identified RNAs or direct biochemical assays which confer specificity to the identified RNAs and associated RNPs in the studied organism. For example, using deletion strains of either individual crRNAs [31] or the whole CRISPR array for performing RIP-seq could help validate direct crRNA-dependent RNA:RNA interactions and identify other potential crRNA-independent interactions with CRISPR-Cas proteins of interest, respectively. In addition, RIP-seq data can be analyzed together with additional available transcriptome data such as differential RNA-sequencing (dRNA-seq) data to distinguish between the enriched primary transcripts marked by a 5′ tri-phosphate (5'PPP) and processed RNAs with a 5′ monophosphate (5′P) group [60, 61]. dRNA-seq has facilitated the identification of transcriptional start sites and small RNAs and helped to decipher processing of CRISPR-derived RNAs in the human pathogen *Streptococcus pyogenes* via tracrRNA [19]. The same approach also helped to uncover distinct crRNA biogenesis mechanisms, leading to the discovery of a new subtype C of Type II CRISPR-Cas systems in *N. meningitidis* and *C. jejuni* [62, 63].

3.6 Outlook

Molecular mechanisms underlying CRISPR-Cas adaptive immunity have been extensively studied over the past decade. While a lot has been uncovered for individual CRISPR systems, networks governing regulation of these systems within the host are yet unknown. Most organisms typically contain a single CRISPR system, while some employ more than one. The interplay between different CRISPR defense systems in varied hosts is crucial for their survival under native conditions. Among all the differences between the CRISPR-Cas classes, types, and subtypes, there lies one commonality: all use RNA-guided ribonucleoprotein complexes. Thus, it is important to know how the RNA guides of these complexes look like and whether RNAs other than system specific crRNAs or tracrRNAs can also be associated with these complexes. A method to study RNP complexes on a transcriptome-wide scale like RIP-seq can help to identify such RNAs. Depending on what is known regarding the CRISPR-Cas protein of interest, other RNA-seq based global pull-down approaches such as CLIP-seq, RIL-seq, or CLASH could also be employed [64–66]. In CLIP-seq, the protein of interest is crosslinked to its interacting RNAs by ultraviolet light in vivo, immunoprecipitated, and bound RNAs are sequenced. This allows for refinement of precise binding sites on RNA as the crosslinking is detected by a change in the nucleotide sequence. The RIL-seq and CLASH approaches capture RNA-RNA pairs by adding a ligation step while the RNAs are in close proximity to the protein of interest post crosslinking and immunoprecipitation [3]. Application of such techniques can help ascertain whether the RNAs are bound directly or indirectly via RNA–RNA interactions to different Cas proteins in varied CRISPR systems and would help shed light on host-mediated regulation and other possible roles for these effector proteins and adaptive immune systems.

4 Notes

1. RIP-seq experiments can be performed in Gram-positive or Gram-negative bacteria to identify RNA substrates of a CRISPR-Cas protein-of-interest (POI). The amount of required cell starting material can vary between 20 and 60 OD_{600} depending on the expression of the POI and chosen growth condition.
2. To keep expression levels of the POI to physiological levels, it is recommended that the POI is chromosomally tagged using a 3xFLAG tag or any other epitope of choice. It is preferred, if a monoclonal antibody is available against the epitope, else the pull-down can be performed using a polyclonal antibody/antiserum. In organisms where genome manipulation is difficult, plasmids can also be used to express the POI. It is important to note that expressing a POI on a plasmid will generate an additional copy of the protein and both genome and plasmid-encoded variants will compete for the same RNAs, leading to an enrichment of plasmid-encoded POI-bound RNAs after sequencing. The copy number of the plasmid should be chosen as to generate levels of the POI in the range of physiological levels and growth kinetics should be monitored to detect any

phenotypic changes upon overexpression of the POI before proceeding with the RIP-seq experiment.

3. Ideally, the POI can be tagged either at its N- or C-terminus. It is advisable to check for expression and activity of the POI post epitope tagging at either end. Once this is verified, strains harboring the epitope-tagged POI should be compared with the untagged WT strain to rule out potential growth defects.
4. If possible, western blotting should be performed following growth curve analysis of the untagged and epitope-tagged strain. Samples collected at different growth phases (lag, early-log, mid-log, late-log/early stationary, and stationary phase) and analyzed on a western blot will help to identify the ideal growth condition, where the expression of the POI is strong enough for immunoprecipitation. Also, initial screening with a western blot helps to confirm the predicted size (kDa) of the tagged POI.
5. The volume of resuspension Buffer A can vary depending on the pellet. This step is performed to wash the cells off the remaining medium and any metabolic waste products generated by the cells, which might affect the subsequent steps of immunoprecipitation.
6. It is important to adjust the volume of the Lysis buffer depending on the amount of bacterial culture pelleted. The volume of the Lysis buffer should be just enough to resuspend the pellet but not dilute it. This should be standardized based on the model organism and the pellet collected. Lysis buffer volumes could impact the overall lysis. In case of volumes exceeding 1 mL, the volume must be equally distributed over two to three fast-prep tubes. After lysis and centrifugation, cleared lysates belonging to the same samples can be pooled together for immunoprecipitation.
7. Bacterial cell lysis using a FastPrep-24 machine or other physical disruption methods should be standardized based on the model organism. Before collecting pellets for RIP-seq, parameters such as the amount of bacterial culture and lysis method need to be standardized. This can be done by visualizing the epitope-tagged POI present post-lysis using a western blot. For Gram-positive cells, this might mean increasing the number of cycles and cycle-times on the FastPrep machine. Also, it is important to maintain native conditions for immunoprecipitation of RNA substrates and samples must be kept on ice throughout the protocol. A larger pellet volume does not necessarily lead to better lysis and more protein starting material for immunoprecipitation.
8. The volume of aliquots collected for RNA/protein fractions will vary depending on the starting volume of the Lysis buffer,

culture pellet and the volume of the cleared lysate recovered post-lysis.

9. To aliquot the beads from its stock, it is recommended to use an autoclaved bottom-cropped 1000 μL tip. To bottom-crop, simply cut 2–3 mm from the bottom of the 1000 μL tip using sterile scissors.
10. The incubation time of the lysate with the anti-FLAG and protein A-sepharose beads can be standardized for each POI. It should be kept in mind that longer incubation times might be detrimental and thus lead to nonspecific binding which might not reflect physiologically relevant RNA:RNA or RNA: protein interactions.
11. It is recommended to store the rest of the SN2 fraction until quality control of the experiment is completed. If the POI is not enriched in the eluate fraction as visualized using western blot, the leftover SN2 fraction can help ascertain whether the POI was bound to the beads or was lost during washing steps of the immunoprecipitation protocol.
12. Care must be taken while performing the washing steps. Any loss of beads is directly proportional to the loss of bound RNPs, thus affecting the analysis.
13. TRIZOL can also be used to recover RNA at this step, following the instructions provided by the manufacturer. It is advisable to add GlycoBlue™ blue to facilitate RNA precipitation, as the RNA yields obtained from immunoprecipitation can be quite low. GlycoBlue™ acts as a carrier, thus aiding in RNA recovery and visualization of the pellet after centrifugation.
14. Using a combination of a 200 μL tip on top of a 1000 μL tip aids in complete removal of the supernatant, without disturbing the pellet. This is a crucial step and any loss of pellets should be avoided.
15. As the protein pellet might be difficult to see among the beads, it is advisable to be cautious while washing and care must be taken to prevent any loss of beads or the protein pellet.
16. Performing a western blot to confirm successful immunoprecipitation of the POI is strongly recommended. This allows to ascertain if the POI was sufficiently enriched in the eluate from the epitope-tagged strain versus the untagged strain. This step also aids in standardization of the coIP protocol by visualizing the tagged protein in other fractions collected (lysate, SN1, SN2, and wash) during the course of the experiment. Performing a northern blot is dependent on the knowledge of previously identified RNAs (acting as positive controls) associated with the POI and can also be excluded if such information is not at hand. In any case, the western blot should

confirm enrichment of the POI prior to proceeding with cDNA library preparation of co-purified transcripts.

Acknowledgments

We would like to thank the members of the Sharma lab for their contributions towards adapting and further developing the RIP-seq protocol for use in Epsilonproteobacteria over the years and Thorsten Bischler from the Core Unit Systems Medicine, University of Würzburg for help with RIP-seq data analysis. We thank Elisabetta Fiore and Mona Alzheimer for critical comments on this book chapter. This work was supported by DFG (Deutsche Forschungsgemeinschaft) grant SH 580/9-1 to C.M.S. within the priority program SPP 2141 "Much more than defence: the multiple functions and facets of CRISPR-Cas."

References

1. Beckmann BM, Castello A, Medenbach J (2016) The expanding universe of ribonucleoproteins: of novel RNA-binding proteins and unconventional interactions. Pflugers Arch 468:1029–1040. https://doi.org/10.1007/s00424-016-1819-4
2. Babitzke P, Lai Y-J, Renda AJ, Romeo T (2019) Posttranscription initiation control of gene expression mediated by bacterial RNA-binding proteins. Annu Rev Microbiol 73:43–67. https://doi.org/10.1146/annurev-micro-020518-115907
3. Saliba A-E, Santos SC, Vogel J (2017) New RNA-seq approaches for the study of bacterial pathogens. Curr Opin Microbiol 35:78–87. https://doi.org/10.1016/j.mib.2017.01.001
4. Hör J, Gorski SA, Vogel J (2018) Bacterial RNA biology on a genome scale. Mol Cell 70:785–799. https://doi.org/10.1016/j.molcel.2017.12.023
5. Koonin EV, Makarova KS, Zhang F (2017) Diversity, classification and evolution of CRISPR-Cas systems. Curr Opin Microbiol 37:67–78. https://doi.org/10.1016/j.mib.2017.05.008
6. Marraffini LA, Sontheimer EJ (2010) CRISPR interference: RNA-directed adaptive immunity in bacteria and archaea. Nat Rev Genet 11:181–190. https://doi.org/10.1038/nrg2749
7. van der Oost J, Jore MM, Westra ER et al (2009) CRISPR-based adaptive and heritable immunity in prokaryotes. Trends Biochem Sci 34:401–407. https://doi.org/10.1016/j.tibs.2009.05.002
8. Hille F, Richter H, Wong SP et al (2018) The biology of CRISPR-Cas: backward and forward. Cell 172:1239–1259. https://doi.org/10.1016/j.cell.2017.11.032
9. Wright AV, Nuñez JK, Doudna JA (2016) Biology and applications of CRISPR systems: harnessing nature's toolbox for genome engineering. Cell 164:29–44. https://doi.org/10.1016/j.cell.2015.12.035
10. Mohanraju P, Makarova KS, Zetsche B et al (2016) Diverse evolutionary roots and mechanistic variations of the CRISPR-Cas systems. Science 353:aad5147. https://doi.org/10.1126/science.aad5147
11. Makarova KS, Wolf YI, Iranzo J et al (2020) Evolutionary classification of CRISPR-Cas systems: a burst of class 2 and derived variants. Nat Rev Microbiol 18:67–83. https://doi.org/10.1038/s41579-019-0299-x
12. Barrangou R, Fremaux C, Deveau H et al (2007) CRISPR provides acquired resistance against viruses in prokaryotes. Science 315:1709–1712. https://doi.org/10.1126/science.1138140
13. Marraffini LA, Sontheimer EJ (2008) CRISPR interference limits horizontal gene transfer in staphylococci by targeting DNA. Science 322:1843–1845. https://doi.org/10.1126/science.1165771
14. Terns MP (2018) CRISPR-based technologies: impact of RNA-targeting systems. Mol Cell

72:404–412. https://doi.org/10.1016/j.molcel.2018.09.018

15. Wang F, Wang L, Zou X et al (2019) Advances in CRISPR-Cas systems for RNA targeting, tracking and editing. Biotechnol Adv 37:708–729. https://doi.org/10.1016/j.biotechadv.2019.03.016
16. Jinek M, Chylinski K, Fonfara I et al (2012) A programmable dual-RNA-guided DNA endonuclease in adaptive bacterial immunity. Science 337:816–821. https://doi.org/10.1126/science.1225829
17. Gasiunas G, Barrangou R, Horvath P, Siksnys V (2012) Cas9-crRNA ribonucleoprotein complex mediates specific DNA cleavage for adaptive immunity in bacteria. Proc Natl Acad Sci U S A 109:E2579–E2586. https://doi.org/10.1073/pnas.1208507109
18. Jiang F, Doudna JA (2017) CRISPR-Cas9 structures and mechanisms. Annu Rev Biophys 46:505–529. https://doi.org/10.1146/annurev-biophys-062215-010822
19. Deltcheva E, Chylinski K, Sharma CM et al (2011) CRISPR RNA maturation by trans-encoded small RNA and host factor RNase III. Nature 471:602–607. https://doi.org/10.1038/nature09886
20. Anders C, Niewoehner O, Duerst A, Jinek M (2014) Structural basis of PAM-dependent target DNA recognition by the Cas9 endonuclease. Nature 513:569–573. https://doi.org/10.1038/nature13579
21. Westra ER, Buckling A, Fineran PC (2014) CRISPR-Cas systems: beyond adaptive immunity. Nat Rev Microbiol 12:317–326. https://doi.org/10.1038/nrmicro3241
22. Louwen R, Horst-Kreft D, de Boer AG et al (2013) A novel link between Campylobacter jejuni bacteriophage defence, virulence and Guillain-Barré syndrome. Eur J Clin Microbiol Infect Dis 32:207–226. https://doi.org/10.1007/s10096-012-1733-4
23. Hale CR, Majumdar S, Elmore J et al (2012) Essential features and rational design of CRISPR RNAs that function with the Cas RAMP module complex to cleave RNAs. Mol Cell 45:292–302. https://doi.org/10.1016/j.molcel.2011.10.023
24. Zegans ME, Wagner JC, Cady KC et al (2009) Interaction between bacteriophage DMS3 and host CRISPR region inhibits group behaviors of Pseudomonas aeruginosa. J Bacteriol 191:210–219. https://doi.org/10.1128/JB.00797-08
25. Mandin P, Repoila F, Vergassola M et al (2007) Identification of new noncoding RNAs in Listeria monocytogenes and prediction of mRNA targets. Nucleic Acids Res 35:962–974. https://doi.org/10.1093/nar/gkl1096
26. Vercoe RB, Chang JT, Dy RL et al (2013) Cytotoxic chromosomal targeting by CRISPR/Cas systems can reshape bacterial genomes and expel or remodel pathogenicity islands. PLoS Genet 9:e1003454. https://doi.org/10.1371/journal.pgen.1003454
27. Babu M, Beloglazova N, Flick R et al (2011) A dual function of the CRISPR-Cas system in bacterial antivirus immunity and DNA repair. Mol Microbiol 79:484–502. https://doi.org/10.1111/j.1365-2958.2010.07465.x
28. Viswanathan P, Murphy K, Julien B et al (2007) Regulation of dev, an operon that includes genes essential for Myxococcus xanthus development and CRISPR-associated genes and repeats. J Bacteriol 189:3738–3750. https://doi.org/10.1128/JB.00187-07
29. Gunderson FF, Cianciotto NP (2013) The CRISPR-associated gene cas2 of Legionella pneumophila is required for intracellular infection of amoebae. MBio 4:e00074–e00013. https://doi.org/10.1128/mBio.00074-13
30. Ratner HK, Escalera-Maurer A, Le Rhun A et al (2019) Catalytically active cas9 mediates transcriptional interference to facilitate bacterial virulence. Mol Cell 75:498–510.e5. https://doi.org/10.1016/j.molcel.2019.05.029
31. Dugar G, Leenay RT, Eisenbart SK et al (2018) CRISPR RNA-dependent binding and cleavage of endogenous RNAs by the Campylobacter jejuni Cas9. Mol Cell 69:893–905.e7. https://doi.org/10.1016/j.molcel.2018.01.032
32. Strutt SC, Torrez RM, Kaya E et al (2018) RNA-dependent RNA targeting by CRISPR-Cas9. elife. https://doi.org/10.7554/eLife.32724
33. Rousseau BA, Hou Z, Gramelspacher MJ, Zhang Y (2018) Programmable RNA cleavage and recognition by a natural CRISPR-Cas9 system from Neisseria meningitidis. Mol Cell 69:906–914.e4. https://doi.org/10.1016/j.molcel.2018.01.025
34. O'Connell MR, Oakes BL, Sternberg SH et al (2014) Programmable RNA recognition and cleavage by CRISPR/Cas9. Nature 516:263–266. https://doi.org/10.1038/nature13769
35. Nelles DA, Fang MY, O'Connell MR et al (2016) Programmable RNA tracking in live cells with CRISPR/Cas9. Cell 165:488–496. https://doi.org/10.1016/j.cell.2016.02.054
36. Zhang A, Wassarman KM, Rosenow C et al (2003) Global analysis of small RNA and

mRNA targets of Hfq. Mol Microbiol 50:1111–1124. https://doi.org/10.1046/j.1365-2958.2003.03734.x

37. Christiansen JK, Nielsen JS, Ebersbach T et al (2006) Identification of small Hfq-binding RNAs in Listeria monocytogenes. RNA 12:1383–1396. https://doi.org/10.1261/rna.49706
38. Sonnleitner E, Sorger-Domenigg T, Madej MJ et al (2008) Detection of small RNAs in Pseudomonas aeruginosa by RNomics and structure-based bioinformatic tools. Microbiology 154:3175–3187. https://doi.org/10.1099/mic.0.2008/019703-0
39. Sittka A, Lucchini S, Papenfort K et al (2008) Deep sequencing analysis of small noncoding RNA and mRNA targets of the global post-transcriptional regulator, Hfq. PLoS Genet 4: e1000163. https://doi.org/10.1371/journal.pgen.1000163
40. Dugar G, Svensson SL, Bischler T et al (2016) The CsrA-FliW network controls polar localization of the dual-function flagellin mRNA in Campylobacter jejuni. Nat Commun 7:11667. https://doi.org/10.1038/ncomms11667
41. Rieder R, Reinhardt R, Sharma C, Vogel J (2012) Experimental tools to identify RNA-protein interactions in Helicobacter pylori. RNA Biol 9:520–531. https://doi.org/10.4161/rna.20331
42. Bilusic I, Popitsch N, Rescheneder P et al (2014) Revisiting the coding potential of the E. coli genome through Hfq co-immunoprecipitation. RNA Biol 11:641–654. https://doi.org/10.4161/rna.29299
43. Göpel Y, Papenfort K, Reichenbach B et al (2013) Targeted decay of a regulatory small RNA by an adaptor protein for RNase E and counteraction by an anti-adaptor RNA. Genes Dev 27:552–564. https://doi.org/10.1101/gad.210112.112
44. Heidrich N, Bauriedl S, Schoen C (2019) Investigating RNA-Protein interactions in Neisseria meningitidis by RIP-Seq analysis. Methods Mol Biol 1969:33–49. https://doi.org/10.1007/978-1-4939-9202-7_3
45. Heidrich N, Bauriedl S, Barquist L et al (2017) The primary transcriptome of Neisseria meningitidis and its interaction with the RNA chaperone Hfq. Nucleic Acids Res 45:6147–6167. https://doi.org/10.1093/nar/gkx168
46. Michaux C, Holmqvist E, Vasicek E et al (2017) RNA target profiles direct the discovery of virulence functions for the cold-shock proteins CspC and CspE. Proc Natl Acad Sci U S A 114:6824–6829. https://doi.org/10.1073/pnas.1620772114
47. Gerovac M, El Mouali Y, Kuper J et al (2020) Global discovery of bacterial RNA-binding proteins by RNase-sensitive gradient profiles reports a new FinO domain protein. RNA 26:1448–1463. https://doi.org/10.1261/rna.076992.120
48. Zhang J, Rouillon C, Kerou M et al (2012) Structure and mechanism of the CMR complex for CRISPR-mediated antiviral immunity. Mol Cell 45:303–313. https://doi.org/10.1016/j.molcel.2011.12.013
49. Brunelle JL, Green R (2014) One-dimensional SDS-polyacrylamide gel electrophoresis (1D SDS-PAGE). Methods Enzymol 541:151–159. https://doi.org/10.1016/B978-0-12-420119-4.00012-4
50. Brown T, Mackey K, Du T (2004) Analysis of RNA by northern and slot blot hybridization. Curr Protoc Mol Biol Chapter 4:Unit 4.9. https://doi.org/10.1002/0471142727.mb0409s67
51. He SL, Green R (2013) Northern blotting. Methods Enzymol 530:75–87. https://doi.org/10.1016/B978-0-12-420037-1.00003-8
52. Robinson JT, Thorvaldsdóttir H, Winckler W et al (2011) Integrative genomics viewer. Nat Biotechnol 29:24–26. https://doi.org/10.1038/nbt.1754
53. Freese NH, Norris DC, Loraine AE (2016) Integrated genome browser: visual analytics platform for genomics. Bioinformatics 32:2089–2095. https://doi.org/10.1093/bioinformatics/btw069
54. Bailey TL, Boden M, Buske FA et al (2009) MEME SUITE: tools for motif discovery and searching. Nucleic Acids Res 37:W202–W208. https://doi.org/10.1093/nar/gkp335
55. Love MI, Huber W, Anders S (2014) Moderated estimation of fold change and dispersion for RNA-seq data with DESeq2. Genome Biol 15:550. https://doi.org/10.1186/s13059-014-0550-8
56. Li Y, Zhao DY, Greenblatt JF, Zhang Z (2013) RIPSeeker: a statistical package for identifying protein-associated transcripts from RIP-seq experiments. Nucleic Acids Res 41:e94. https://doi.org/10.1093/nar/gkt142
57. Uren PJ, Bahrami-Samani E, Burns SC et al (2012) Site identification in high-throughput RNA-protein interaction data. Bioinformatics 28:3013–3020. https://doi.org/10.1093/bioinformatics/bts569
58. Mann M, Wright PR, Backofen R (2017) IntaRNA 2.0: enhanced and customizable

prediction of RNA-RNA interactions. Nucleic Acids Res 45:W435–W439. https://doi.org/10.1093/nar/gkx279

59. Zadeh JN, Steenberg CD, Bois JS et al (2011) NUPACK: Analysis and design of nucleic acid systems. J Comput Chem 32:170–173. https://doi.org/10.1002/jcc.21596
60. Sharma CM, Hoffmann S, Darfeuille F et al (2010) The primary transcriptome of the major human pathogen Helicobacter pylori. Nature 464:250–255. https://doi.org/10.1038/nature08756
61. Heidrich N, Dugar G, Vogel J, Sharma CM (2015) Investigating CRISPR RNA biogenesis and function using RNA-seq. Methods Mol Biol 1311:1–21. https://doi.org/10.1007/978-1-4939-2687-9_1
62. Zhang Y, Heidrich N, Ampattu BJ et al (2013) Processing-independent CRISPR RNAs limit natural transformation in Neisseria meningitidis. Mol Cell 50:488–503. https://doi.org/10.1016/j.molcel.2013.05.001
63. Dugar G, Herbig A, Förstner KU et al (2013) High-resolution transcriptome maps reveal strain-specific regulatory features of multiple Campylobacter jejuni isolates. PLoS Genet 9:e1003495. https://doi.org/10.1371/journal.pgen.1003495
64. Melamed S, Peer A, Faigenbaum-Romm R et al (2016) Global Mapping of Small RNA-Target Interactions in Bacteria. Mol Cell 63:884–897. https://doi.org/10.1016/j.molcel.2016.07.026
65. Waters SA, McAteer SP, Kudla G et al (2017) Small RNA interactome of pathogenic E. coli revealed through crosslinking of RNase E. EMBO J 36:374–387. https://doi.org/10.15252/embj.201694639
66. Holmqvist E, Wright PR, Li L et al (2016) Global RNA recognition patterns of post-transcriptional regulators Hfq and CsrA revealed by UV crosslinking in vivo. EMBO J 35:991–1011. https://doi.org/10.15252/embj.201593360
67. Jiao C, Sharma S, Dugar G, et al (2021) Non-canonical crRNAs derived from host transcripts enable multiplexable RNA detection by Cas9. Science 372:941–948. https://doi.org/10.1126/science.abe7106

Chapter 7

Rapidly Characterizing CRISPR-Cas13 Nucleases Using Cell-Free Transcription-Translation Systems

Katharina G. Wandera and Chase L. Beisel

Abstract

Cell-free transcription-translation (TXTL) systems produce RNAs and proteins from added DNA. By coupling their production to a biochemical assay, these biomolecules can be rapidly and scalably characterized without the need for purification or cell culturing. Here, we describe how TXTL can be applied to characterize Cas13 nucleases from Type VI CRISPR-Cas systems. These nucleases employ guide RNAs to recognize complementary RNA targets, leading to the nonspecific collateral cleavage of nearby RNAs. In turn, RNA targeting by Cas13 has been exploited for numerous applications, including in vitro diagnostics, programmable gene silencing in eukaryotes, and sequence-specific antimicrobials. As part of the described method, we detail how to set up TXTL assays to measure on-target and collateral RNA cleavage by Cas13 as well as how to assay for putative anti-CRISPR proteins. Overall, the method should be useful for the characterization of Type VI CRISPR-Cas systems and their use in ranging applications.

Key words Acrs, Cas13a, Cas13b, Cell-free system, CRISPR-Cas, Lysate, gRNA, TXTL

1 Introduction

Cell-free transcription-translation (TXTL) systems represent cellular lysates or solutions of purified components that recapitulate transcription and translation in a test tube. By adding linear or circular DNA to the TXTL mix, RNA and protein are made in a matter of minutes to hours [1, 2]. Reactions can also be conducted in as little as a few microliters, allowing large numbers of TXTL reactions to be conducted in parallel with microtiter plates. The reaction conditions can also be tightly controlled to maximize protein production or metabolic activity. While TXTL has been used for protein production or as diagnostic platforms, one emerging application of TXTL is characterizing functional RNAs and proteins [3–6]. In contrast to in vitro assays and cell-based assays, TXTL does not require any purification beyond the source DNA, and no culturing is required. By removing cell growth and

Erik Dassi (ed.), *Post-Transcriptional Gene Regulation*, Methods in Molecular Biology, vol. 2404,
https://doi.org/10.1007/978-1-0716-1851-6_7,

viability, biomolecules that might prove toxic to the cell can also be tested. As part of an assay, the RNAs and/or proteins are produced from the added DNA, and their activity is monitored based on how the assay is configured. For example, gene regulators can be assayed by including a fluorescent reporter encoded in separate DNA targeted by the regulator. By measuring fluorescence over time, a quantitative and dynamic readout can be obtained that can garner different insights into the synthesis and activity of the assessed biomolecule [7, 8]. Separately, libraries of DNA can be added and then re-extracted for next-generation sequencing, such as when comprehensively elucidating DNA binding motifs [9, 10].

One expanding use of TXTL-based assays is the characterization of CRISPR-Cas systems and technologies [11, 12]. CRISPR-Cas systems represent prokaryotic defense systems whose Cas nucleases are directed by their guide RNAs (gRNAs) to cleave complementary nucleic acids. These same nucleases and gRNAs have been co-opted as programmable tools for a wide range of applications spanning gene editing and gene regulation to diagnostics and antimicrobials [13–16]. Most of these applications have relied on the DNA-targeting nucleases Cas9 and Cas12a, where our group has developed different TXTL assays to characterize DNA cleavage by different orthologs [11], factors influencing gRNA production and activity [10], and protospacer-adjacent motifs (PAMs) recognized by each nuclease as part of target interrogation [9, 17, 18]. Beyond these two nucleases, the recently discovered Cas13 nucleases have been garnering increasing attention [19–21]. These nucleases have the unique ability to target RNA. Furthermore, target recognition activates the HEPN ribonuclease domain that resides outside of the target site, leading to cleavage of any RNA it encounters. Collateral RNA cleavage has been the centerpiece of Cas13-based in vitro diagnostics [22, 23]. Finally, Cas13 has been recently used as programmable antimicrobial by driving widespread RNA degradation in bacterial cells expressing an antibiotic resistance marker [24]. In contrast, in eukaryotic cells collateral RNA activity appears to be mostly limited to target RNA, allowing the use of Cas13 for programmable gene silencing similar to RNA interference [19, 25]. These applications have led to the development of different tools for gRNA design tailored to different applications [26, 27]. Beyond these applications, nature boasts a rich diversity of Cas13 nucleases, while anti-CRISPR proteins (Acrs) specifically against these nucleases have been reported that could serve to better control their activity [28, 29]. Therefore, there is a need to accelerate the characterization of Cas13 nucleases and their associated gRNAs. Here, we report a set of methods based on TXTL for this purpose (Fig. 1). We demonstrate how to configure a TXTL assay to measure on-target and collateral RNA cleavage using Cas13 nucleases from two different CRISPR subtypes, and we show how the assay can be implemented to screen for putative Acrs [30]. These methods should facilitate the further exploration and use of these unique nucleases as CRISPR technologies.

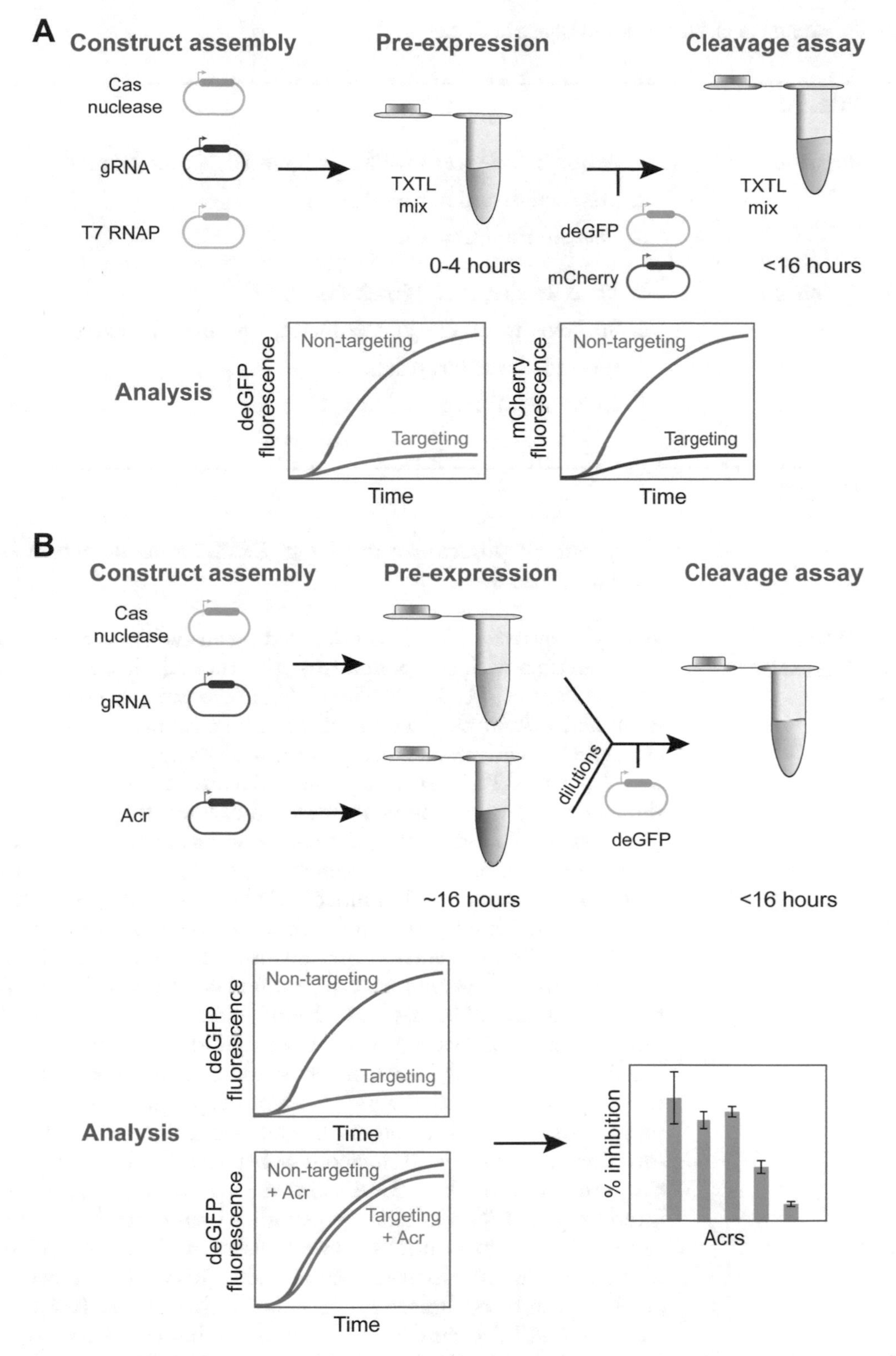

Fig. 1 Assessing RNA cleavage by Cas13 in TXTL. (**a**) General setup for assessing on-target and collateral RNA cleavage. deGFP serves as the readout of on-target cleavage because its mRNA is targeted by the expressed gRNA, whereas mCherry serves as the read of collateral cleavage because its mRNA is not targeted by the expressed gRNA. (**b**) General setup for assessing putative anti-CRISPR proteins against Cas13 nucleases. Inhibition is calculated by running samples with the targeting and non-targeting gRNAs with or without the anti-CRISPR protein

2 Materials

2.1 Reagents

1. Arbor Biosciences myTXTL Sigma 70 Master Mix kit.
2. DNA clean and concentrator kit.
3. Plasmid midiprep kit.

2.2 Equipment

1. Plate Reader, e.g., Biotek Synergy H1.
2. Spectrophotometer/fluorometer, e.g., from DeNovix.
3. 96-well V bottom plate.
4. Cover mat for 96-well plate.

3 Methods

Carry out all procedures involving TXTL lysate on ice, unless otherwise specified.

3.1 Design of the Expression Constructs

We regularly use a TXTL system based on an *Escherichia coli* lysate supplemented with components to drive transcription and translation (*see* **Note 1**) [1, 31]. While this lysate can produce high protein levels and sustain production for up to ~16 h, many other lysates are available and may be suitable for a given application [1, 32–34]. The added DNA encoding the Cas13 nuclease, the gRNA, and the reporter plasmid can be in the form of linear or plasmid DNA (*see* **Note 2**). In either case, the construct has to allow for strong expression in *E. coli* under exponential growth, as these conditions reflect the state of the cells immediately prior to lysate preparation. Therefore, to achieve strong transcription, it is important to choose a suitable promoter, such as the constitutive promoter J23108. A list of suitable σ^{70} promoters can be found elsewhere [35, 36]. For the constructs used here, we use the inducible T7 promoter. As this polymerase is not part of the lysate, an additional construct encoding the T7 RNA polymerase must be added. A terminator to halt transcription is encoded after the gene, e.g., the synthetic t500 terminator that acts through rho-independent termination. We introduce at least 15–20 nucleotides between the last repeat and the terminator of the gRNA construct to avoid the terminator interfering with Cas13 binding. Another important factor in the success of protein synthesis is the translation efficiency, which is determined by the ribosome-binding site (RBS). The translation efficiency can be optimized using online tools, e.g., the RBS calculator [37]. All proteins used here have additionally been codon-optimized for *E. coli* to achieve high translation efficiency.

To illustrate these methods, we have included two different examples using Cas13 nucleases from *Leptotrichia shahii* (LsCas13a) and *Bergeyella zoohelcum* (BzCas13b) and their respective gRNAs. All plasmids are listed in Table 1 and are partially available on Addgene. To clarify which strand is selected when choosing the target, we show a detailed example with gRNA-1 of BzCas13b (Fig. 2). It is important to note here that Cas13 targets RNA and not DNA (Fig. 3); therefore, the spacer must be complementary to the coding strand of the target plasmid. In total, we show the results from testing gRNAs targeting different locations in the deGFP mRNA in comparison to a non-targeting gRNA for both LsCas13a and BzCas13b (Fig. 4). We also show how varying the amount of added DNA can tune targeting activity (Fig. 5), allowing this activity to be tuned to alter the sensitivity of the assay.

When choosing an appropriate spacer, it is important to consider that some Cas13 nucleases require a specific protospacer flanking sequence (PFS) [38, 39]. This sequence is part of the target mRNA and, for some nucleases, necessary for the activation of Cas13 interference. In total, the PFS requirements of Cas13 are not very strict, with a PFS of 3′ A, U or C in the case of LsCas13a [19]. The sequence is also complementary to the flanking portion of the repeat, where extensive complementarity between both regions was shown to block immunity [40]. It is important to note here that target recognition by the RNP complex leads to collateral RNA cleavage, which does not require a specific PFS or complementarity to the gRNA. We measure cleavage of the gRNA target as well as collateral cleavage by measuring the fluorescence of two fluorescence reporters. The mRNA of the deGFP reporter plasmid is specifically targeted by the RNP complex. The mRNA of the second reporter plasmid (mCherry) is not directly targeted and instead degraded through collateral RNA cleavage activity (Fig. 6).

We also extended the TXTL assay beyond measuring on-target RNA cleavage by characterizing putative Acrs against Cas13 nucleases (Fig. 7). AcrVIA4 and AcrVIA7 showed weak inhibition of GFP expression (10.3% and 21.3%, respectively, compared to no Acr in the presence of the non-targeting gRNA) [28, 29]. In the following, we provide a step-by-step description for the design of constructs expressing the nuclease, a gRNA, and Acrs.

3.1.1 Nuclease

1. Determine amino acid (aa) sequence of the nuclease of interest, e.g., by using the NCBI database or the CRISPR-Cas Finder [41]. If it is a well-established CRISPR-Cas system, plasmids can also be available through Addgene (Table 1).
2. Codon-optimize the aa sequence using any of the available tools, e.g., from IDT [42].
3. Clone the codon-optimized sequence under the control of a strong promoter, such as the constitutive promoter J23108 or the inducible T7 promoter (*see* **Notes 3** and **4**).

Table 1
List of all plasmids used in this manuscript, including Addgene number, if available

Lab Number	Components	Source	Benchling Link	Addgene Number
CBS-0011	T7 RNA polymerase	[1]	https://benchling.com/s/seq-8eQXtPpoGA4m28uE2GL2	–
CBS-0338	deGFP (target)	[1]	https://benchling.com/s/seq-IXOQKCe1NW2rtQlZ5u0u	–
CBS-0466	pLsCas13a-gRNA-NT	This study	https://benchling.com/s/seq-ZtDisAT5pPWTUKcH5b1Q	164857
CBS-0606	mmCherry	[1]	https://benchling.com/s/seq-TZCzv6TK9v73bDUuNMos	–
CBS-1238	LsCas13a	[10]	https://benchling.com/s/seq-pFCvwO7NtNCdZlBTdeeX	–
CBS-1316	pBzCas13b-gRNA-1	This study	https://benchling.com/s/seq-JuDzDSaPfFxw8DcYFyee	164858
CBS-1317	pBzCas13b-gRNA-2	This study	https://benchling.com/s/seq-iikmX9hQhl8gi20hUAdU	164859
CBS-1318	pBzCas13b-gRNA-3	This study	https://benchling.com/s/seq-n8HrinpBd6B3G8e7Noju	164860
CBS-1319	pBzCas13b-gRNA-NT	This study	https://benchling.com/s/seq-Oy5PH76yIAIiRyfnvutv	164861
CBS-1639	pBzCas13b	This study	https://benchling.com/s/seq-UczifYTPkx11Om88fObq	164862
CBS-2261	pAcrVIA4	This study	https://benchling.com/s/seq-EgN2uWisQPZPsOtbbtw1	164863
CBS-2264	pAcrVIA7	This study	https://benchling.com/s/seq-cBfC6nT7zNJCIXvI1rCA	164864
CBS-2447	pLsCas13a-gRNA-1	This study	https://benchling.com/s/seq-E4pVebW8Uz21jiX4EoCr	164865
CBS-2448	pLsCas13a-gRNA-2	This study	https://benchling.com/s/seq-RScdWGoan3nrAteVAmkB	164866
CBS-2449	pLsCas13a-gRNA-3	This study	https://benchling.com/s/seq-1bN3uPGU7qOFOfadEG7y	164867

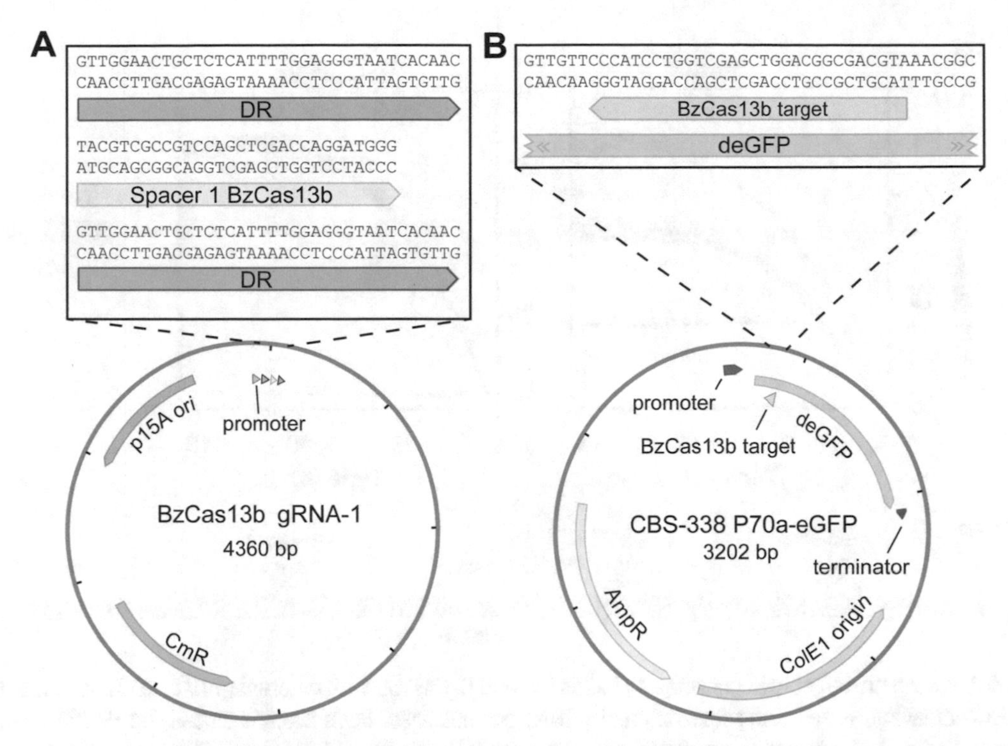

Fig. 2 Plasmid constructs encoding a gRNA and targeted reporter. (**a**) Plasmid map showing gRNA-1 used by BzCas13b and designed to target the deGFP mRNA. In this example, the gRNA is encoded as a single-spacer CRISPR array. Tick marks indicate increments of 500 bp within the plasmid. (**b**) Plasmid map showing the target location in the *deGFP* gene. The guide sequence is designed to base pair with the transcribed mRNA encoding deGFP

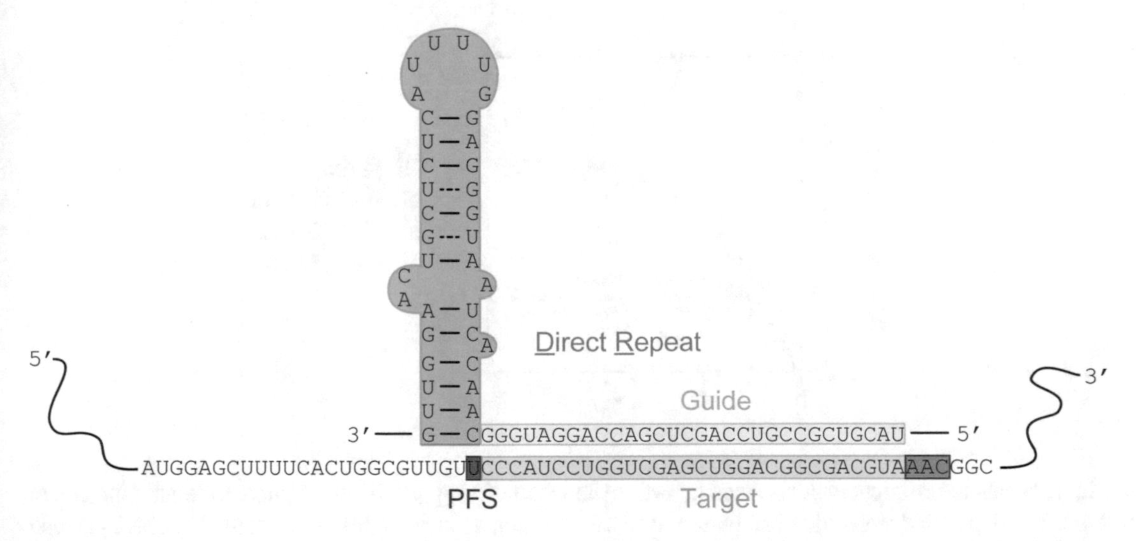

Fig. 3 The processed gRNA base pairing with the target mRNA. The gRNA and target sequences match those displayed in Fig. 2. The key regions are indicated, including the processed direct repeat, the guide, the PFS, and the target

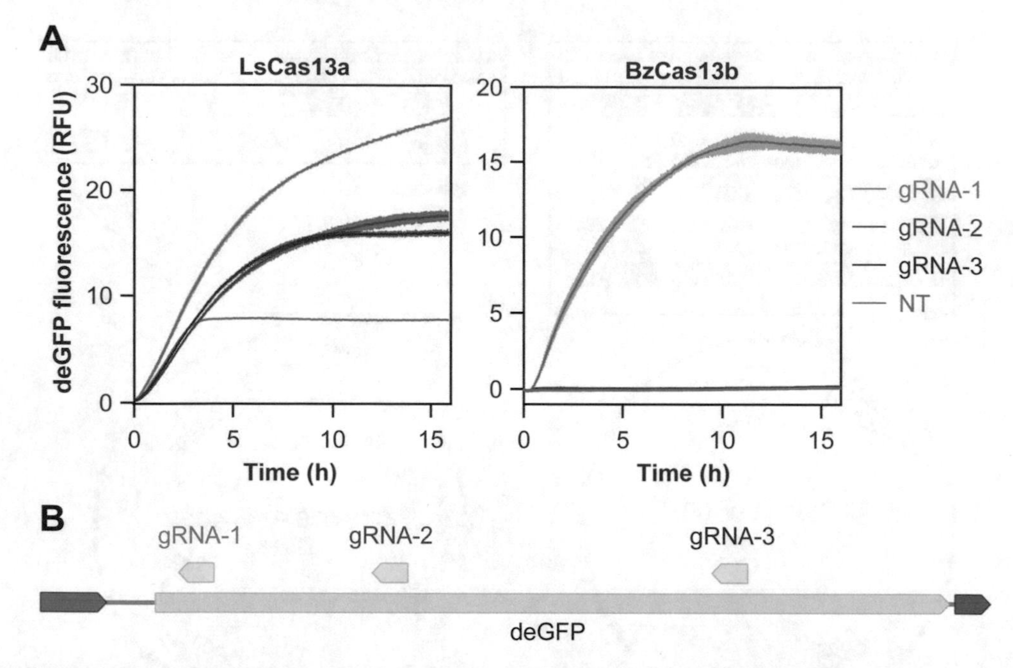

Fig. 4 Assaying on-target RNA cleavage by LsCas13a and BzCas13b with different gRNAs. (**a**) Time courses of the RNA cleavage assay using four different gRNAs per nuclease. Each assay included the deGFP reporter plasmid and a Cas13 nuclease pre-expressed with a targeting or non-targeting (NT) gRNA. The nuclease and gRNA were pre-expressed for four hours before the deGFP plasmid was added. The fluorescence was measured for 16 h at 29 °C in a plate reader. Lines and shaded regions represent the mean and standard deviation of duplicates. (**b**) Target locations within the *deGFP* gene. The targets of the gRNAs are 32 nts for LsCas13a and 30 nts for BzCas13b based on the length of their respective spacers

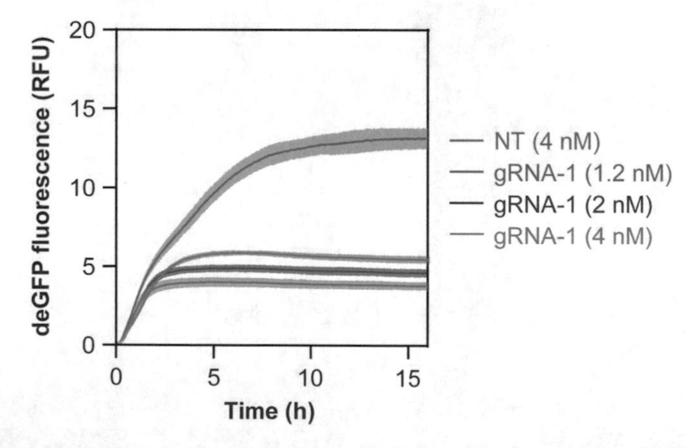

Fig. 5 Assaying on-target RNA cleavage by BzCas13b when varying gRNA-1 production. Different amounts of the gRNA-1 plasmid were added as part of the TXTL reaction as indicated. The non-targeting gRNA plasmid (NT) was added to yield a final concentration of 4 nM. The nuclease, gRNA, and deGFP plasmids were added at the same time. The fluorescence was measured for 16 h at 29 °C in a plate reader. Lines and shaded regions represent the mean and standard deviation of quadruplicates

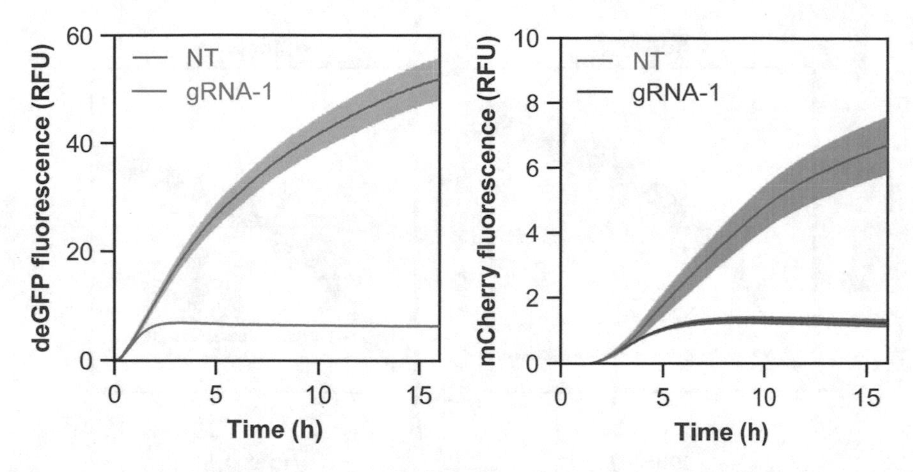

Fig. 6 Assessing on-targeting and collateral RNA cleavage by LsCas13a. Time courses of the RNA cleavage assay using both the targeted deGFP reporter plasmid and a non-targeted mCherry reporter plasmid. Each assay included the LsCas13 nuclease pre-expressed with gRNA-1 or the non-targeting (NT) gRNA. The nuclease and gRNA were pre-expressed for 16 h before being combined with the deGFP and mCherry plasmids in fresh TXTL mix. The fluorescence was measured for 16 h at 29 °C in a plate reader. Lines and shaded regions represent the mean and standard deviation of quadruplicates

3.1.2 gRNA

1. Determine the direct repeat and the length of the spacer in your CRISPR-Cas13 array of interest, such as by using CRISPR-Cas Finder [41].
2. Choose a targeted plasmid (e.g., deGFP plasmid named pTXTL-P70a(2)-deGFP HP available through Arbor Biosciences). Any gene can be used as a target, as long as it is transcribed in TXTL.
3. Design the target sequence (also called the protospacer) based on the targeted plasmid and the PFS of the nuclease.
   (a) Find a PFS in the gene of interest, e.g., *GFP*, and use the nucleotides next to it as a protospacer. Keep in mind that Cas13 targets RNA and not DNA. Therefore only target within the transcribed region of the gene. If the start and end of the transcript are unclear, select targets within the open reading frame of the gene.
   (b) Assess the secondary structure of your chosen protospacer, e.g., with NUPACK [43]. Avoid extensive secondary structures, due to the strong preference of Cas13 for ssRNA [19, 27].
   (c) Clone the spacer sequence between two direct repeats and under the control of a strong promoter, e.g., the constitutive promoter J23108 or the inducible T7 promoter. *See* Fig. 2 for a detailed example. Depending on which nuclease is used, it is enough to encode just one repeat next to the spacer. In this case, it has to be clear if the repeat needs

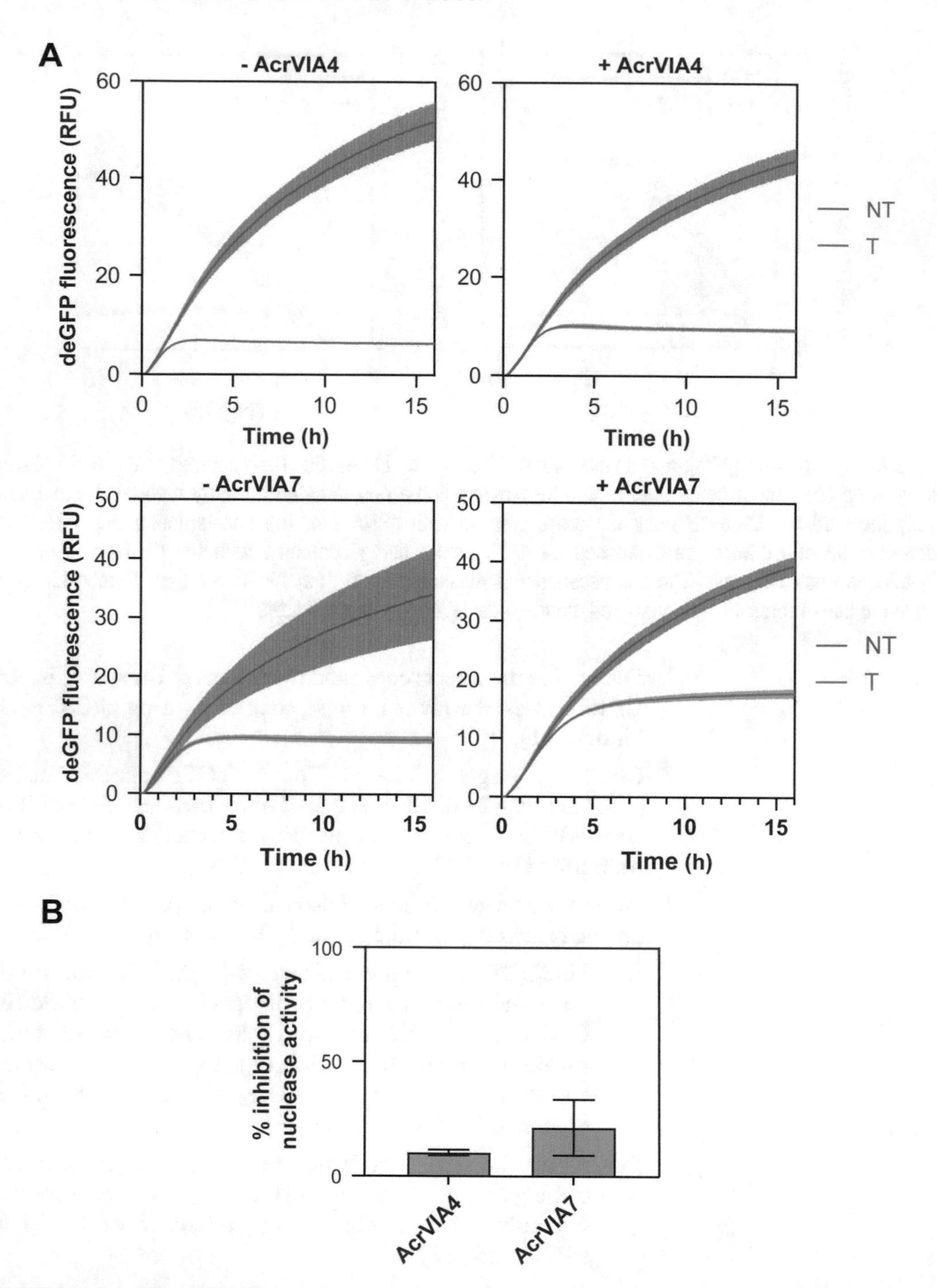

Fig. 7 Assessing the inhibitory activity of Acrs against LsCas13a. (**a**) Time courses of the RNA cleavage assay with two different Acrs. Each assay included the deGFP reporter plasmid and a Cas13 nuclease pre-expressed with gRNA-1 or non-targeting (NT) gRNA, and pre-expressed AcrVIA4 or AcrVIA7. These pre-expression mixes were then combined with the deGFP plasmid in fresh TXTL mix. The fluorescence was measured for 16 h at 29 °C in a plate reader. Lines and shaded regions represent the mean and standard deviation of quadruplicates. (**b**) Calculated inhibition. The extent of inhibition of RNA cleavage by LsCas13a was calculated using the end-point measurements

to be upstream or downstream of the spacer [44]. For these constructs, the length of a given guide can be determined experimentally similar to prior work (*see* ref. **19** as one example), although the natural spacer length can again be used. Here, we used spacers of 30 nts and 32 nts for BzCas13b and LsCas13a, respectively.

(d) To avoid steric hindrance between the gRNA and nuclease during binding, at least 15–20 nucleotides should lie between the last repeat and the terminator of the gRNA construct.

3.1.3 Acrs

1. Determine the amino acid sequence of the protein of interest, e.g., either an already published Acr or a putative Acr.
2. Codon-optimize the aa sequence using any of the available tools, e.g., from IDT [42].
3. Include a strong promoter for expression in *E. coli*, either constitutive or inducible, a strong ribosome-binding site (RBS), the coding region of the putative Acr, and a rho-independent terminator (*see* **Notes 3** and **4**).
4. The coding region of the *acr* gene can be replaced with any putative *acr*.

3.2 Purification of DNA

The DNA used here is dissolved in nuclease-free water.

3.2.1 Plasmid DNA

1. Culture *E. coli* cells harboring the plasmid for TXTL in LB medium with the respective antibiotic in an overnight culture.
2. Isolate plasmid DNA with a plasmid midiprep kit (*see* **Note 5**).
3. Perform a second DNA purification step using a DNA clean and concentrator kit.
4. Determine the DNA concentration using a spectrophotometer/fluorometer, e.g., by DeNovix.
5. Set the initial DNA concentration for each construct according to the amount of DNA needed for the respective experiment, e.g., 48 nM for the Acr constructs and make the appropriate DNA stock.

3.2.2 Linear DNA

Linear DNA can be used instead of plasmid DNA (*see* **Note 2**).

1. Linear constructs can for example be ordered as gBlocks from IDT or genes of interest (GOI) can be linearized from a plasmid via PCR.
2. Design primers flanking the GOI including promoter and terminator regions, e.g., with the NCBI primer design tool [45].

3. Perform a standard PCR to multiply the fragment. Depending on the amount of DNA that is needed, multiple reactions can be performed and combined afterwards.
4. Clean the PCR product with a standard DNA clean-up kit.
5. Set the initial DNA concentration for each construct according to the amount of DNA needed for the respective experiment, e.g., 48 nM for the Acr constructs and make the appropriate DNA stock.

3.3 TXTL Pre-expression and RNA Cleavage Assay to Observe On-target and Collateral RNA Cleavage Using Fluorescence Reporters

To enable formation of an active ribonucleoprotein complex (RNP) between Cas13 and a gRNA, both Cas13 and the gRNA are pre-expressed together in a single TXTL reaction by adding the DNA encoding nuclease and gRNA and the T7 RNA polymerase, if necessary (*see* **Note 4**). After a first incubation step, the targeted deGFP reporter plasmid is added and fluorescence can be measured over time. If desired, a plasmid encoding another fluorescent reporter, e.g., mCherry, lacking the target sequence can be added to observe collateral damage.

1. Thaw the appropriate volume of TXTL on ice. Upon preparation of a master mix, take into account that about 10% more reaction volume than actually needed should be prepared due to inaccuracies during pipetting.
2. Add the appropriate amount of DNA encoding the nuclease, gRNA, and T7 RNA polymerase (*see* **Note 4**) to the TXTL mix, *see* Table 2. Mix gently by vortexing and spin it down in a tabletop centrifuge afterwards (*see* **Note 6**).
3. Incubate the mix for 4 h at 29 °C in a thermocycler or incubator (*see* **Notes 7** and **8**).
4. Add the appropriate amount of DNA encoding the deGFP reporter plasmid to the TXTL mix (*see* **Note 9**).
5. *Optional:* Add the appropriate amount of DNA encoding the mCherry reporter plasmid (final concentration of 1 nM) to the TXTL mix (*see* **Notes 10** and **11**).
6. Load 5-μl duplicates of each reaction in the bottom of a 96-well V bottom plate. Also load 5-μl duplicates of control reactions (*see* Table 2 and **Note 11**).
7. Seal the plate with a cover mat to prevent evaporation over the course of the reaction.
8. Place the sealed plate in a plate reader to measure GFP fluorescence (Ex 485 nm, Em 528 nm, *see* **Note 10**). The plate reader should be pre-warmed to 29 °C (*see* **Note 7**).
9. Incubate the reactions for 16 h at 29 °C (*see* **Note 7**) and measure fluorescence every 3 min (Ex 485 nm, Em 528 nm).

Table 2
Components for expression of Cas13, a gRNA, and a fluorescence reporter plasmid in TXTL to observe cleavage of a targeted RNA. The table states the reaction volume needed for one experiment. The master mix for this reaction was prepared containing either the targeting or non-targeting gRNA plasmid. The construct encoding the T7 RNA polymerase is added if any of the components are under the control of a T7 promoter. If that is not the case, then additional water is added in place of the T7 RNA polymerase plasmid

Component	Volume (μl)	Initial concentration (nM)	Final concentration (nM)	Negative control (μl)
TXTL	9	–	–	9
Cas13 plasmid	1	12–48	1–4	–
gRNA plasmid	1	12–48	1–4	–
T7 RNA polymerase	0.5	4.8	0.2	–
Water	0	–	–	3
deGFP plasmid	0.5	24	1	–
mCherry plasmid			*1*	–

3.4 TXTL Pre-expression and RNA Cleavage Assay to Assess Inhibition by Acrs

To enable formation of an active Cas13-gRNA complex, the RNP complex is pre-expressed in TXTL by adding the DNA encoding nuclease, gRNA, and the T7 RNA polymerase. The reaction is incubated at 29 °C for 16 h to ensure maximal generation of the complex (*see* **Note 7**). The putative Acr is pre-expressed in a separate reaction to avoid nonspecific inhibitory effects on the biogenesis of the RNP complex [11]. Of course the assay can also be changed to observe potential inhibition of the biogenesis by only pre-expressing the Acr, but not the nuclease and gRNA (*see* **Note 8**). After incubation, a targeted deGFP reporter plasmid together with the pre-expressed RNP complex and Acr is added to fresh TXTL mix and fluorescence can be measured over time. In this setup, like in the one above, a plasmid encoding another fluorescent reporter, e.g., mCherry, can be added to observe collateral RNA damage.

3.4.1 Pre-expression of Nuclease and gRNA

1. Thaw the appropriate volume of TXTL on ice. Upon preparation of a master mix, take into account that about 10% more reaction volume than actually needed should be prepared due to inaccuracies during pipetting.
2. Add the appropriate amount of DNA encoding the nuclease and gRNA to the TXTL mix, as listed in Table 3. Mix the solution gently by vortexing and spin it down afterwards (*see* **Notes 6** and **9**).

Table 3
Components for pre-expression of Cas13 and a gRNA in TXTL. The table states the reaction volume needed for one experiment. The master mix for this reaction was prepared containing either the targeting or non-targeting gRNA plasmid

Component	Volume (μl)	Initial concentration (nM)	Final concentration (nM)
TXTL	9	–	–
Cas13 plasmid	1	12–48	1–4
gRNA plasmid	1	12–48	1–4
Water	1	–	–

Table 4
Components for pre-expression of an Acr in TXTL. The table states the reaction volume needed for one experiment. The master mix for this reaction was prepared containing all components besides the Acr plasmid

Component	Volume (μl)	Initial concentration (nM)	Final concentration (nM)
TXTL	9	–	–
Acr plasmid	1	48	4
T7 RNA polymerase	1	2.4	0.2
Water	1	–	–

3. Place the 12-μl reactions in a thermocycler or incubator and incubate them for 16 h at 29 °C (*see* **Note 7**).

3.4.2 Pre-expression of Acrs

1. Thaw the appropriate volume of TXTL on ice. Upon preparation of a master mix, take into account that about 10% more reaction volume than actually needed should be prepared due to inaccuracies during pipetting.
2. Add the appropriate amount of DNA encoding the Acr to the TXTL mix, as listed in Table 4. Mix gently by vortexing and spin it down afterwards (*see* **Notes 6** and **9**).
3. Place the 12-μl reactions in a thermocycler or incubator and incubate them for 16 h at 29 °C (*see* **Note 7**).

3.4.3 RNA Cleavage Assay

1. Thaw the appropriate volume of TXTL on ice.
2. Prepare the reaction by adding the components listed in Table 5 to a fresh PCR tube. Mix gently by vortexing and spin it down afterwards (*see* **Notes 6** and **9**).
3. Load 5-μl duplicates of each reaction in the bottom of a 96-well V bottom plate. Be careful not to introduce any bubbles.
4. Seal the plate with a cover mat to prevent evaporation over the course of the reaction.

Table 5
RNA cleavage assay in TXTL to observe potential inhibitory activity of putative Acrs. The mRNA resulting from the plasmid encoding deGFP is targeted by the Cas13:gRNA complex. Upon inhibitory activity of the Acr, target RNA cleavage will not take place or be reduced. If desired, mCherry can be added to observe collateral RNA cleavage and, in the case of expressed Acrs that inhibit Cas13, the inhibition of both target RNA cleavage and collateral activity

Component	Volume (μl)	Initial concentration (nM)	Final concentration (nM)	Negative control (μl)
TXTL	9	–	–	9
Pre-expressed Cas13 and gRNA	1	–	–	–
Pre-expressed Acr	1	–	–	–
deGFP plasmid	1	12	1	–
Water	0	–	–	3

5. Place the sealed plate in a plate reader to measure GFP fluorescence (Ex 485 nm, Em 528 nm). The plate reader should be pre-warmed to 29 °C (*see* **Notes 7** and **10**).
6. Incubate the reactions for 16 h at 29 °C (*see* **Note 7**) and measure fluorescence every 3 min (Ex 485 nm, Em 528 nm).

3.5 Data Processing

The plate reader can export the fluorescence data as an Excel spreadsheet after the run is finished. Here, we explain how to process the fluorescence data to evaluate to what degree the Cas13 nuclease altered deGFP and mCherry fluorescence levels and the extent to which the Acr inhibited RNA cleavage.

1. Export the data in an Excel spreadsheet after the plate reader run is finished. It should include time points, and fluorescence intensity values for each well and each time point.
2. Subtract the background fluorescence. Background fluorescence was measured for each plate reader separately using a TXTL reaction containing 3 μl of water and no DNA (*see* **Note 11**).
3. Calculate the average of the replicates and visualize the data on a graph by plotting the fluorescence over time.
4. If the experiment was performed in triplicates or even quadruplicates, the standard deviations for the replicates can be calculated and used to show error bars for each time point in the graph. Another possibility is to calculate the standard deviation and error by replicates conducted in separate runs.
5. Calculate the fold-reduction for the reporter construct using the ratio of deGFP concentrations after 16 h of the reaction containing non-targeting gRNA over the reaction containing targeting gRNA. The same can be done for the fluorescence values of mCherry (*see* **Note 12**).

6. Percent inhibition of RNA cleavage by the nuclease can be calculated using the following formula:

$$\%\text{Inhibition of nuclease activity} = 100\% * \left(\frac{\text{GFPt,Acr}/\text{GFPnt,Acr} - \text{GFPt-}/\text{GFPnt-}}{1 - \text{GFPt-}/\text{GFPnt-}}\right)$$

4 Notes

1. Here, we use the commercially available myTXTL mix from Arbor Biosciences. Other commercial sources and instructions on how to produce the mix in the lab are available [31].
2. If linear DNA is used, the TXTL reaction requires GamS protein (2 μM in the final reaction) or linear DNA encoding Chi sites that block rapid degradation of linear DNA by RecBCD [2, 46].
3. Some proteins can have a toxic effect in bacteria. While this is technically no problem in TXTL, since it is a cell-free system, it can be a problem when propagating the plasmid in *E. coli*. Therefore, it can be advantageous to clone nucleases or putative Acrs under the control of an inducible promoter to avoid potential toxic effects.
4. If an inducible promoter is used, the respective inducers have to be added to the reaction, e.g., in the form of DNA in the case of the T7 promoter, where a plasmid encoding the T7 RNA polymerase is needed. Other inducible or constitutive promoters can be used, a list can be found online [31, 35, 36].
5. We have found that preparing plasmid DNA using midiprep leads to more consistent experimental results, therefore we only used midiprep kits for DNA preparation.
6. Do not vortex on full speed, half speed is sufficient (equals about 1400 rpm). Spin down briefly in a tabletop centrifuge.
7. We conduct all reactions at 29 °C, the optimal temperature for deGFP production, but the temperature can be varied between 25 °C and 42 °C [36].
8. TXTL mix is active for up to 16 h. For short pre-expression times between 0 and 4 h, the mix can be subsequently used in the cleavage assays. Pre-expression times can be varied between 0 and 4 h. This can be tested in a time course experiment and adapted to the respective experiment. If longer pre-expression times are performed, it is necessary to use fresh TXTL mix for the cleavage assay. Both methods are described in the Methods section as separate experiments.

9. Do not exceed a maximum volume of 12 μl; instead, adjust concentrations of stock DNA if necessary. If necessary, fill the reaction with water to reach 12 μl of volume.
10. Other fluorescent reporter plasmids can be used in this experimental setup. If more than one fluorescent reporter is used, consider the values for excitation and emission, so they don't overlap. Here, we used deGFP (Ex 485 nm, Em 528 nm) and mCherry (Ex 587 nm, Em 610 nm).
11. Background fluorescence was measured for each plate reader separately using a TXTL reaction containing 9 μl of TXTL mix and 3 μl of water but no added DNA (Table 2). Positive controls were added for each fluorescence reporter, containing only TXTL mix, the reporter plasmid, and filled up to 12 μl with water.
12. Other ways of analyzing the data are possible, but not described in detail here. If multiple replicates, e.g., quadruplicates, are performed, outliers can be determined by Grubb's analysis. Another possibility is to determine the time point at which NT and T deviate from each other.

Acknowledgments

We thank Dr. Chunyu Liao for providing the plasmid expressing the *Leptotrichia shahii* Cas13a, Ms. Fani Ttofali for providing the plasmid expressing the non-targeting gRNA for LsCas13a, and Ms. Franziska Wimmer for providing feedback on the protocol.

Funding: *This work was supported by the DARPA Safe Genes program (HR0011-17-2-0042 to C.L.B.).*

References

1. Garamella J, Marshall R, Rustad M, Noireaux V (2016) The all E. coli TX-TL toolbox 2.0: a platform for cell-free synthetic biology. ACS Synth Biol 5:344–355
2. Marshall R, Maxwell CS, Collins SP et al (2017) Short DNA containing χ sites enhances DNA stability and gene expression in E. coli cell-free transcription-translation systems. Biotechnol Bioeng 114:2137–2141
3. Silverman AD, Karim AS, Jewett MC (2020) Cell-free gene expression: an expanded repertoire of applications. Nat Rev Genet 21:151–170
4. Tinafar A, Jaenes K, Pardee K (2019) Synthetic biology goes cell-free. BMC Biol 17
5. Takahashi MK, Tan X, Dy AJ et al (2018) A low-cost paper-based synthetic biology platform for analyzing gut microbiota and host biomarkers. Nat Commun 9:3347
6. Swartz JR (2018) Expanding biological applications using cell-free metabolic engineering: an overview. Metab Eng 50:156–172
7. Agrawal DK, Marshall R, Noireaux V, Sontag ED (2019) In vitro implementation of robust gene regulation in a synthetic biomolecular integral controller. Nat Commun 10:5760
8. Marshall R, Noireaux V (2018) Synthetic biology with an All E. coli TXTL system: quantitative characterization of regulatory elements and gene circuits. Methods Mol Biol 1772:61–93
9. Maxwell CS, Jacobsen T, Marshall R et al (2018) A detailed cell-free transcription-translation-based assay to decipher CRISPR

protospacer-adjacent motifs. Methods 143:48–57

10. Liao C, Ttofali F, Slotkowski RA et al (2019) Modular one-pot assembly of CRISPR arrays enables library generation and reveals factors influencing crRNA biogenesis. Nat Commun 10:2948
11. Marshall R, Maxwell CS, Collins SP et al (2018) Rapid and scalable characterization of crispr technologies using an E. coli cell-free transcription-translation system. Mol Cell 69:146–157.e3
12. Marshall R, Beisel CL, Noireaux V (2020) Rapid testing of CRISPR nucleases and guide RNAs in an E. coli cell-free transcription-translation system. STAR Protocol 1:100003
13. Knott GJ, Doudna JA (2018) CRISPR-Cas guides the future of genetic engineering. Science 361:866–869
14. Champer J, Buchman A, Akbari OS (2016) Cheating evolution: engineering gene drives to manipulate the fate of wild populations. Nat Rev Genet 17:146–159
15. Simeonov DR, Marson A (2019) CRISPR-based tools in immunity. Annu Rev Immunol 37:571–597
16. Wu WY, Lebbink JHG, Kanaar R et al (2018) Genome editing by natural and engineered CRISPR-associated nucleases. Nat Chem Biol 14:642–651
17. Jacobsen T, Ttofali F, Liao C et al (2020) Characterization of Cas12a nucleases reveals diverse PAM profiles between closely-related orthologs. Nucleic Acids Res 48:5624–5638
18. Leenay RT, Maksimchuk KR, Slotkowski RA et al (2016) Identifying and visualizing functional PAM diversity across CRISPR-Cas systems. Mol Cell 62:137–147
19. Abudayyeh OO, Gootenberg JS, Konermann S et al (2016) C2c2 is a single-component programmable RNA-guided RNA-targeting CRISPR effector. Science 353:aaf5573
20. Shmakov S, Abudayyeh OO, Makarova KS et al (2015) Discovery and functional characterization of diverse class 2 CRISPR-Cas systems. Mol Cell 60:385–397
21. Yan WX, Chong S, Zhang H et al (2018) Cas13d is a compact RNA-targeting type VI CRISPR effector positively modulated by a WYL-domain-containing accessory protein. Mol Cell 70:327–339.e5
22. Mustafa MI, Makhawi AM (2020) SHERLOCK and DETECTR: CRISPR-Cas systems as potential rapid diagnostic tools for emerging infectious diseases. J Clin Microbiol 59: e00745-20. https://doi.org/10.1128/JCM.00745-20
23. Gootenberg JS, Abudayyeh OO, Lee JW et al (2017) Nucleic acid detection with CRISPR-Cas13a/C2c2. Science 356:438–442
24. Kiga K, Tan X-E, Ibarra-Chávez R et al (2020) Development of CRISPR-Cas13a-based antimicrobials capable of sequence-specific killing of target bacteria. Nat Commun 11:2934
25. Abudayyeh OO, Gootenberg JS, Essletzbichler P et al (2017) RNA targeting with CRISPR-Cas13. Nature 550:280–284
26. Metsky HC, Welch NL, Haradhvala NJ et al Efficient design of maximally active and specific nucleic acid diagnostics for thousands of viruses
27. Bandaru S, Tsuji MH, Shimizu Y et al (2020) Structure-based design of gRNA for Cas13. Sci Rep 10:11610
28. Lin P, Qin S, Pu Q et al (2020) CRISPR-Cas13 inhibitors block RNA editing in bacteria and mammalian cells. Mol Cell 78:850–861.e5
29. Meeske AJ, Jia N, Cassel AK et al (2020) A phage-encoded anti-CRISPR enables complete evasion of type VI-A CRISPR-Cas immunity. Science 369:54–59
30. Wandera KG, Collins SP, Wimmer F et al (2020) An enhanced assay to characterize anti-CRISPR proteins using a cell-free transcription-translation system. Methods 172:42–50
31. Sun ZZ, Hayes CA, Shin J et al (2013) Protocols for implementing an *Escherichia coli* based TX-TL cell-free expression system for synthetic biology J Vis Exp e50762
32. Dondapati SK, Stech M, Zemella A, Kubick S (2020) Cell-free protein synthesis: a promising option for future drug development. BioDrugs 34:327–348
33. Johnston WA, Alexandrov K (2014) Production of eukaryotic cell-free lysate from Leishmania tarentolae. Methods Mol Biol 1118:1–15
34. Thoring L, Dondapati SK, Stech M et al (2017) High-yield production of "difficult-to-express" proteins in a continuous exchange cell-free system based on CHO cell lysates. Sci Rep 7:11710
35. Promoters/Catalog/Anderson—parts.igem.org. http://parts.igem.org/Promoters/Catalog/Anderson. Accessed 26 Mar 2020
36. Shin J, Noireaux V (2010) Efficient cell-free expression with the endogenous E. coli RNA polymerase and sigma factor 70. J Biol Eng 4:8

37. Salis HM (2011) The ribosome binding site calculator. Methods Enzymol 498:19–42
38. Smargon AA, Cox DBT, Pyzocha NK et al (2017) Cas13b is a type VI-B CRISPR-associated RNA-guided RNase differentially regulated by accessory proteins Csx27 and Csx28. Mol Cell 65:618–630.e7
39. Cox DBT, Gootenberg JS, Abudayyeh OO et al (2017) RNA editing with CRISPR-Cas13. Science 358:1019–1027
40. Meeske AJ, Marraffini LA (2018) RNA guide complementarity prevents self-targeting in type VI CRISPR systems. Mol Cell 71:791–801.e3
41. CRISPR-CAS++. https://crisprcas.i2bc.paris-saclay.fr/CrisprCasFinder/Index. Accessed 27 Mar 2020
42. Website. https://eu.idtdna.com/codonopt. Accessed 3 Dec 2020
43. Zadeh JN, Steenberg CD, Bois JS et al (2011) NUPACK: analysis and design of nucleic acid systems. J Comput Chem 32:170–173
44. O'Connell MR (2019) Molecular mechanisms of RNA targeting by Cas13-containing Type VI CRISPR–Cas systems. J Mol Biol 431:66–87
45. Primer designing tool. https://www.ncbi.nlm.nih.gov/tools/primer-blast/. Accessed 8 Dec 2020
46. Sitaraman K, Esposito D, Klarmann G et al (2004) A novel cell-free protein synthesis system. J Biotechnol 110:257–263

Part III

Interactomics

Chapter 8

Studying RNP Composition with RIP

Annalisa Rossi and Alberto Inga

Abstract

RNA is never left alone throughout its life cycle. Together with proteins, RNAs form membraneless organelles, called ribonucleoprotein particles (RNPs) where these two types of macromolecules strongly influence each other's functions and destinies. RNA immunoprecipitation is still one of the favorite techniques which allows to simultaneously study both the RNA and protein composition of the RNP complex.

Key words RNA-binding protein, Immunoprecipitation, Ribonucleoprotein

1 Introduction

Post-transcriptional mechanisms regulating gene expression depend on a combined action of RNA-binding proteins (RBP) and RNAs. Soon after the start of transcription RNA molecules associate with a distinct class of RBPs, giving rise to macromolecular, functional units called ribonucleoprotein particles (RNP), which control either gene expression or the functions of the components present in the complex. In this contest, RBPs crucially regulate key steps of RNA metabolism, including splicing, polyadenylation, capping, export, localization, editing, translation, and turnover. In addition, RNAs present in the RNP can also regulate interactions, localization, and functions of the RBPs [1–4].

RBPs interact with different affinities with RNA. They are characterized by the presence of one or more RNA-binding domains (RBDs), containing 60–100 residues that usually associate with RNA in a structure and sequence-dependent manner. To date, more than 40 RBDs have been identified, the most common class being the RNA recognition motif found in >50% of the RBPs. Other RBDs include the K-homology (KH) domain, the Larp module, zinc fingers, double-stranded RNA-binding motifs,

Erik Dassi (ed.), *Post-Transcriptional Gene Regulation*, Methods in Molecular Biology, vol. 2404,
https://doi.org/10.1007/978-1-0716-1851-6_8,

DEAD-box domains, and the Piwi/Argonaute/Zwille (PAZ) domain [5–8].

Because the RNP composition determines the fate of any given mRNA in virtually all phases of its life cycle, it is perhaps not surprising that changes in RNP composition or function are increasingly recognized as causes of human disease. Alterations in the expression and localization of RBPs can influence the expression levels of oncogenes and tumor-suppressor genes and can lead to diverse cancer-related cellular phenotypes such as proliferation, angiogenesis, and invasion [9–11]. Further, misregulated assembly of RNPs in the central nervous system often leads to neurodegenerative diseases such as amyotrophic lateral sclerosis and spinal muscular atrophy [12–14].

A detailed insight into the composition of these highly dynamic RNPs, their assembly, and remodeling is pivotal for understanding how they control gene expression. Substantial advances have been made towards this goal through recent technologies in the analysis of RNAs and RNPs.

Here we describe one of the most basic method, the RNA immunoprecipitation (RIP), which relies on the immunoprecipitation of RNP complexes typically achieved through the use of an antibody targeting one of the RBP in the RNP complex, with subsequent isolation, analysis, and identification of RNAs bound to the RBP. Such identification can be obtained by qRT-PCR in the case of known targets or by RNA-sequencing in the case of unknown targets or when a more comprehensive, potentially unbiased analysis is needed. The main advantages of the RIP protocol are that it gives the possibility of isolating RNP complexes, allowing identification of components other than mRNAs, such as other regulatory or RNA-processing proteins, as well as small noncoding RNAs. In addition, RIP is an ideal method for studying remodeling of RNPs during the dynamic processes of post-transcriptional gene expression [15].

The protocol we present here is based on the RIP protocol published by Keene and colleagues but with few modifications. All steps of the procedure from harvesting cells to RNA extraction are given. In addition, we provide several hints as a guide for optimizing specific steps of the experiment [16].

As an example to illustrate the protocol, we describe how we performed a RIP of the DHX30 RNA helicase and its associated RNAs starting from HCT116 cells.

2 Materials

2.1 Tissue Cell Culture Components

1. HCT116 is a human colorectal carcinoma cell line and was maintained in RPMI (Gibco) media supplemented with 10% FBS, antibiotics (100 units/mL penicillin plus 100 mg/mL streptomycin), and 2 mM L-glutamine (*see* **Note 1**).

2.2 Sample Collection Components

1. Phosphate-buffered saline (PBS) for cell culture (pH 7.4): 10× PBS contains potassium phosphate monobasic (KH_2PO_4) 1.440 g/L, sodium chloride (NaCl) 90 g/L, sodium phosphate dibasic (Na_2HPO_4 -$7H_2O$) 7.950 g/L.
2. Polysome lysis buffer (PLB): 100 mM Potassium chloride (KCl), 5 mM magnesium chloride ($MgCl_2$), 10 mM HEPES (pH 7.0), 0.5% Nonidet P-40 (NP40), 1 mM dithiothreitol (DTT), 100 U/mL RNase Out, 1× complete protease inhibitor cocktail (Merk). To prepare 5 mL PLB add 50 μL of 1 M HEPES (pH 7.0), 500 μL of 1 M KCl, 25 μL of 1 M $MgCl_2$, and 100 μL of 25% NP40 to 4207.5 mL of nuclease-free H_2O. Then add 5 μL of 1 M DTT, 12.5 μL RNase Out, and 100 μL of 50× protease inhibitor cocktail (*see* **Note 2**).

2.3 RNP Immunoprecipitation Components

1. Protein A/G dynabeads: this choice depends on the species and IgG type of the antibody to be conjugated.
2. Antibodies: antibody recognizing an RBP of interest, and an isotype-matched control antibody (here: rabbit anti-DHX30 (Bethyl A302-218A), and Normal Rabbit anti-IgG (Merck-Millipore 12-370)).
3. NT2 buffer: 50 mM Tris–HCl (pH 7.4), 150 mM NaCl, 1 mM $MgCl_2$, 0.05% NP40.

   To make 50 mL of NT2 buffer, add 2.5 mL of 1 M Tris (pH 7.4), 1.5 mL of 5 M NaCl, 50 μL of 1 M $MgCl_2$, and 100 μL of 25% NP40 to 45.85 mL nuclease-free H_2O.
4. RNasin.
5. Dithiothreitol (DTT).
6. TRIZOL.
7. Chloroform, isopropanol, and glycogen.

3 Methods

3.1 Tissue Culture and RNP Lysate Collection

1. Maintain HCT116 cells in the RPMI cell growth medium at 37 °C and 5% CO_2. For a complete experiment, two 15 cm dishes at 70–80% of confluence will be needed.
2. Decant the growth medium, place the dish on ice, wash the adherent cells with ice-cold PBS, collect the cells off the dish using a scraper, and transfer them into a pre-chilled 50 mL conical tube. Spin down the cells at 200 × *g* at 4 °C for 5 min. Wash the cell pellet with cold PBS, spin down again, and aspirate the supernatant as much as possible, without disturbing the cell pellet.
3. Resuspend the cell pellet in 1 mL of PLB buffer by pipetting up and down. Allow the lysate to chill on ice for 5 min before immediately freezing and storing the pellet at −80 °C until you proceed with the experiment (*see* **Note 3**).

3.2 Antibody Coating of Protein A/G Beads

For the immunoprecipitation of DHX30 protein, we use magnetic protein A dynabeads and coat them manually with rabbit anti-DHX30 antibody or normal rabbit IgG, according to the manufacturer's descriptions.

1. Vortex the dynabeads for 30 s and transfer 25 μL of beads per sample into two nuclease-free low-retention 1.5 mL tubes. One will be used for the RBP of interest and the other one for the control RIP.
2. Wash the beads two times with 500 μL of NT2. Collect the beads using a magnetic stand and resuspend beads by pipetting up and down with low-affinity filter tips.
3. Dissolve 3–5 μg of specific antibody or isotype-matched IgG in 100 μL of NT2 buffer supplemented with protease inhibitors, mix by pipetting up and down, and add it to the beads (*see* **Note 4**).
4. Incubate the beads at 4 °C on a rotating wheel at 20 rpm for at least 2 h. (Alternatively, incubate at 4 °C on a rotating wheel, overnight, the day before the immunoprecipitation).
5. Immediately before usage, wash the antibody-coated beads two times with an ice-cold NT2 buffer supplemented with RNase and protease inhibitors to remove unbound antibody and potential contaminants.
6. After the final wash, resuspend the designated amount of antibody-coated beads per sample in 720 μL of ice-cold NT2 buffer. Add to each sample/tube 40 U/mL of RNase inhibitors, 1× protease inhibitors, and DTT to a final concentration of 1 mM, and keep them on ice until the lysate samples are ready to proceed with the IP.

3.3 Immunoprecipitation Reaction and RNA Extraction

1. Thaw the cell lysate (**step 3** of Subheading 3.1) on ice.
2. Centrifuge the lysate at 4 °C at 20,000 × *g* for 30 min to clear the sample.
3. Transfer the cleared lysate into a new pre-chilled microfuge tube and store it on ice. Be careful not to disturb the pellet that is formed. At least 800 μL of lysate will be collected.
4. Take two 40 μL aliquots of the cleared lysate of each sample and store them on ice. The first one represents total cellular RNA as input for the RIP and will be needed for a subsequent qRT-PCR or RNA-seq library preparation. The total RNA input is used to calculate and quantify the enrichment of specifically RBP-bound transcripts. The second aliquot is used for control western blotting to verify the input RBP amount. Add 80 μL of NT2 buffer to both aliquots. This matches the total RNA and protein concentration to the subsequent immunoprecipitation steps.

5. For each sample wash 10–15 μL of protein A/G magnetic beads as in **step 2** of Subheading 3.2. Add 360 μL of lysate to the washed beads and proceed with the pre-clearing step incubating 1 h on a rotating wheel, at 4 °C. This step reduces background signal (*see* **Note 5**).
6. Use the magnetic stand to collect the beads and transfer the pre-cleared lysate to the beads coated with either the antibody recognizing the RBP and the control IgG from **step 6** of Subheading 3.2 (*see* **Note 6**).
7. Slowly rotate the IP sample at 4 °C tumbling end over end for 4 h or overnight (*see* **Note 7**).
8. Use the magnetic stand to collect the beads and transfer the supernatant into a new tube. Take a 40 μL aliquot of the supernatant and store it on ice for a subsequent control western blotting. This aliquot serves as a control for estimating the IP efficiency.
9. Wash the beads five times with 1 mL of ice-cold NT2 buffer by pipetting up and down with low-affinity filter tips. At the last washing step, dissolve the beads in 400 μL of the washing buffer and take 40 μL of the beads slurry and store it on ice for IP WB control. This sample, together with the input and supernatant, serves to control for effective RBP immunoprecipitation and potential loss of RBP during the washing steps (*see* **Note 8**).
10. Release the RNA from the RNP complexes by adding 1 mL of TRIZOL reagent directly to the tube (*see* **Note 9**).
11. Isolate RNA following the manufacturer's descriptions from either TRIZOL or by using the column-based Direct-Zol RNA MiniPrep kit.
12. Resuspend or elute the RNA in up to 15 μL RNase-free water for subsequent RNA quantification. RNA can be stored at 80 °C for months (*see* **Note 10**).

3.4 RNP Immunoprecipitation Controls

As mentioned before, in the case the RNA targets of the RBP are known, RIP technique can be suitable to analyze the composition and remodeling of its RNP in particular experimental conditions. Instead, if the targets of the RBP are still unknown, after RIP it is possible to proceed with RNA-seq library preparation. In both cases it is important to perform control experiments to assess enrichment of both the RBP and RNA components of the RNP being immunoprecipitated.

1. Perform SDS–PAGE followed by Western blotting running the collected lysate samples, i.e., input, unbound, and IPs and incubating the membrane with the antibody of your RBP of interest and control antibody (*see* Fig. 1a and **Note 11**).

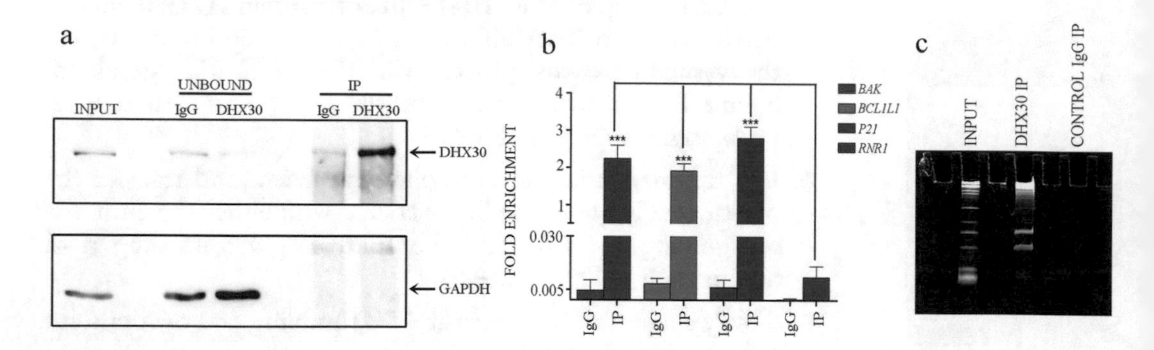

Fig. 1 An example of RIP results. (**a**) The efficiency and specificity of immunoprecipitation of DHX30 antibody was verified by western blot, comparing unbound fractions and immunoprecipitates obtained with the specific antibody or the control IgG. (**b**) RNA immunoprecipitation (RIP) assays using a DHX30 antibody in HCT116 cells. Data are plotted as the percentage of input fold enrichment relative to the signal obtained for each mRNA examined in the control immunoprecipitation (IgG). Bars plot the average and the standard deviation of three biological replicates. *P*-value was calculated comparing the amount of each specific RNA with the amount of *RNR1*, a negative control; ***$p < 0.001$, Student's t test. (**c**) Specific RNAs are enriched in the IP DHX30 eluates. DHX30-specific antibody or control IgG was used to immunoprecipitate DHX30 containing RNPs. RNAs were isolated and run onto 9.7% Urea-Denaturing PAGE. The gel was then stained with Sybr Gold (Thermo-Fisher) (Modified from Rizzotto et al. Cell Reports 2020, PMID 32234473)

2. If you have information about candidate transcripts bound to your RBP, perform qRT-PCR on the RNA samples (total RNA input, IP, and isotype-matched IP control) (*see* Fig. 1b and **Note 12**).
3. Instead, if no information about the potential targets is available and therefore it is not possible to proceed with control qRT-PCR, it is suggested to perform a denaturing RNA Urea-PAGE running input and IPs. In this way the best conditions for the RNA immunoprecipitation can be found and checked before proceeding with library preparation and RNA-seq (*see* Fig. 1c, **Note 13** and [17]).

4 Notes

1. Studying the RNP composition in a physiological context is highly recommended. However, IP-grade antibodies are not always commercially available for a specific RBP. In this case it is necessary to transfect and ectopically express an epitope-tagged RBP, and immunoprecipitate it using the antibody against the tag.
2. RNA samples must be handled cautiously to protect them from degradation caused by nucleases and heat. Always wear gloves. Keep samples at 4 °C as much as possible. Clean the workbench with nuclease-inhibitor solutions like RNase AWAY or

equivalent. All solutions and buffers used for handling RNA samples should be prepared with pure and nuclease-free water and processed with nuclease-free filter tips in nuclease-free low-retention reaction tubes.

3. If a cycle of freezing and thawing is necessary to complete the cell lysis and to avoid post-lysis reassortment, on the contrary, one more cycle can determine the disruption of the complexes and degradation of the components.
4. It is recommended to set the amount of the antibody before starting with the RIP experiment. An excessive amount can increase a nonspecific signal, on the contrary using a too small amount can lead to a loss of signal.
5. RNAs have a certain grade of nonspecific affinity with protein A/G magnetic beads. It is strongly recommended to perform a pre-clearing step. However, this procedure can dramatically reduce the capture of low abundance RNA species.
6. Keene and other colleagues suggest to dilute PBL lysates ten times in NT2 buffer following these conditions: 850 μL of NT2, 100 μL of PBL lysates, 20 mM EDTA, 1 mM DTT, 40 U/mL of RNase inhibitors, 1× protease inhibitors [16]. However, at least for DHX30 immunoprecipitation, we observed better results by diluting PBL lysates only three times in NT2 buffer.
7. Generally, 4 h are sufficient for the binding and it is recommended to avoid RNA degradation. However, when the RBP or the RNAs being investigated are expressed at low levels, overnight incubation is suggested.
8. Washing step is critical to reduce background signal. It is necessary to test the conditions for the specific RNP being investigated. To reduce background, it may be necessary to use more stringent wash conditions, such as adding from 0.5 M to 3 M urea, 0.1% or less SDS, or deoxycholate to the NT2 wash buffer. All tubes should be kept on ice as much as possible while working quickly during the washing process to reduce degradation.
9. Use of glycogen and overnight precipitation at −20 °C can improve the efficiency of the RNA recovery and extraction.
10. The volume of resuspension/elution can be adjusted to the expected yield. Of course, this will affect RNA concentrations for subsequent procedures. For accurate RNA quantification of the RIP samples use sensitive methods (e.g., Qubit RNA IQ Assay for the Qubit 4.0 Fluorometer, ThermoFisher).
11. This western blot control is necessary to verify if the RBP of interest was sufficiently immunoprecipitated from the sample. Expect to see a depleted signal in your supernatant and

enrichment in your IP/bead fraction. However, complete RBP depletion in the supernatant fraction is not required. For the nonspecific isotype control IP, there should not be a signal in the IP/bead fraction for your RBP of interest. While IP conditions are being optimized, it is recommended to include an aliquot of the IP wash step to assess RBP loss due to too harsh buffer conditions.

12. Quantify the RNA samples and reverse transcribe approximately 1–10 ng of each RNA sample into cDNA. Verify the enrichment of candidate-bound transcript using qRT-PCR and include several controls of potentially non-bound transcripts. It is critical to observe enrichment of the candidate-bound transcript in the RNP IP sample over input and isotype-matched control IP sample.

13. Even if sensitive RNA quantification methods are now available, before proceeding with expensive library preparation and sequencing, it is suggested to perform a denaturing RNA Urea-PAGE. Using an 8–9% gel and staining with Sybr Gold expect to see a faint or no signal in the control IgG and an appreciable signal in the IP.

References

1. Martin KC, Ephrussi A (2009) Review mRNA localization: gene expression in the spatial dimension. Cell 136:719–730. https://doi.org/10.1016/j.cell.2009.01.044
2. Müller-Mcnicoll M, Neugebauer KM (2013) How cells get the message: dynamic assembly and function of mRNA-protein complexes. Nat Rev Genet 14:275–287. https://doi.org/10.1038/nrg3434
3. Gehring NH, Wahle E, Fischer U (2017) Deciphering the mRNP code: RNA-bound determinants of post-transcriptional gene regulation. Trends Biochem Sci 42:369–382. https://doi.org/10.1016/j.tibs.2017.02.004
4. Mitchell SF, Parker R (2014) Principles and properties of eukaryotic mRNPs. Mol Cell 54:547–558. https://doi.org/10.1016/j.molcel.2014.04.033
5. Lunde BM, Moore C, Varani G (2007) RNA-binding proteins: modular design for efficient function. Nat Rev Mol Cell Biol 8:479–490. https://doi.org/10.1038/nrm2178
6. Ascano M, Hafner M, Cekan P, Gerstberger S, Tuschl T (2012) Identification of RNA-protein interaction networks using PAR-CLIP. Wiley Interdiscip Rev RNA 3:159–177. https://doi.org/10.1002/wrna.1103
7. Corley M, Burns MC, Yeo GW (2020) How RNA-binding proteins interact with RNA: molecules and mechanisms. Mol Cell 78:9–29. https://doi.org/10.1016/j.molcel.2020.03.011
8. Hentze MW, Castello A, Schwarzl T, Preiss T (2018) A brave new world of RNA-binding proteins. Nat Rev Mol Cell Biol 19:327–341. https://doi.org/10.1038/nrm.2017.130
9. Fidaleo M, De Paola E, Paronetto MP (2016) The RNA helicase a in malignant transformation. Oncotarget 7(19):28711–28723
10. Pereira B, Billaud M, Almeida R (2017) RNA-binding proteins in cancer: old players and new actors. Trends Cancer 3:506–528. https://doi.org/10.1016/j.trecan.2017.05.003
11. Elcheva IA, Spiegelman VS (2020) Targeting RNA-binding proteins in acute and chronic leukemia. Leukemia 35(2):360–376. https://doi.org/10.1038/s41375-020-01066-4
12. Shukla S, Parker R (2016) Hypo- and hyper-assembly diseases of RNA–protein complexes. Trends Mol Med 22:615–628. https://doi.org/10.1016/j.molmed.2016.05.005
13. Thelen MP, Kye MJ (2020) The role of RNA binding proteins for local mRNA translation: implications in neurological disorders. Front

Mol Biosci 6:161. https://doi.org/10.3389/fmolb.2019.00161

14. Advani VM, Ivanov P (2020) Stress granule subtypes: an emerging link to neurodegeneration. Cell Mol Life Sci 77:4827–4845. https://doi.org/10.1007/s00018-020-03565-0
15. Morris AR, Mukherjee N, Keene JD (2010) Systematic analysis of posttranscriptional gene expression. Wiley Interdiscip Rev Syst Biol Med 2:162–180. https://doi.org/10.1002/wsbm.54
16. Keene JD, Komisarow JM, Friedersdorf MB (2006) RIP-Chip: the isolation and identification of mRNAs, microRNAs and protein components of ribonucleoprotein complexes from cell extracts. Nat Protoc 1:302–307. https://doi.org/10.1038/nprot.2006.47
17. Rossi A, Moro A, Tebaldi T, Cornella N, Gasperini L, Lunelli L, Quattrone A, Viero G, Macchi P (2017) Identification and dynamic changes of RNAs isolated from RALY-containing ribonucleoprotein complexes. Nucleic Acids Res 45:6775–6792. https://doi.org/10.1093/nar/gkx235

Chapter 9

PAR-CLIP: A Method for Transcriptome-Wide Identification of RNA Binding Protein Interaction Sites

Charles Danan, Sudhir Manickavel, and Markus Hafner

Abstract

During *p*ost-*t*ranscriptional *g*ene *r*egulation (PTGR), *R*NA *b*inding *p*roteins (RBPs) interact with all classes of RNA to control RNA maturation, stability, transport, and translation. Here, we describe *P*hoto*a*ctivatable-*R*ibonucleoside-Enhanced Crosslinking and *I*mmunoprecipitation (PAR-CLIP), a transcriptome-scale method for identifying RBP binding sites on target RNAs with nucleotide-level resolution. This method is readily applicable to any protein directly contacting RNA, including RBPs that are predicted to bind in a sequence- or structure-dependent manner at discrete *R*NA *r*ecognition *e*lements (RREs), and those that are thought to bind transiently, such as RNA polymerases or helicases.

Key words RNA binding protein (RBP), RNA, Photoactivatable-Ribonucleoside-Enhanced Crosslinking and Immunoprecipitation (PAR-CLIP), Crosslinking and Immunoprecipitation (CLIP), Post-transcriptional gene regulation (PTGR), RNA recognition element (RRE), Non-coding RNA, mRNA, Binding site

1 Introduction

All classes of RNA are subject to *p*ost-*t*ranscriptional *g*ene *r*egulation (PTGR), including splicing, 5′- and 3′-end-modification, editing, transport, translation, and degradation [1–3]. These processes are critical for the regulation of protein-coding *m*essenger *R*NA (mRNA), as well as for the biogenesis and function of *n*on-*c*oding *R*NAs (ncRNAs, e.g., ribosomal RNA, microRNA, small interfering RNA, etc.), which themselves have a wide range of gene-regulatory functions [4]. PTGR is coordinated by the actions of *r*ibo*n*ucleo*p*roteins (RNPs), protein-RNA complexes comprised of one or more *R*NA *b*inding *p*roteins (RBPs) and associated coding or non-coding RNAs.

Charles Danan and Sudhir Manickavel contributed equally to this work.

Erik Dassi (ed.), *Post-Transcriptional Gene Regulation*, Methods in Molecular Biology, vol. 2404,
https://doi.org/10.1007/978-1-0716-1851-6_9,

The fundamental importance of PTGR is reflected in analyses of abundance, expression patterns, and evolutionary conservation of RBPs. In human cell lines and tissues, approximately 20% of the protein-coding transcriptome is comprised of RBPs, making RBPs more abundant than most other classes of proteins. The low tissue-specificity and deep evolutionary conservation of most RBP families suggest that many PTGR processes are ancient and equally essential for all cells [4]. Dysregulation of PTGR is observed in a wide variety of human pathologies, ranging from musculoskeletal and autoimmune disorders, to neurodegenerative disease, to essentially all forms of cancer [5–7].

Dissection of PTGR networks requires the careful characterization of the molecular interactions of RBPs with their RNA ligands and other binding partners, but this effort is complicated by the vast size of PTGR networks. In humans, there are approximately 1500 proteins containing identified *R*NA binding domains (RBDs), and over 20,000 protein-coding mRNAs in addition to the thousands of diverse non-coding RNAs [8]. Each RBP binds at defined sequence and structural elements termed *R*NA *r*ecognition *e*lements (RREs). However, RREs are short and partially degenerate, confounding reliable computational predictions and sparking the need for experimental methods to comprehensively identify RREs on a transcriptome-wide scale [9].

Traditionally, RREs were characterized individually in a reductive process; sequences from known RNA targets were analyzed and then putative RREs were biochemically validated. Characterization of RNPs on a transcriptome-wide scale first became possible using RNP Immunoprecipitation (RIP) followed by comprehensive identification and quantification of recovered RNAs by microarray or next-generation sequencing analysis (RIP-Chip or RIP-seq) [10]. However, RIP methods are limited to the analysis of kinetically stable interactions. Furthermore, the RRE needs to be inferred computationally from the sequence of the long recovered RNAs, which is only successful for RREs with high information content [11, 12].

*C*ross*l*inking and *I*mmuno*p*recipitation (CLIP) approaches use UV light to covalently crosslink RBPs with their RNA targets at the site of interaction. The covalent bond between the RBP and target RNAs allows for limited RNase digestions to trim the RNA to the footprint protected by the RBP, as well as additional stringent purification steps after IP, including denaturing polyacrylamide gel electrophoresis and blotting onto nitrocellulose membranes. The recovered RNA segments can then be sequenced using next-generation sequencing technologies to reveal target transcripts and RREs on a transcriptome-wide scale [13, 14].

Here we provide a step-by-step protocol for *P*hoto*a*ctivatable-*R*ibonucleoside-Enhanced CLIP (PAR-CLIP) (Fig. 1). In PAR-CLIP, photoactivatable ribonucleosides—4-thiouridine

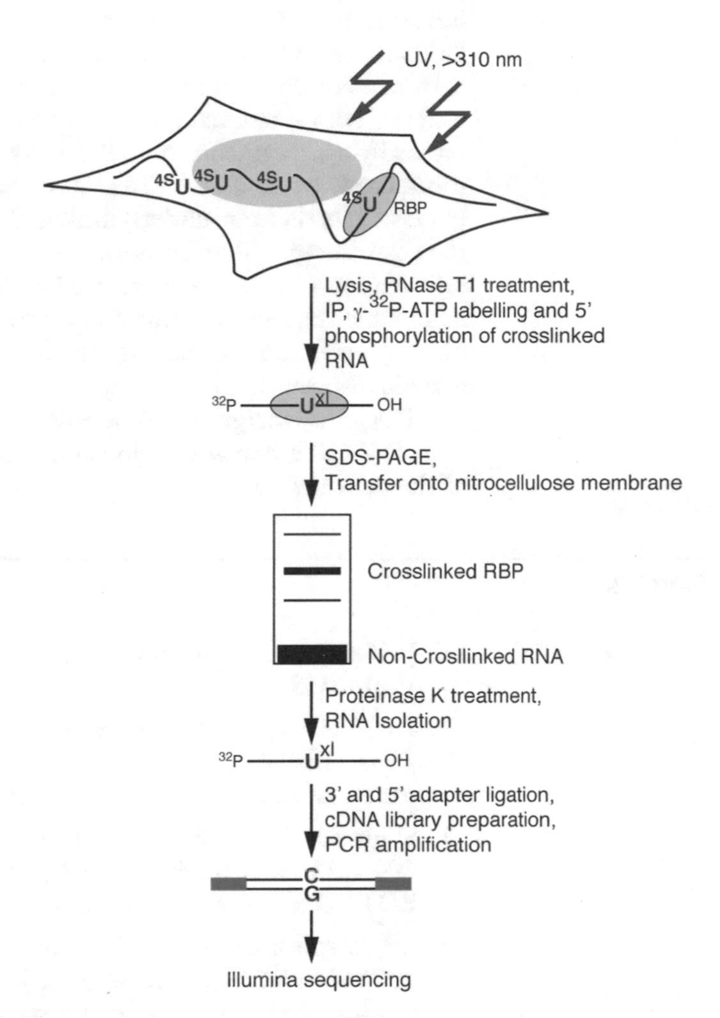

Fig. 1 Outline of the PAR-CLIP methodology. PAR-CLIP begins with incorporation of photoactivatable thioribonucleosides into nascent transcripts followed by crosslinking with long-wavelength > 310 nm UV. Crosslinked RNA-RBP complexes are isolated by immunoprecipitation and further purified by SDS-PAGE. After recovery from the purified radioactive band, the RNA is carried through a small RNA cDNA library preparation protocol for sequencing. Reverse transcription of crosslinked RNA with incorporated photoactivatable thioribonucleosides, followed by PCR amplification, leads to a characteristic mutation (T-to-C when using 4SU and G-to-A when using 6SG) that is used to identify the RNA recognition elements

(4SU), or more rarely, 6-thioguaniosine (6SG)—are incorporated into nascent RNA transcripts. The labeled RNAs are excited in living cells with UVA or UVB light (>310 nm) and yield photo-adducts with interacting RBPs. Besides an increased crosslinking efficiency compared to 254 nm CLIP, a key feature of PAR-CLIP is a characteristic mutation (T-to-C for 4SU and G-to-A for 6SG) introduced during reverse transcription at the position of cross-linking. This mutation pinpoints the sites of RNA-RBP interaction with nucleotide resolution. And, more importantly, it enables the user to computationally remove the ubiquitous background of co-purifying fragments of cellular RNAs that otherwise may be misinterpreted as signal [15]. The resulting detailed interaction maps will further our understanding of the mechanisms underlying the pathologic dysregulation of PTGR components. This information can also be integrated with emerging data from other large-scale sequencing efforts to interrogate whether variations in binding sites contribute to phenotypic variations or complex genetic disease.

The following guide covers all experimental steps of PAR-CLIP and cDNA library construction and touches on a number of aspects of the data analysis.

2 Materials

1. 1 M 4-Thiouridine (4SU) stock solution: 260.27 mg 4SU in 1 ml DMSO.
2. 1× NP40 lysis buffer: 50 mM HEPES, pH 7.5, 150 mM KCl, 2 mM EDTA, 1 mM NaF, 0.5% (v/v) NP40, 0.5 mM DTT, complete EDTA-free protease inhibitory cocktail (Roche).
3. High-salt wash buffer: 50 mM HEPES-KOH, pH 7.5, 500 mM KCl, 0.05% (v/v) NP40, 0.5 mM DTT, complete EDTA-free protease inhibitor cocktail (Roche).
4. Dephosphorylation Buffer: 50 mM Tris–HCl, pH 7.9, 100 mM NaCl, 10 mM $MgCl_2$, 1 mM DTT.
5. Polynucleotide Kinase (PNK) Buffer without DTT: 50 mM Tris–HCl, pH 7.5, 50 mM NaCl, 10 mM $MgCl_2$.
6. PNK Buffer with DTT: 50 mM Tris–HCl, pH 7.5, 50 mM NaCl, 10 mM $MgCl_2$, 5 mM DTT.
7. SDS-PAGE Loading Buffer: 10% glycerol (v/v), 50 mM Tris–HCl, pH 6.8, 2 mM EDTA, 2% SDS (w/v), 100 mM DTT, 0.1% bromophenol blue.
8. 1× Transfer Buffer with Methanol: 1× NuPAGE Transfer Buffer, 20% MeOH.
9. 2× Proteinase K Buffer: 100 mM Tris–HCl, pH 7.5150 mM NaCl, 12.5 mM EDTA, 2% (w/v) SDS.
10. Acidic Phenol/Chloroform/IAA: 25 ml acidic phenol, 24 ml chloroform, 1 ml isoamyl alcohol, pH 4.2.

11. 10× RNA Ligase Buffer without ATP: 0.5 M Tris–HCl, pH 7.6, 0.1 M $MgCl_2$, 0.1 M 2-mercaptoethanol, 1 mg/ml acetylated BSA.
12. 10× RNA Ligase Buffer with ATP: 0.5 M Tris–HCl, pH 7.6, 0.1 M $MgCl_2$, 0.1 M 2-mercaptoethanol, 1 mg/ml acetylated BSA, 2 mM ATP.
13. Formamide Gel Loading Dye: 50 mM EDTA, 0.05% (w/v) bromophenol blue in formamide.
14. SuperScript IV Reverse Transcriptase Kit including 5× buffer and enzyme (Invitrogen).
15. 10× dNTP Solution: 2 mM dATP, 2 mM dCTP, 2 mM dGTP, 2 mM dTTP.
16. 10× PCR Buffer: 100 mM Tris–HCl, pH 8.0, 500 mM KCl, 1% Triton-X-100, 20 mM $MgCl_2$, 10 mM 2-mercaptoethanol.
17. Dynabeads Protein G: Invitrogen.
18. 15 ml Centrifuge Tubes.
19. 1.5 ml DNA LoBind Tubes: Eppendorf.
20. RNase T1 (1000 U/μl): Fermentas.
21. Calf Intestinal Alkaline Phosphatase (10,000 U/ml): New England Biolabs (NEB).
22. T4 Polynucleotide Kinase (10,000 U/ml): NEB.
23. γ-^{32}P-ATP, 10 mCi/ml, 1.6 μM: Perkin Elmer.
24. NuPAGE Novex 4–12% BT Midi 1.0 gel: Invitrogen.
25. 20× NuPAGE MOPS running buffer: Invitrogen.
26. Protein Size Marker.
27. 20× NuPAGE Transfer Buffer: Invitrogen.
28. 0.45 μm Nitrocellulose Membrane.
29. Proteinase K (Powder): Roche.
30. GlycoBlue, 10 mg/ml: Ambion.
31. Truncated and mutated RNA Ligase 2, T4 Rnl2 (1-249) K227Q, 1 mg/ml: NEB; plasmid for recombinant expression can also be obtained at addgene.org.
32. T4 RNA Ligase (10 U/μl): Thermo Scientific.
33. SuperScript III Reverse Transcriptase: Invitrogen.
34. Taq DNA Polymerase, 5 U/μl.
35. MinElute Gel Extraction Kit: Qiagen.
36. Pre-adenylated, barcoded 3′ Adapter (DNA):

    AppNNTGACTGTGGAATTCTCGGGTGCCAAGG.

    Note: underlined sequence represents a barcode sequence that can be changed in order to allow multiplexed sequencing.

37. 5′ Adapter (RNA): GUUCAGAGUUCUACAGUCCGAC GAUC.
38. RT Primer: GCCTTGGCACCCGAGAATTCCA.
39. 5′ Primer: AATGATACGGCGACCACCGACAGGTTCA GAGTTCTACAGTCCGA.
40. 3′ Primer:
    CAAGCAGAAGACGGCATACGAGAT CGTGAT GTGA CTGGAGTTCCTTGGCACCCGAGAATTCCA.
    Note: underlined sequence represents an Illumina barcode sequence that can be changed in order to allow multiplexed sequencing.
41. RNA Size Marker, 19 nt, 5′ phosphorylated: 5′ p CGUACGCG GUUUAAACGA.
42. RNA Size Marker, 35 nt, 5′ phosphorylated:
    5′ p CUCAUCUUGGUCGUACGUACGCGGAAUA GUUUAAACUGU.

3 Methods

Before beginning PAR-CLIP, please *see* **Notes 1–7** for essential preparatory steps.

3.1 Preparation of UV-Crosslinked RNPs

3.1.1 Expanding Cells

1. Expand cells in appropriate growth medium in 15-cm plates. As a starting point, we recommend using a number of cells that will result in 1.5–3 ml of wet cell pellet. For HEK293 cells, approximately 50–100 × 10^6 cells will result from 5 to 10 15-cm plates. Grow cells to approximately 80% confluency.
2. Sixten hours before crosslinking, add 4SU to a final concentration of 100 μM (1:1000 v/v of a 1 M 4SU stock solution) directly to the cell culture medium (*see* **Note 8**).

3.1.2 UV-Crosslinking for Adherent Cells

1. Aspirate or pour off media from plates (*see* **Note 9**).
2. Irradiate cells uncovered with a dose of 0.15 J/cm^2 of >310 nm UV light in a Spectrolinker XL-1500 (Spectronics Corporation) equipped with >310 nm light bulbs or similar device.
3. Cover cells in 1 ml PBS and scrape cells off with a rubber policeman. Transfer the cell suspension to 50 ml centrifugation tubes and collect by centrifugation at 500 × *g* at 4 °C for 5 min. Discard the supernatant.

Stopping point: If you do not want to continue directly with cell lysis and immunoprecipitation, snap freeze the cell pellet in liquid nitrogen and store at −80 °C. Cell pellets can be stored for at least 12 months.

3.1.3 UV-Crosslinking for Cells Grown in Suspension

1. Collect cells by centrifugation at 500 × *g* at 4 °C for 5 min. Aspirate or pour off media.
2. Take up cells in 10 ml PBS and transfer onto one 15-cm cell culture plate.
3. Irradiate uncovered with a dose of 0.2 J/cm^2 of >310 nm UV light in a Spectrolinker XL-1500 (Spectronics Corporation) equipped with >310 nm light bulbs or similar device.
4. Transfer cells into a 50 ml centrifugation tube and collect by centrifugation at 500 × *g* for 5 min at 4 °C and discard the supernatant.

Stopping point: If you do not want to continue directly with cell lysis and immunoprecipitation, snap freeze the cell pellet in liquid nitrogen and store at −80 °C. Cell pellets can be stored for at least 12 months.

3.1.4 Cell Lysis and RNase T1 Digest

1. Take up crosslinked cell pellet in 3 volumes of 1× NP40 lysis buffer and incubate on ice for 10 min in a 15 ml centrifuge tube.
2. Clear cell lysate by centrifugation at 13,000 × *g* at 4 °C for 15 min. In the meantime, begin to prepare the magnetic beads (*see* Sect. 3.2.1).
3. Transfer supernatant to a new 15 ml centrifuge tube. Discard the pellet.
4. Add RNase T1 to a final concentration of 1 U/μl and incubate at 22 °C for 15 min. Cool reaction subsequently for 5 min on ice before proceeding (*see* **Note 10**).

   Keep a 100 μl aliquot of cell lysate and store at −20 °C to control for RBP expression in Sect. 3.2.4.

3.2 Immunoprecipitation and Recovery of Crosslinked Target RNA Fragments

See **Note 11** for guidelines on handling and washing of magnetic beads.

3.2.1 Preparation of Magnetic Beads

1. Transfer 20 μl of Protein G magnetic beads per ml of cell lysate to a 1.5 ml microcentrifuge tube (for a typical experiment, 120–200 μl of beads) (*see* **Note 12**).
2. Wash beads twice in 1 ml of PBS.
3. Resuspend beads in twice the volume of PBS relative to the original bead volume aliquoted.
4. Add minimum 0.25 mg of antibody per ml original bead volume and incubate on a rotating wheel for 40 min at room temperature.

5. Wash beads twice in 1 ml of PBS to remove unbound antibody.
6. Resuspend beads in one original bead volume of PBS.

3.2.2 Immunoprecipitation (IP), Second RNase T1 Digestion, and Dephosphorylation

1. Add 20 μl of freshly prepared antibody-conjugated magnetic beads per 1 ml of partial RNase T1-treated cell lysate (from Sect. 3.1.4, **step 4**) and incubate in 15 ml centrifuge tubes on a rotating wheel for 1 h at 4 °C.
2. Collect magnetic beads on a magnetic particle collector for 15 ml centrifuge tubes.
3. Keep a 100 μl aliquot of supernatant and store at −20 °C to control for RBP depletion in Sect. 3.2.4. Discard the remaining supernatant.
4. Add 1 ml of 1× NP40 lysis buffer to the centrifugation tube and transfer the suspension to a 1.5 ml microcentrifuge tube (*see* **Note 13**).
5. Wash beads twice in 1 ml of 1× NP40 lysis buffer.
6. Take up cells in one original bead volume of 1× NP40 lysis buffer.
7. Add RNase T1 to a final concentration of 1 U/μl and incubate the bead suspension at 22 °C for 15 min. Cool subsequently on ice for 5 min.
8. Wash beads twice in 1 ml of 1× NP40 lysis buffer (*see* **Note 14**).
9. Wash beads twice in 400 μl of dephosphorylation buffer.
10. Resuspend beads in 1 original bead volume of dephosphorylation buffer.
11. Add calf intestinal alkaline phosphatase to a final concentration of 0.5 U/μl and incubate the suspension at 37 °C for 10 min.
12. Wash beads twice in 1 ml of 1× NP40 lysis buffer.
13. Wash beads twice in 1 ml of polynucleotide kinase (PNK) buffer without DTT (*see* **Note 15**).
14. Resuspend beads in 1 original bead volume of PNK buffer with DTT.

3.2.3 Radiolabeling of RNA Segments Crosslinked to Immunoprecipitated Proteins

1. To the bead suspension described above, add T4 polynucleotide kinase to 1 U/μl and γ-^{32}P-ATP to a final concentration of 0.5 μCi/μl (1.6 μM ATP) in one original bead volume. Incubate at 37 °C for 30 min.
2. Add non-radioactive ATP to obtain a final concentration of 100 μM and incubate at 37 °C for another 5 min.
3. Wash magnetic beads five times with 800 μl of PNK buffer without DTT. Store a 100 μl aliquot of radioactive wash waste for use as radioactive markers in future steps.

4. Resuspend the beads in 70 μl of SDS-PAGE loading buffer and incubate for 5 min in a heat block at 95 °C to denature and release the immunoprecipitated RNPs. Vortex and centrifuge briefly.
5. Remove the magnetic beads on the magnetic separator and transfer the supernatant (i.e., radiolabeled RNP immunoprecipitate) to a clean 1.5 ml microcentrifuge tube.

Stopping point: The sample can be stored at −20 °C for a prolonged period of time. However, the half-life of ^{32}P is 14.5 days, and we therefore recommend continuing with the protocol within 2 weeks.

3.2.4 SDS Polyacrylamide Gel Electrophoresis, Transfer, and Recovery of RNA from Nitrocellulose Membrane

1. Prepare a 4–12% Bis-Tris polyacrylamide gel. We recommend using the first half of the gel for separation of the radiolabeled RNP IP and the second half for immunoblotting to control for RBP expression and IP efficiency. In the first half of the gel, load 40 μl of the radiolabeled RNP IP per well. Each RNP IP sample should be loaded adjacent to a ladder and there should be at least one lane distance between different samples. In the second half of the gel, load 10 μl of cell lysate (from Sect. 3.1.4, **step 4**), 10 μl supernatant (from Sect. 3.2.2, **step 3**), and 2 μl of the radiolabeled IP.
2. Run the gel at 200 V for 40 min.
3. Using semi-dry blotting, transfer proteins onto a 0.45 μm nitrocellulose membrane in 1× transfer buffer at 2 mA/cm^2 current for 1 h.
4. Using a scalpel or razor blade, split the nitrocellulose membrane in two, separating the RNP IP samples from the samples for immunoblotting. Proceed to **step 5** with the RNP IP samples. With the lanes for testing IP and RBP expression, perform a Western blot to probe for your RBP or RBP-tag.
5. Label 3 corners of the membrane and each band of the protein length marker with 1 μl of radioactive wash waste from Section 2C, **step 3**. Wrap the membrane in plastic film (e.g., Saran wrap) to avoid contamination of the phosphorimager screen.
6. Expose the membrane to a blanked phosphorimager screen for 1 h at room temperature and visualize on a phosphorimager. If the radioactivity of the recovered RNP is weak, you can expose the membrane for longer.

   Stopping point: The membrane can be stored at −20 °C for a prolonged period of time.
7. Print the image from the phosphorimager onto an overhead projector transparency film; make sure the image is scaled to

100% for printing. Align the transparency film printout on top of the membrane using the labeled corners for orientation.

8. Cutting through the transparency and the membrane directly beneath, excise the bands on the nitrocellulose membrane that correspond to the expected size of the RBP.
9. Cut the nitrocellulose excisions further by slicing them into ~5 smaller pieces. Transfer the pieces into a 1.5 ml low adhesion tube (e.g., siliconized or DNA LoBind tubes).

3.2.5 Proteinase K Digestion

1. Add 400 μl of 1x Proteinase K buffer to the nitrocellulose pieces followed by the addition of approximately 2 mg Proteinase K. Vortex, briefly centrifuge, and incubate at 55 °C for 1 h 30 min.
2. Extract the RNA by addition of 2 volumes of acidic phenol/chloroform/IAA (25:24:1, pH 4.0) directly to the Proteinase K digestion. Vortex for 15 s and centrifuge at >14,000 × *g* at 4 °C for 5 min. Remove the aqueous phase without disturbing the organic phase or interphase, and transfer the aqueous phase to a new 1.5 ml low adhesion microcentrifuge tube. If the organic or interphase is accidentally disturbed, centrifuge the sample again and reattempt.
3. Add 1 volume of chloroform to the recovered aqueous phase to remove residual phenol. Vortex for 15 s and centrifuge at >14,000 × *g* at 4 °C for 5 min. Remove the aqueous phase without disturbing the organic phase, and transfer the aqueous phase to a new 1.5 ml low adhesion microcentrifuge tube.
4. To the isolated aqueous phase, add 1/10 volume of 3 M NaCl, 1 μl 15 mg/ml GlycoBlue, and 3 volumes of 100% ethanol. Mix thoroughly by inverting the tube at least five times and incubate at −20 °C or −80 °C for 20 min. Proceed to cDNA library preparation.

Stopping point: If kept at −20 °C, RNA can be safely stored for several months as an ethanol precipitate.

3.3 cDNA Library Preparation and Deep Sequencing

The following section describes the standard small RNA cDNA library preparation protocol described for cloning of small regulatory RNAs, found in Reference 16. Before generating the small cDNA libraries following the steps described below, we strongly recommend reading this protocol. The main differences in the procedure described here are: (a) the use of a non-barcoded 3′ adapter, (b) no spike-in of radioactive RNA size markers, and (c) no spike-in of calibrator oligoribonucleotides.

See **Note 16** for general guidelines for the cDNA library preparation.

3.3.1 3′ Adapter Ligation

1. Prepare 5′-^{32}P-labeled RNA size marker cocktail. Use of the size markers will control for successful ligation and indicate the length of the bands that need to be cut out of the gel.
2. Pre-run the gel for 30 min at 30 W using 1× TBE buffer. While the gel is pre-running, move on to step 1.3.
3. Radiolabel the size markers individually in a 10 μl reaction containing 1 μM RNA, 10 U T4 polynucleotide kinase, and 50 μCi γ-^{32}P-ATP at 37 °C for 15 min.
4. Quench the reactions from **step 1** by adding 10 μl of denaturing formamide gel loading solution to each reaction.
5. Denature the RNA by incubating the tubes for 1 min at 90 °C.
6. Load each sample into one well of the 15% denaturing polyacrylamide gel. In order to avoid cross-contamination, make sure to space the size markers with a minimum 2-well distance from each other.
7. Run the gel for 50 min at 30 W using 1× TBE buffer, until the bromophenol blue dye is close to the bottom of the gel.
8. Dismantle the gel, leaving it mounted on one glass plate. Using a scalpel or razor blade, cut crosses of approx. 1 cm length in 3 corners of the gel. Into these crosses, pipette 1 μl of the radioactive waste (stored in Section 3.2.3, **step 3**) to facilitate alignment of the gel to the phosphorimager paper printout. Wrap the gel in plastic film (e.g., Saran wrap) to avoid contamination of the phosphorimager screen.
9. Expose the gel for 5 min to a phosphorimager screen at −20 °C.
10. Align the gel on top of a printout scaled to 100% according to the position of the three spots of radioactive waste. Cut out the radioactive bands corresponding to the length marker.
11. Place the gel slices in 1.5 ml low adhesion microcentrifuge tubes and cover in 0.3 M NaCl (>300 μl). Elute the ligation product into the NaCl using constant agitation at 4 °C overnight (a rotating wheel works well).
12. The following day, take off the supernatant, add 1 μl 15 mg/ml GlycoBlue, mix well, and follow with addition of 3 volumes of 100% ethanol. Mix thoroughly by inverting the tube at least five times and incubate at −20 °C or −80 °C for 20 min.
13. Centrifuge the precipitated RNA at >14,000 × *g* at 4 °C for 30 min.
14. Remove the supernatant completely without disturbing the pellets. Air-dry the pellets for 10 min.
15. Resuspend the pellets in 10 μl water and combine the solutions to obtain the concentrated size marker cocktail.

16. Transfer 1 μl of this cocktail to a new low adhesion tube and dilute it 1:50 in water to obtain the diluted size marker cocktail. Mix by pipetting up and down several times.
17. 10 μl of this diluted size marker cocktail will be used in **step 3**. Store the remaining diluted and concentrated size marker cocktail at −20 °C for future PAR-CLIPs. One preparation of concentrated size marker cocktail can be used for multiple experiments. When diluting the size marker cocktail in future experiments, take into account the 14.5 day half-life of ^{32}P.
18. Spin sample from Sect. 3.2.5, **step 4** at >14,000 × *g* at 4 °C for 20 min. A blue pellet should be visible at the bottom of the tube.
19. Remove the supernatant completely without disturbing the pellet. Air-dry the pellet for 10 min.
20. Resuspend the pellet in 10 μl water.
21. Prepare the following reaction mixture for ligation of the adenylated 3′ adapter, multiplying the volumes by number of ligation reactions to be performed plus 2 extra volumes to include the diluted radioactive RNA size marker cocktail and to account for pipetting error:
    (a) 2 μl 10× RNA ligase buffer without ATP.
    (b) 6 μl 50% aqueous DMSO.
    (c) 1 μl 100 μM adenylated 3′ adapter oligonucleotide.
22. Add 9 μl of the reaction mixture to each sample, including the 10 μl of diluted radioactive RNA size marker cocktail.
23. Denature the RNA by incubating the tubes for 1 min at 90 °C. Immediately place the tubes on ice and incubate for 2 min.
24. Add 1 μl of Rnl2(1-249)K227Q ligase (1 μg/μl), swirl gently with your pipette tip, and incubate the tubes overnight on ice at 4 °C.
25. Prepare a 20-well, 15% denaturing polyacrylamide gel (15 cm wide, 17 cm long, 0.5 mm thick; 25 ml gel solution).
26. Pre-run the gel for 30 min at 30 W using 1× TBE buffer.
27. Add 20 μl of formamide gel loading solution to each 3′ adapter ligation reaction.
28. Denature the RNA by incubating the tubes for 1 min at 90 °C.
29. Load each sample into one well of the 15% denaturing polyacrylamide gel. In order to avoid cross-contamination, make sure to space different samples appropriately; we recommend a two-well distance.

30. Split the marker reaction, loading one half on opposite ends of the gel to frame the PAR-CLIP samples. Once again, avoid cross-contamination by keeping a two-well distance between samples and markers.
31. Run the gel for 45 min at 30 W using 1× TBE buffer, until the bromophenol blue dye is close to the bottom of the gel.
32. Dismantle the gel, leaving it mounted on one glass plate. Using a scalpel or razor blade, cut crosses of approximately 1 cm length in 3 corners of the gel. Into these crosses, pipette 1 μl of the radioactive waste (stored in Section 2C, **step 3**) to facilitate alignment of the gel to the phosphorimager paper printout. Wrap the gel in plastic film (e.g., Saran wrap) to avoid contamination of the phosphorimager screen.
33. Expose the gel for at least 1 h to a phosphorimager screen, keeping the cassette at −20 °C to prevent diffusion of RNA. If the radioactivity of the recovered RNA is weak, you can expose the gel overnight at −20 °C.
34. Align the gel on top of a printout scaled to 100% according to the position of the three spots of radioactive waste. The 3-′-ligated 19 and 35 nt markers should be visible on the print-out, possibly with two additional lower bands representing unligated 19 and 35 nt marker. Using the ligated markers as guides, cut out sample RNA of 19 to 35 nt length, ligated to the 3′ adapter. Cut out the ligated markers as well (*see* **Note 17**).
35. Place the gel slices in separate 1.5 ml low adhesion microcentrifuge tubes and cover in 0.3 M NaCl (>300 μl). Elute the ligation product into the NaCl using constant agitation at 4 °C overnight (a rotating wheel works well).
36. The following day, take off the supernatant, add 1 μl 15 mg/ml GlycoBlue, mix well, and follow with addition of 3 volumes of 100% ethanol. Mix thoroughly by inverting the tube at least five times and incubate at −20 °C or −80 °C for 20 min.

Stopping point: If kept at −20 °C, RNA can be safely stored for several months as an ethanol precipitate.

3.3.2 5′ Adapter Ligation

1. Centrifuge the precipitated RNA at >14,000 × *g* at 4 °C for 30 min. A blue pellet should be visible at the bottom of the tubes.
2. Remove the supernatant completely without disturbing the pellet. Air-dry the pellet for 10 min.
3. Resuspend the pellet in 9 μl water.
4. Prepare the following reaction mixture for ligation of the 5′ adapter, multiplying the volumes by number of ligation

reactions to be performed plus 2 extra volumes to include the RNA size markers and to account for pipetting errors:

(a) 2 μl 10× RNA ligase buffer with ATP.

(b) 6 μl 50% aqueous DMSO.

(c) 1 μl 100 μM 5′ adapter oligonucleotide.

5. Add 9 μl of the reaction mixture to each sample, including the 3′-ligated radioactive RNA size markers.
6. Denature the RNA by incubating the tubes for 1 min at 90 °C. Immediately place the tubes on ice and incubate for 2 min.
7. Add 2 μl T4 RNA ligase, swirl gently with your pipette tip, and incubate for 1 h at 37 °C. While the samples are incubating, prepare the polyacrylamide gel.
8. Prepare a 20-well, 12% denaturing polyacrylamide gel (15 cm wide, 17 cm long, 0.5 mm thick; 25 ml gel solution).
9. Pre-run the gel for 30 min at 30 W using 1× TBE buffer.
10. Add 20 μl of formamide gel loading solution to each 5′ adapter ligation reaction.
11. Denature the RNA by incubating the tubes for 1 min at 90 °C.
12. Load the gel as described in Sect. 3.3.1, **steps 29** and **30**, and run for 45 min at 30 W using 1× TBE buffer, until the bromophenol blue dye is close to the bottom of the gel.
13. Image the gel as described in Sect. 3.3.1, **steps 32** and **33**, and excise the new ligation product (*see* **Note 18**).
14. Place the gel slices in 1.5 ml low adhesion microcentrifuge tubes and cover in 0.3 M NaCl (>300 μl). Elute the ligation product into the NaCl using constant agitation at 4 °C overnight (a rotating wheel works well).
15. The following day, take off the supernatant, add 1 μl 15 mg/ml GlycoBlue, mix well, and follow with addition of 3 volumes of 100% ethanol. Mix thoroughly by inverting the tube at least five times and incubate at −20 °C or −80 °C for 20 min.

Stopping point: If kept at −20 °C, RNA can be safely stored for several months as an ethanol precipitate.

3.3.3 Reverse Transcription

1. Centrifuge the precipitated RNA at >14,000 × *g* at 4 °C for 30 min. A blue pellet should be visible at the bottom of the tubes.
2. Remove the supernatant completely without disturbing the pellet and allow the pellet to air-dry for 10 min.
3. Resuspend the pellet in 4.6 μl water and transfer to a thermocycler tube.

4. Prepare the following reaction mixture for reverse transcription, multiplying the volumes by number of reverse transcription reactions to be performed plus 1 extra volume to account for pipetting errors:
   (a) 1.5 μl 0.1 M DTT.
   (b) 3 μl 5× first-strand synthesis buffer.
   (c) 4.2 μl 10× dNTPs.
   (d) 1 μl 100 μM RT primer.
5. Before addition of the reaction mixture, denature the RNA by incubating the tubes for 30 s at 90 °C in a thermocycler, and then hold at 50 °C.
6. Add 9.7 μl of the reaction mix to each sample and incubate for 3 min at 50 °C. Add 0.75 μl of Superscript III reverse transcriptase, mix gently by flicking the tube twice and incubate for 2 h at 50 °C.
7. Add 85 μl water and mix well.

   Stopping point: cDNA can be stored indefinitely at −20 °C.

3.3.4 PCR Amplification

1. Prepare the following mix multiplied by the number of samples:
   (a) 40 μl 10× PCR buffer.
   (b) 40 μl 10× dNTPs.
   (c) 2 μl 100 μM 5′ primer.
   (d) 2 μl 100 μM 3′ primer.
   (e) 272 μl water.

   89 μl of the reaction mix will be used in a pilot PCR reaction to determine the optimal number of PCR cycles for amplification, and the remaining mixture will be used for a large-scale PCR.
2. To 89 μl of the reaction mix add 10 μl from the cDNA solution and 1 μl of Taq polymerase.
3. Perform a standard 100 μl, 30 cycle PCR with the following conditions: 45 s at 94 °C, 85 s at 50 °C, 60 s at 72 °C.
4. Beginning with the 12th cycle and ending with the 30th cycle, remove a 10 μl aliquot from each PCR reaction every 3 cycles (i.e., at cycles 12, 15, 18, etc.).
5. Analyze the 10 μl aliquots on a 2.5% agarose gel alongside a 25 bp ladder. The expected PCR product should appear between 95 and 110 bp. When ligated and amplified with the correct primers, the 19 and 35 nt markers appear at 95 and 110 bp, respectively. Often, a lower band appears at 72 bp corresponding to the direct ligation products of the 3′ and 5′

adapters. Define the optimal cycle number for cDNA amplification, which should be within the exponential amplification phase of the PCR, approximately 5 cycles away from reaching the saturation level of PCR amplification (*see* **Note 19**).

6. Using the remaining PCR cocktail, perform three 100 μl PCR reactions with the optimal cycle number identified above.
7. Combine the individual 100 μl reactions and precipitate with 3 volumes of 100% ethanol.
8. Take up the pellet in 60 μl 1× DNA loading dye.
9. Run the sample on two wells of a 2.5% agarose gel alongside a 25 bp ladder.
10. Visualize the DNA on a UV transilluminator and excise the gel piece containing cDNA between 85 and 120 bp of length.
11. Extract the DNA using the Qiagen MinElute Gel Extraction Kit, following the instructions of the manufacturer. Use 30 μl elution buffer to recover the DNA.
12. Submit 10 μl of the purified cDNA to Illumina sequencing. We recommend using 50 cycle single-end sequencing on a HiSeq, NextSeq, NovaSeq, or equivalent machine.

Optional: Determination of Incorporation Levels of 4-Thiouridine into Total RNA

When using a cell line for the first time for PAR-CLIP, it might be helpful to assess its capability to take up 4SU.

1. Supplement growth medium with 100 μM of 4SU 16 h prior to harvest, provide regular media to one control plate.
2. The following day, harvest cells using a cell scraper and spin down at 500 × *g* for 5 min at 4 °C.
3. Remove supernatant and resuspend the pellet in 3 volumes of TrIzol reagent (Sigma), follow the manufacturer's instructions.
4. Further purify total RNA using Qiagen RNAeasy according to the manufacturer's instructions (*see* **Note 20**).
5. Digest and dephosphorylate total RNA to single nucleosides by incubating 40 μg of purified total RNA for 16 h at 37 °C with 0.4 U bacterial alkaline phosphatase (e.g., Worthington Biochemical) and 0.09 U snake venom phosphodiesterase (e.g., Worthington Biochemical) in a 30 μl volume.
6. As a reference standard, use a synthetic 4SU labeled RNA (previously we used CGUACGCGGAAUACUUCGA(4SU)U), which is subjected to complete enzymatic digestion.
7. Separate the resulting mixtures of ribonucleosides by HPLC on a Supelco Discovery C18 (bonded phase silica 5 μM particle, 250 × 4.6 mm) reverse phase column (Bellefont). HPLC buffers are 0.1 M triethylammonium acetate (TEAA) in 3% acetonitrile (A) and 90% acetonitrile in 0.1 M TEAA (B).

8. Use an isocratic gradient: 0% B for 15 min, 0–10% B for 20 min, 10–100% B for 30 min.
9. Clean HPLC column with a 5 min 100% wash between runs.

3.4 PAR-CLIP Analysis

With current depths of Illumina sequencing reaching >200 million sequence reads per sample, PAR-CLIP data analysis requires sophisticated approaches to identify binding sites [17]. Several bio-computational pipelines for PAR-CLIP data analysis have been made available, including PARalyzer [18], PIPE-CLIP [19], WavclusteR [20], doRina [21], CLIPZ [22], Starbase [23], miR-TarCLIP [24], Piranha [25], and dCLIP [26]. After initial analysis, you may calculate the common sequence motifs of the RRE using one of the several programs initially developed for the analysis of transcription-factor binding sites on DNA, including MEME [27], MDScan [28], PhyloGibbs [29], cERMIT [30], and Gimsan [31].

Generally, the analysis of the sequence reads begins by alignment to the genome, allowing for at least one error (substitution, insertion, or deletion) to capture crosslinked reads with crosslinking-induced mutations. Next, overlapping sequence reads are grouped, taking into account the frequency of crosslinking-induced mutations. To allow insights into the RBP's binding preferences, these groups of overlapping sequence reads can then be mapped against the transcriptome to annotate and categorize them as derived from 5′ untranslated region (UTR), coding sequence (CDS), 3′UTR, introns, rRNA, long non-coding RNAs, tRNAs, and so forth.

The frequency of the T-to-C mutations (or G-to-A mutations when using 6SG) allows ranking of groups to predict those interactions with the highest functional impact. In addition, it may be useful to provide a limited set of high-confidence interaction sites as input into motif-finding programs to facilitate the detection of the underlying RRE. Some of the analysis pipelines, such as PARalyzer, take advantage of the frequency and distribution of crosslinking-induced mutations to predict the shortest possible region of interaction between RBP and RNA that harbors the RRE.

CLIP-based approaches provide a genome-wide view of the protein-RNA interaction sites and routinely identify tens of thousands of interaction sites in the transcriptome. However, additional experimentation—as well as clear ranking of binding sites—is necessary to relate RNA binding to phenotypes arising from knockout, overexpression, or mutation of the RBP. For example, the effect of RNA binding on transcript stability and alternative splicing can be assayed using microarray analysis and RNA sequencing analysis. Quantitative proteomics (SILAC, iTRAQ) and ribosome profiling are increasingly available as methods to assess translational regulation by RBPs [32]. Analysis of RBPs involved in RNA transport and other processes may require the development of more specialized assays.

4 Notes

1. For NP40 Lysis Buffer prepare a stock of 5× buffer without DTT and protease inhibitors. Add DTT and protease inhibitor to 1× buffer directly before use.
2. Not every antibody will retain its binding ability in 500 mM KCl—adjust the salt concentration accordingly for the high-salt wash buffer. If in doubt use lysis buffer for washing. Also add DTT and protease inhibitor directly to high-salt wash buffer before experiment.
3. This protocol describes the procedure for analysis of endogenously expressed, or recombinant constitutively expressed, or inducibly expressed RBPs. The PAR-CLIP protocol will work with any cell line expressing detectable levels of RBP as long as there is an efficient antibody for *immunoprecipitation* (IP). However, some antibody quality testing is necessary before beginning PAR-CLIP. If using an antibody that specifically recognizes your RBP-of-interest, perform stringent quality testing of the IP with your antibody *before* attempting PAR-CLIP. We recommend transiently transfecting the cells with a vector for expression of the protein of interest with an N-terminal or C-terminal epitope fusion, such as FLAG, MYC, or HA. Follow transfection with IP using an RBP-specific antibody and Western blotting for the epitope tag using reliable commercial antibodies. At this time you should also test for the maximal monovalent salt concentration compatible with your IP. Increasing salt concentration will result in fewer co-purifying proteins and RNAs but can also lead to loss of bound RNP. For reference, the FLAG antibody tolerates up to 500 mM KCl.
4. Guidelines for the use of *4-thiouridine* (4SU) may need to be adapted for use in the desired cell lines or model organisms; the concentration of 4SU and the length of UV-light exposure in this protocol were optimized for HEK293 cells. For other cell lines, the user may want to determine the optimal, non-toxic 4SU concentration and labeling time. In cell lines or model organisms with weak 4SU uptake, it may be necessary to enhance or introduce expression of nucleoside transporters, such as *uracil phophoribosyltransferase* (UPRT) [33, 34]. We have also included an optional section at the end of the PAR-CLIP procedure for determining the incorporation of 4SU into total RNA. The energy dose of UV light necessary for crosslinking may vary due to differing transparency of the sample compared to mammalian cells grown in monolayers. For example, cells plated as dense suspensions, yeast, and worms exhibit higher opacity [35].

5. We recommend use of positive and negative controls, particularly when performing the pilot PAR-CLIP experiments. An appropriate negative control could comprise the use of IgG isotype control as a substitute for the RBP antibody; this will allow the user to visualize fragments of abundant cellular RNAs, as well as RNPs co-purifying through nonspecific interactions with antibodies and magnetic beads (Thermo Scientific MA1-10407). For a positive control, plasmids encoding FLAG/HA-tagged RBPs previously characterized by PAR-CLIP are available on www.addgene.org.
6. The on-bead RNase T1 digestion described in Section 2B, **step 7** should be optimized for your individual RBP. Each RBP binding footprint provides a different level of protection from RNase T1, resulting in shorter or longer RNA fragments after the RNA is isolated by Proteinase K digestion. RNA fragments between 19 and 35 nt are ideal for small cDNA library preparation. Fragments shorter than 19 nt have a higher probability of mismapping compared to longer reads. Fragments longer than 35 nt cannot be fully sequenced by standard 50 base single-end sequencing. The concentration of RNase T1 suggested in this protocol may be too high for certain RBPs, resulting in RNA less than 19 nt long. To determine the correct concentration of RNaseT1 for the on-bead RNaseT1 digestion, perform the PAR-CLIP protocol through Section 2. When you reach Section 2B, **step 7**, perform a set of separate digestions with RNaseT1 concentrations ranging from 0 to 100 U/μl. After Section 2 is complete, analyze the resulting RNA on a denaturing polyacrylamide gel as described in Section 3A, **steps 10–16**. If the majority of RNA is below 19 nt, over-digestion has occurred and the RNase T1 concentration must be reduced appropriately.
7. Use low adhesion microcentrifuge tubes (e.g., siliconized or DNA LoBind tubes) for all manipulations of the small RNAs after the Proteinase K digestion. The minute amounts of small RNAs to be recovered after RNA isolation will readily adsorb to the walls of standard tubes.
8. It is also possible to use 100 μM of 6-thioguanosine (6SG) as the photoactivatable ribonucleoside. 6SG has a lower crosslinking efficiency compared to 4SU and will result in a G-to-A mutation instead of a T-to-C mutation at the crosslinking site.
9. A thin film of remaining media helps prevent cells from drying, and does not interfere with crosslinking.
10. Incubation at room temperature is also sufficient if there are no means of incubation at 22 °C.
11. Guidelines for working with magnetic beads:

(a) Before pipetting beads from the source container, always mix thoroughly by shaking or vortexing.

(b) To prevent drying and loss of function, do not leave beads uncovered for prolonged periods of time.

Step-by-step for washing magnetic beads:

(a) Place the beads in suspension on a magnetic separator and let stand for 1 min or until solution clears.

(b) Carefully remove buffer from the tube without disturbing the beads.

(c) Add buffer to the tube while the tube is on the magnetic separator.

(d) Remove the tube from the magnetic separator and resuspend the beads either by flicking, shaking, or vortexing. To prevent loss of beads, we do not recommend mixing by pipetting.

(e) Briefly centrifuge the tube to collect beads caught on the tube cap.

(f) Place the beads in suspension on a magnetic separator and let stand for 1 min or until solution clears.

(g) Remove the supernatant and resuspend in the appropriate buffer, or repeat **steps 1–5** for additional wash steps.

12. For small volumes of lysate do not use less than 35 μl of magnetic beads to account for minor loss during handling.

13. Make sure that you do not exceed the maximum salt concentration at which the antibody recognizes its antigen.

14. Optional: Reduce IP background by performing a high-salt wash. Replace 1× NP40 lysis buffer with 1× high-salt wash buffer. Only perform this step if you are confident you will not exceed the maximum salt concentration at which the antibody recognizes its antigen.

15. To avoid bead damage, do not expose magnetic beads to high DTT concentration for prolonged time.

16. Take care to avoid contamination of the minute amounts of RNA, e.g., with RNases. Use RNAse-free water and store RNA at −20 or −80 °C. Previously prepared cDNA libraries may contaminate lab surfaces and equipment and will readily amplify during PCR. Use sterile filter tips wherever possible.

17. Avoid recovering RNA <19 nt, which have the potential to complicate the subsequent bioinformatics analysis, as they have a higher probability of mismapping compared to longer reads.

18. Optional: Excise the markers and keep them as controls for the reverse transcription and PCR.

19. As the PCR reaction approaches saturation of PCR product, reagents within the reaction become limiting, leading to selective amplification of certain transcripts over others.
20. It is important to add 0.1 mM dithiothreitol (DTT) to wash buffers and subsequent enzymatic steps to prevent oxidization of 4SU during RNA isolation and analysis.

References

1. Sonenberg N, Hinnebusch AG (2009) Regulation of translation initiation in eukaryotes: mechanisms and biological targets. Cell 136:731–745
2. Moore MJ, Proudfoot NJ (2009) Pre-mRNA processing reaches back to transcription and ahead to translation. Cell 136:688–700
3. Martin KC, Ephrussi A (2009) mRNA localization: gene expression in the spatial dimension. Cell 136:719–730
4. Cech TR, Steitz JA (2014) The noncoding RNA revolution-trashing old rules to forge new ones. Cell 157:77–94
5. Lukong KE, Chang K-W, Khandjian EW, Richard S (2008) RNA-binding proteins in human genetic disease. Trends Genet 24:416–425
6. Cooper TA, Wan L, Dreyfuss G (2009) RNA and disease. Cell 136:777–793
7. Castello A, Fischer B, Hentze MW, Preiss T (2013) RNA-binding proteins in Mendelian disease. Trends Genet 29:318–327
8. Gerstberger S, Hafner M, Tuschl T (2014) A census of human RNA-binding proteins. Nat Rev Genet 15:829–845
9. König J, Zarnack K, Luscombe NM, Ule J (2011) Protein-RNA interactions: new genomic technologies and perspectives. Nat Publ Group 13:77–83
10. Tenenbaum SA, Carson CC, Lager PJ, Keene JD (2000) Identifying mRNA subsets in messenger ribonucleoprotein complexes by using cDNA arrays. Proc Natl Acad Sci U S A 97:14085–14090
11. Gilbert C, Svejstrup JQ (2006) RNA immunoprecipitation for determining RNA-protein associations in vivo. Curr Protoc Mol Biol, Chapter 27, Unit 27.4–27.4.11
12. Gerber AP, Luschnig S, Krasnow MA et al (2006) Genome-wide identification of mRNAs associated with the translational regulator PUMILIO in Drosophila melanogaster. Proc Natl Acad Sci U S A 103:4487–4492
13. López de Silanes I, Zhan M, Lal A et al (2004) Identification of a target RNA motif for RNA-binding protein HuR. Proc Natl Acad Sci U S A 101:2987–2992
14. Maes OC, Chertkow HM, Wang E, Schipper HM (2009) MicroRNA: implications for Alzheimer disease and other human CNS disorders. Curr Genomics 10:154–168
15. Hafner M, Landthaler M, Burger L et al (2010) Transcriptome-wide identification of RNA-binding protein and microRNA target sites by PAR-CLIP. Cell 141:129–141
16. Hafner M, Renwick N, Farazi TA et al (2012) Barcoded cDNA library preparation for small RNA profiling by next-generation sequencing. Methods 58:164–170
17. Ascano M, Hafner M, Cekan P et al (2011) Identification of RNA-protein interaction networks using PAR-CLIP. WIREs RNA 3:159–177
18. Corcorabin DL, Georgiev S, Mukherjee N et al (2011) PARalyzer: definition of RNAding sites from PAR-CLIP short-read sequence data. Genome Biol 12:R79
19. Chen B, Yun J, Kim MS et al (2014) PIPE-CLIP: a comprehensive online tool for CLIP-seq data analysis. Genome Biol 15:1–10
20. Sievers C, Schlumpf T, Sawarkar R, Comoglio F, Paro R (2012) Mixture models and wavelet transforms reveal high confidence RNA-protein interaction sites in MOV10 PAR-CLIP data Nucl. Acids Res 40(2):160
21. Anders G, Mackowiak SD, Jens M et al (2012) doRiNA: a database of RNA interactions in post-transcriptional regulation. Nucleic Acids Res 40:D180–D186
22. Khorshid M, Rodak C, Zavolan M (2011) CLIPZ: a database and analysis environment for experimentally determined binding sites of RNA-binding proteins. Nucleic Acids Res 39: D245–D252
23. Yang JH, Li JH, Shao P et al (2011) starBase: a database for exploring microRNA-mRNA interaction maps from Argonaute CLIP-Seq and Degradome-Seq data. Nucleic Acids Res 39:D202–D209

24. Chou CH, Lin FM, Chou MT et al (2013) A computational approach for identifying microRNA-target interactions using high-throughput CLIP and PAR-CLIP sequencing. BMC Genomics 14(Suppl 1):S2
25. Uren PJ, Bahrami-Samani E, Burns SC et al (2012) Site identification in high-throughput RNA-protein interaction data. Bioinformatics 28:3013–3020
26. Wang T, Xie Y, Xiao G (2014) dCLIP: a computational approach for comparative CLIP-seq analyses. Genome Biol 15:R11
27. Bailey TL (2002) Discovering novel sequence motifs with MEME. Curr Protoc Bioinformatics. Chapter 2, Unit 2.4–2.4.35
28. Liu XS, Brutlag DL, Liu JS (2002) An algorithm for finding protein–DNA binding sites with applications to chromatin-immunoprecipitation microarray experiments. Nat Biotechnol 20:835–839
29. Siddharthan R, Siggia ED, van Nimwegen E (2005) PhyloGibbs: a Gibbs sampling motif finder that incorporates phylogeny. PLoS Comp Biol 1:e67
30. Georgiev S, Boyle AP, Jayasurya K et al (2010) Evidence-ranked motif identification. Genome Biol 11:R19
31. Ng P, Keich U (2008) GIMSAN: a Gibbs motif finder with significance analysis. Bioinformatics 24:2256–2257
32. Brewis IA, Brennan P (2010) Proteomics technologies for the global identification and quantification of proteins. Adv Protein Chem Struct Biol 80:1–44
33. Guruharsha KG, Rual JF, Zhai B et al (2011) A protein complex network of *Drosophila melanogaster*. Cell 147:690–703
34. Kucerova L, Poturnajova M, Tyciakova S, Matuskova M (2012) Increased proliferation and chemosensitivity of human mesenchymal stromal cells expressing fusion yeast cytosine deaminase. Stem Cell Res 8:247–258
35. Jungkamp AC, Stoeckius M, Mecenas D et al (2011) In vivo and transcriptome-wide identification of RNA binding protein target sites. Mol Cell 44:828–840

Chapter 10

A Pipeline for Analyzing eCLIP and iCLIP Data with *Htseq-clip* and *DEWSeq*

Sudeep Sahadevan, Thileepan Sekaran, and Thomas Schwarzl

Abstract

Individual-nucleotide crosslinking and immunoprecipitation (iCLIP) sequencing and its derivative enhanced CLIP (eCLIP) sequencing are methods for the transcriptome-wide detection of binding sites of RNA-binding proteins (RBPs). This chapter provides a stepwise tutorial for analyzing iCLIP and eCLIP data with replicates and size-matched input (SMI) controls after read alignment using our open-source tools *htseq-clip* and *DEWSeq*. This includes the preparation of gene annotation, extraction, and preprocessing of truncation sites and the detection of significantly enriched binding sites using a sliding window based approach suitable for different binding modes of RBPs.

Key words Enhanced individual-nucleotide crosslinking and Immunoprecipitation, iCLIP, eCLIP, seCLIP, Next-generation sequencing, RNA-binding proteins, RBP binding-site detection

1 Introduction

In recent years, high-throughput mass spectrometry technologies identified thousands of novel RNA-binding proteins (RBPs) [1]. Crosslinking and immunoprecipitation (CLIP) combined with next-generation sequencing (NGS) techniques allowed the transcriptome-wide detection of a protein's binding sites. UV light is used to induce a crosslink between protein and its target RNA in very close proximity. Individual-nucleotide CLIP (iCLIP) and its derivatives use the tendency of reverse transcriptases to stop at the crosslink sites for binding-site detection [2]. Enhanced CLIP (eCLIP) is a successor of iCLIP and is based on the same principles. However, alongside an updated chemistry, most importantly, eCLIP introduced a size-matched input (SMI) control [3]. This special control is crucial for controlling many underlying biases of CLIP protocols, especially for RBPs without strong affinity to binding motifs or with binding positions that are hard to detect

Erik Dassi (ed.), *Post-Transcriptional Gene Regulation*, Methods in Molecular Biology, vol. 2404,
https://doi.org/10.1007/978-1-0716-1851-6_10,

due to the low yield of crosslinking and the very high background of CLIP protocols.

In this chapter, we provide a workflow for analyzing i/eCLIP data which tests for significant enrichment of crosslink sites of the sample over SMI controls taking replicate information into consideration when testing. We present two open-source packages developed for this purpose: Firstly, *htseq-clip*, a Python package for the extraction of crosslink sites from truncations and counting them into windows using custom gene annotations. Secondly, *DEWSeq*, a R/Bioconductor package for detecting the binding sites using a sliding window approach, for testing the immunoprecipitated (IP) i/eCLIP samples versus SMI controls.

1.1 Properties of e/iCLIP Data

iCLIP and its derivatives rely on reverse transcriptases (RTs) stopping at the crosslink site between RNA and polypeptides remaining from the digested protein. It is believed that this happens >80% of the time, depending on protein and type of RTs (contrary to PAR-CLIP [4] or HITS-CLIP [5], where RTs with a higher likelihood of skipping the crosslink site are preferred). Note that truncation and crosslink sites are often used in literature interchangeably in this context.

The reverse transcribed cDNA is sequenced with high-throughput NGS machines. The crosslink sites are positions adjacent to the start of the sequenced read (Fig. 1). A crosslink site can be extracted from the truncation as single nucleotide genomic positions with an offset using *htseq-clip*. Depending on the protein, the binding sites have different distributions of truncation sites around their binding sites, which is caused by, but not limited to effects such as:

1. Primary sequence position specific binding (Binding to primary sequence motifs or precise locations like the exon junction complex does).
2. Binding preference to RNA structure.
3. Double- or single-stranded RNA binding.
4. RNA modifications.

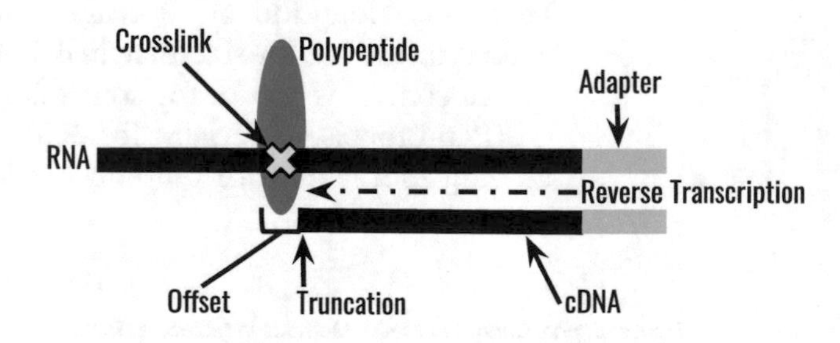

Fig. 1 Truncation at crosslinking sites

5. Efficiency of crosslinking for the protein under investigation (generally <1%).
6. Likelihood of reverse transcriptase stopping at crosslink site.
7. Background RNA contamination.
8. Library size and sequencing depth.

The combination of those effects will only in a few cases lead to perfect peak-shape patterns, exon junction complexes (EJC), and splicing factors being prime examples. In most other cases, the profiles will be more scattered and the background will be much higher. A sliding window approach will be able to handle both cases when testing for significant enrichments of truncation sites in the IP sample over the SMI control. *DEWSeq* is combining sliding window testing with generalized linear models, Bayesian fitting and significant testing from *DESeq2* [6], a well-established package for differential gene expression in NGS data, based on generalized linear models (GLMs). A major advantage of this approach is the possibility to detect short as well as longer stretches of enriched crosslink sites and that the replicate information is used for significance testing.

1.2 Controls

For finding significant binding in IP samples, one needs the right controls to compare against. In CLIP-type experiments, there are different types of controls such as non-specific antibody IgG controls, no crosslink (noCL) control, and empty beads control for unspecific binding. Size-matched input controls (SMIs) control for non-specific background effects induced by crosslinking and other biases induced by sample preparation steps [3]. It was shown that SMI can successfully be used to control for the CLIP-induced biases [3]. Our workflow therefore requires SMI controls, ideally in replicates. noCL, IgG, and empty bead controls should be only in the evaluation of called binding sites.

1.3 From Raw to Aligned Reads

Depending on the CLIP protocol type, libraries are either sequenced in single-read or paired-end mode containing unique molecular identifiers (UMIs) and multiplexing sequencing indices on predefined positions (e.g., first 10 nt of read 1 might be UMIs, next 4 nucleotides could be indices for used multiplexing). Those nucleotides have to be extracted from the raw sequences, samples have to be demultiplexed and PCR duplicates removed using UMI information. Due to the individuality of preprocessing, we will simply discuss those steps. We recommend using tools which can account for sequencing errors in the indices and UMI sequences, like the *Je Suite* [7] for example.

In the next step, the reads need to be trimmed for sequencing adapters with, e.g., *cutadapt* [8]. Attention: The original eCLIP protocol includes two adapter trimming steps to help with the multi-ligation problems of the chemistry [3]. We strongly recommend to stick to the corresponding data preprocessing protocols (e.g., eCLIP_analysis_SOP_v2.2.pdf from encodeproject.org) for this.

After discarding low quality reads and general quality control with, e.g., *FASTQC* and *multiQC*, reads need to be aligned with a splice-aware aligner like *NovoAlign* (http://www.novocraft.com/products/novoalign) or *STAR* [9], followed by PCR duplicate removal using UMI, if not already done [7]. The resulting .bam alignment files as well as gene annotation are the inputs for our data analysis *htseq-clip* and *DEWSeq* workflow.

2 Data Analysis

2.1 System Requirements

A machine with at least 32 GB of RAM running Linux/Unix or MacOS.

2.2 Data Requirements

For the analysis, we require the following files:

1. .bam alignment files, after PCR duplication removal with UMIs, in replicates with SMI controls.
2. an annotation file in .gff3 format, e.g. from GENCODE (http://gencodegenes.org).

2.3 From Aligned Reads to Counts with Htseq-Clip

2.3.1 Installation Requirements

htseq-clip requires the following:

1. Python programming environment (version $\geq$3.5).
2. Installation of the Python package *HTSeq*.

   Please note that *HTSeq* requires the *pysam* package for processing alignment files, which currently only works on Linux and OSX operating systems.

2.3.2 Quick Installation

Once all the requirements are fully met, *htseq-clip* can be installed as:

```
$ pip install htseq-clip
```

2.3.3 Conda Environment

It is recommended to use the *conda* package management system for managing multiple Python environments and to resolve dependencies relatively easily. Please install *conda* on your machine following the guidelines provided in the *conda* website. After the successful installation of *conda*, *htseq-clip* can be installed using the following steps:

1. Create a new "htseq-clip" conda environment and activate the environment:

```
(base) $ conda create -y -n htseq-clip
(base) $ conda activate htseq-clip
```

2. Install *HTSeq* and *pysam*:

```
(htseq-clip) $ conda install -c bioconda pysam
(htseq-clip) $ conda install -c bioconda htseq
```

3. Finally, *htseq-clip* can be installed:

```
(htseq-clip) $ pip install htseq-clip
```

2.3.4 Version Used in this Chapter

The *htseq-clip* version used in chapter can be installed as:

```
$ pip install htseq-clip==2.2.0b0
```

2.3.5 Help

Help pages for *htseq-clip* are available at https://htseq-clip.readthedocs.io/.

2.3.6 Overview

htseq-clip requires alignment files in .bam format and corresponding gene annotation file in .gff3 format. The workflow displayed in Fig. 2 can be grouped into three steps:

1. Preparing annotation, this step includes: flattening of the gene annotations to unique annotations for chromosomal positions, creating sliding windows from flattened annotation and creating a mapping file.
2. Extract crosslink sites and summarize to counts per sliding windows.
3. Create a count matrix for further analysis with *DEWSeq*.

2.4 A Minimal Reproducible Example

We demonstrate the use of *htseq-clip* on SLBP (stem loop binding protein) eCLIP data generated by the ENCODE consortium using K562 human cell lines [3]. The experimental design used by the consortium consists of two IP samples compared against one SMI sample. More replicates would be desirable, 2 samples vs. 1 control is the minimal requirement for *DEWSeq*.

The .bam alignment files were generated by the consortium by aligning the reads after processing to hg38 human genome assembly using *STAR* aligner [3]. For additional details of the read

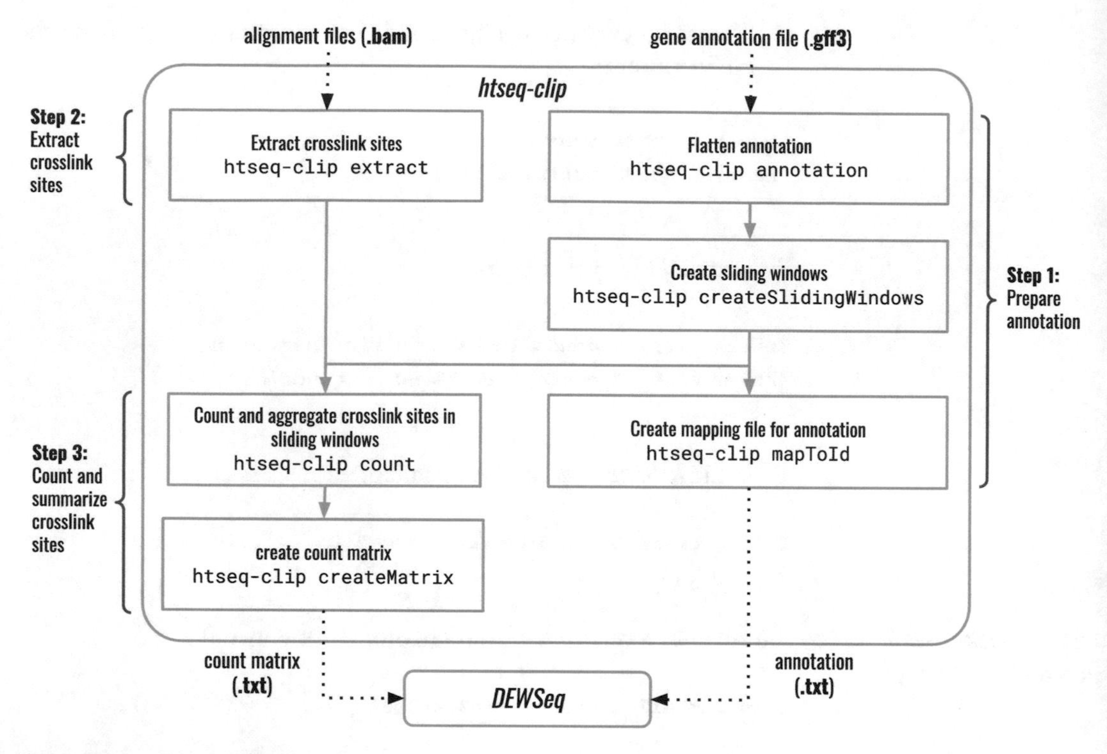

Fig. 2 htseq-clip workflow

Table 1
SLBP eCLIP data description and source

Experiment accession	Bam file accession	Description	Experiment URL
ENCSR483NOP	ENCFF218ZEI ENCFF511HSJ	IP sample 1 IP sample 2	https://www.encodeproject.org/experiments/ENCSR483NOP/
ENCSR950WBG	ENCFF879UID	SMI sample	https://www.encodeproject.org/experiments/ENCSR950WBG/

processing steps used by the consortium, please refer to the links in Table 1.

2.4.1 First Steps

Please note that all the commands below were executed in a Linux shell. Create a new directory called "SLBP_analysis."

```
$ mkdir -p SLBP_analysis
```

We will create the following subdirectories for file organization:

1. *annotation*: annotation files directory,
2. *bam*: bam file directory,
3. *sites*: directory for extracted crosslink (truncation) sites from command extract,
4. *counts*: directory for summarized counts from commands count and createMatrix.

```
$ cd SLBP_analysis
$ mkdir -p annotation bam sites counts
```

2.4.2 Download BAM Files

```
$ cd bam
$ wget https://www.encodeproject.org/files/ENCFF218ZEI/@@download/ENCFF218ZEI.bam
$ wget https://www.encodeproject.org/files/ENCFF511HSJ/@@download/ENCFF511HSJ.bam
$ wget https://www.encodeproject.org/files/ENCFF879UID/@@download/ENCFF879UID.bam
```

2.4.3 Download Annotation

Download human hg38 annotation in .gff3 format. This example uses GENCODE release 27.

```
$ cd ../annotation
$ wget ftp://ftp.ebi.ac.uk/pub/databases/gencode/Gencode_human/release_27/gencode.v27.annotation.gff3.gz
```

The rest of this example analysis will follow the same order as it is shown in Fig. 2.

2.4.4 Prepare Annotation

This section describes how to prepare annotation files required for the analysis with *htseq-clip*. First, activate the "htseq-clip" *conda* environment to be able to access *htseq-clip*:

```
$ conda activate htseq-clip
```

2.4.5 Flatten Annotation

Flatten a given annotation file in .gff3 format to BED6 format as follows:

```
$ cd /path/to/SLBP_analysis/annotation
$ htseq-clip annotation -g gencode.v27.annotation.gff3 -o gencode.v27.annotation.bed
```

Parameters used:

1. -g/--gff GFF formatted annotation file, supports .gz files.
2. -o/--output Output file name. If the file name is given with .gz suffix, it is gzipped. If no file name is given, output is print to console.

For other optional parameters, please refer to *htseq-clip* documentation available here: https://htseq-clip.readthedocs.io/en/latest/documentation.html#id1

Attention: The default values for optional parameters of **extract** function assumes that the input .gff3 is GENCODE formatted.

2.4.6 Create Sliding Windows

Create sliding windows from the flattened annotation file as:

```
$ htseq-clip createSlidingWindows -i gencode.v27.annotation.
bed -w 50 -s 20 -o SLBP_K562_w50s20.txt
```

Parameters used:

1. -i/--input Flattened annotation file, *see* Subheading 2.4.5.
2. -w/--windowSize Window size in number of base pairs for the sliding window (default: 50, Note: if you are unsure, we would recommend 75–100 nt).
3. -s/--windowStep Window step size for sliding window (default: 20).
4. -o/--output Output file name. If the file name is given with .gz suffix, it is gzipped. If no file name is given, output is print to console.

2.4.7 Create Mapping

Use the column "name" from the output of the annotation command or **createSlidingWindows** command to create a mapping file. This file will be used in downstream analysis with *DEWSeq*.

```
$ htseq-clip mapToId -a SLBP_K562_w50s20.txt -o
SLBP_K562_w50s20_annotation.txt.gz
```

Parameters used:

1. -a/--annotation Flattened annotation file from annotation command or sliding window file from createSlidingWindows command.
2. -o/--output Output file name. If the file name is given with .gz suffix, it is gzipped. If no file name is given, output is print to console.

2.5 Extracting and Counting Truncation Sites

This section describes *htseq-clip* commands and steps for extracting and counting crosslink sites from alignment files.

2.5.1 Extract Crosslink Sites

The crosslink sites are extracted from alignment files with an offset of 1 nucleotide from the start position of the second read of the paired-end alignment:

```
$ cd /path/to/SLBP_analysis
$ htseq-clip extract -i bam/ENCFF218ZEI.bam -e 2 -s s -g -1 --
primary -o sites/ENCFF218ZEI.bed
$ htseq-clip extract -i bam/ENCFF511HSJ.bam -e 2 -s s -g -1 --
primary -o sites/ENCFF511HSJ.bed
$ htseq-clip extract -i bam/ENCFF879UID.bam -e 2 -s s -g -1 --
primary -o sites/ENCFF879UID.bed
```

Parameters used:

1. -i/--input Input .bam file.
2. -e/--mate For paired-end sequencing, select the read/mate to extract the crosslink sites from, accepted choices: 1, 2.
3. -s/--site Crosslink site choices, accepted choices: s, i, d, m, e for extracting start, insertion, deletion or middle, end (default: e).
4. -g/--offset Number of nucleotides to offset for crosslink sites (default: 0).
5. --primary Use this flag consider only primary alignments of multimapped reads.
6. -o/--output Output file name. If the file name is given with .gz suffix, it is gzipped. If no file name is given, output is print to console.

For other optional parameters please refer to htseq-clip documentation available here: https://htseq-clip.readthedocs.io/en/latest/documentation.html#extract

2.5.2 Count and Aggregate Sites

Count the number of truncation sites per window using the output files from `extract` command and sliding window files generated using **createSlidingWindows** command:

```
$ htseq-clip count -i sites/ENCFF218ZEI.bed -a annotation/
SLBP_K562_w50s20.txt -o counts/ENCFF218ZEI.csv
$ htseq-clip count -i sites/ENCFF511HSJ.bed -a annotation/
SLBP_K562_w50s20.txt -o counts/ENCFF511HSJ.csv
$ htseq-clip count -i sites/ENCFF879UID.bed -a annotation/
SLBP_K562_w50s20.txt -o counts/ENCFF879UID.csv
```

Parameters used:

1. -i/--input Extracted crosslink sites from extract command.
2. -a/--ann Flattened annotation file from annotation command or sliding window file from createSlidingWindows command.
3. -o/--output Output file name. If the file name is given with .gz suffix, it is gzipped. If no file name is given, output is print to console.

For other optional parameters, please refer to htseq-clip documentation available here: https://htseq-clip.readthedocs.io/en/latest/documentation.html#count

2.5.3 Create Count Matrix

Create an R friendly matrix file from multiple files created by **count** command.

```
$ htseq-clip createMatrix -i counts/ -b ENCFF -o counts/
SLBP_K562_w50s20_counts.txt.gz
```

Parameters used:

1. -i/--inputFolder Folder name with output files from **count** command.
2. -b/--prefix Use files only with this given file name prefix (default: None).
3. -o/--output Output file name. If the file name is given with .gz suffix, it is gzipped. If no file name is given, output is print to console.

For other optional parameters, please refer to htseq-clip documentation available here: https://htseq-clip.readthedocs.io/en/latest/documentation.html#count

2.6 Additional Help

1. Please use the issues page in *htseq-clip* github repository (https://github.com/EMBL-Hentze-group/htseq-clip/issues) to report issues, bugs or for feature requests.
2. A yaml file to recreate the htseq-clip conda environment used in this analysis is available here: https://github.com/EMBL-Hentze-group/DEWSeq_analysis_helpers/blob/master/SLBP_example/htseq-clip.yaml

2.7 Differential Analysis with DEWSeq

2.7.1 Installation of DEWSeq and Required Packages

DEWSeq is the companion R/Bioconductor package for the statistical analysis of the data files generated by *htseq-clip*. We can use *BiocManager* for installation. Please note that all following codes snippets are to be executed in R.

```
> if(!requireNamespace("BiocManager", quietly = TRUE))
 install.packages("BiocManager")
> BiocManager::install("DEWSeq")
```

Following packages are required for our analysis workflow, you can install them with following command:

```
> install.packages(c("IHW", "tidyverse", "data.table", "ggre-
pel"))
```

2.7.2 Loading Libraries

Once all packages are installed, we use the function **require** to load them:

```
> require(DEWSeq)
> require(IHW)
> require(tidyverse)
> require(data.table)
> require(ggrepel)
```

2.7.3 Help

A detailed documentation and an example vignette are available in *DEWSeq* Bioconductor page: https://bioconductor.org/packages/devel/bioc/html/DEWSeq.html

2.7.4 Data Import

In this section, the count matrix generated by *htseq-clip* command **createMatrix** and mapping file generated by **mapToId** command will be used as inputs.

```
> setwd('/path/to/SLBP_analysis')
> count_matrix <- fread('counts/SLBP_K562_w50s20_counts.txt.
gz', stringsAsFactors=FALSE, sep="\t", header=TRUE)
> count_matrix <- column_to_rownames(count_matrix,'uni-
que_id')
```

For details of the sample data in this table, please refer to Table 1 in section "A minimal reproducible example." The sample information table given below is generated based on the table.

```
> col_data <- data.frame(type=c('IP','IP','SMI'), row.names=colnames(count_matrix))
> annotation_file <- 'annotation/SLBP_K562_w50s20_annotation.txt.gz'
```

Next, we create a DESeqDataSet object using the function **DESeqDataSetFromSlidingWindows**:

```
> ddw <- DESeqDataSetFromSlidingWindows(countData=count_matrix, colData=col_data, annotObj=annotation_file, design=~type)
```

2.7.5 Estimate Size Factors

The first step after creating the DESeqDataSet is to estimate the size factors for library size normalization. This determines a stable measure for library sizes which will be used to account for library size differences. The default method of size factor estimation is exactly the same as for *DESeq2*:

```
> ddw <- estimateSizeFactors(ddw)
```

CLIP samples often show high amounts of noncoding RNAs, which can bias the library size estimation. Using the annotation annotated and preprocessed by *htseq-clip*, one can restrict the type of RNAs which are used for library size estimation. For example, if we want to calculate the library sizes based on mRNAs only, which is recommend for most cases, this step can be modified as:

```
> ddw_mRNAs <- ddw[ rowData(ddw)[,"gene_type"] == "protein_coding", ]
> ddw_mRNAs <- estimateSizeFactors(ddw_mRNAs)
> sizeFactors(ddw) <- sizeFactors(ddw_mRNAs)
```

For additional methods of size factor estimation, please refer to the github repository mentioned under the section "Additional detailed analysis."

2.7.6 Prefiltering

Removing windows with very few crosslink sites will benefit further analysis. There are no hard and fast rules to determine an ideal crosslink count threshold. This threshold depends on the experiment being performed, culture conditions, and the protein under investigation. We would encourage users to fix a cut-off threshold based on the protein of interest, and the experience users have in dealing with the datasets from the said protein.

For demonstration purposes, we are using a cut-off threshold of 1, and we further stipulate that there must be at least 2 samples in the data, where the number of crosslink count is greater than 1.

A general approach would be to specify that at least half the number of samples should have the given cut-off threshold per window. But in this case, since the analysis is done only using 3 samples, we chose the condition that for each sliding window, there must be at least 2 samples with greater than 1 crosslink site count to be included in further analysis.

```
> keep <- which(rowSums(counts(ddw)>1)>=2)
> ddw <- ddw[keep,]
> ddw
```

2.7.7 Estimate Dispersion and Model Fit

The estimation of dispersions of counts is also based on *DESeq2*. The following example shows dispersion estimation using local fit.

```
> ddw <- estimateDispersions(ddw, fitType='local', quiet=-
TRUE)
```

We have implemented an additional strategy to decide whether to use parametric or local fit based on the analysis of residuals from fit dispersions. For additional details on this strategy, please refer to the github repositories mentioned under the section "Additional variations."

2.7.8 Differential Expressed Windows Analysis

In the next step, we call significantly enriched windows. For this, *DEWSeq* supports both Wald test and LRT (Likelihood-ratio test) implemented in *DESeq2*. But, the example below demonstrates the use of LRT test, as our experiments with ENCODE datasets (data not shown here) have shown that the results from LRT have more biologically relevant results (in this case we are using presence of primary sequence motifs in enriched windows as a proxy for biological relevance).

```
> ddw <- nbinomLRT(ddw, full = ~type, reduced = ~1)
```

DEWSeq package provides a function called **resultsDEWSeq**, which is a modified version of the `results` function from *DESeq2*.

This function can be used to extract the results after significance testing either using Wald test or using LRT:

```
> resultWindows <- resultsDEWSeq(ddw, contrast = c("type",
"IP", "SMI"),tidy = TRUE) %>% as_tibble
```

2.7.9 Multiple Hypothesis Correction

DEWSeq results can be corrected for FDR (False discovery rate) using Benjamini Hochberg (BH) method as follows:

```
> resultWindows[,'p_adj'] <- p.adjust(resultWindows$pvalue,
method="BH")
```

We recommend using the Bioconductor package *IHW* [10] over BH method for multiple hypothesis correction. Multiple testing correction using IHW can be done as follows:

```
> resultWindows[,"p_adj_IHW"] <- adj_pvalues(ihw(pvalue ~
baseMean, data = resultWindows, alpha = 0.05, nfolds = 10))
```

Please refer to *IHW* vignettes and documentation for additional details.

2.7.10 Combining Windows to Regions

One of the final steps is to merge adjacent differentially binding windows into binding regions. In the example below, results after FDR correction using *IHW*:

```
> resultRegions <- extractRegions(windowRes=resultWindows,
padjCol="p_adj_IHW", padjThresh=0.05, log2FoldChangeThresh=1)
%>% as_tibble
```

This is the final result table from *DEWSeq* analysis returning significantly enriched binding regions. For a description of the columns in this table, please refer to *DEWSeq* documentation available here: https://bioconductor.org/packages/release/bioc/manuals/DEWSeq/man/DEWSeq.pdf

2.7.11 Exporting Results

The enriched windows and binding regions (resultWindows and resultRegions) can be exported to BED format for visualization in genome browsers such as IGV as follows:

```
> toBED(windowRes=resultWindows, regionRes=resultRegions,
padjThresh=0.05, padjCol="p_adj_IHW", fileName="SLBP_re-
gions_w50s20.bed")
```

2.7.12 Visualization: Volcano Plot

Similar to RNA-seq analysis data visualizations, differential binding windows results from DEWSeq can also be visualized using volcano plot. A major difference here is that in DEWSeq results, the results are available only for windows with $\log_2$ fold change ≥ 0.

A customized R script can be used to visualize differentially binding windows results. This R script is available in our github repository: https://github.com/EMBL-Hentze-group/DEWSeq_analysis_helpers/tree/master/Volcano_plot. This script can be sourced and used for generating volcano plot as follows:

```
> source( https://raw.githubusercontent.com/EMBL-
Hentze-group/DEWSeq_analysis_helpers/master/Vol
cano_plot/volcanoplot.R)
> tophits <- resultWindows %>% filter(
p_adj_IHW<=1e-5 & log2FoldChange>=10 )%>%
  select(gene_name,log2FoldChange) %>%
  arrange(-log2FoldChange) %>%
  select(gene_name) %>%
  unlist() %>%
  unique()
  # the set of functions chained above selects the
genes with windows
  # which shows highest differential binding in IP
samples compared to SMI
> volcanoplot(resultWindows, padj_col =
'p_adj_IHW', gene_names = tophits[c(1:3)])
```

2.7.13 Session Info

See Subheading 1.12 “Session info” in file: https://github.com/EMBL-Hentze-group/DEWSeq_analysis_helpers/blob/master/SLBP_example/SLBP_analysis.pdf.

2.7.14 Additional Help

1. Please use the issues page in *DEWSeq* github repository: https://github.com/EMBL-Hentze-group/DEWSeq/issues to report issues, bugs or for feature requests.
2. An Rmarkdown, knitted html and pdf files for this analysis are available in the github repository: https://github.com/EMBL-Hentze-group/DEWSeq_analysis_helpers/tree/master/SLBP_example

3 Further Improvements and Variations of the Analysis

3.1 Advanced Models

DEWSeq uses *DESeq2* for model fitting and significance calling. Although *DEWSeq* was optimized for a one-sided IP vs SMI comparisons, the underlying GLMs allow complex experimental designs and models. For example, one can use blocking factors for factoring in cell line differences:

```
> ddw <- DESeqDataSetFromSlidingWindows(countData=count_matrix, colData=col_data, annotObj=annotation_file, design=~protein + cell_line)
```

If you want to do two-sided tests, we would recommend doing traditional *DESeq2* tests.

3.2 Quality Control

The possibility of doing PCA or other dimension reduction techniques allows for QC which is very welcome given the complexity of CLIP data. For these techniques, please refer to the examples given in *DESeq2* vignette.

3.3 Additional Variations

An example is found as Rmarkdown on our github repository: https://github.com/EMBL-Hentze-group/DEWSeq_analysis_helpers/tree/master/SLBP_example

Please refer to the html or the pdf file in the repository for the following analysis steps:

1. Estimate size factors for only protein coding genes: Section **1.4.1**.
2. Estimate size factors without significant windows: Section **1.4.2**.
3. Decide best fit for the dataset: Section **1.6.1**.

In addition, we have also generated a parameterized Rmarkdown file, which can be used for the complete analysis of an i/eCLIP dataset. This Rmarkdown and additional details can be found in the following github repository: https://github.com/EMBL-Hentze-group/DEWSeq_analysis_helpers/tree/master/Parametrized_Rmd.

This parameterized Rmarkdown has the following features:

1. estimating size factor only for mRNA genes,
2. deciding on the best fit for the dataset,
3. option to choose between LRT and Wald test.

For a full description of the input files and parameters required for this Rmarkdown, please see the README.md file in the repository.

4 Other Tools

Peak calling is a very elegant method to detect binding sites, given that there are peaks identifiable. When browsing aligned bam files with genome browsers like *IGV* [11] and peaks can be identified visually, a peak caller which is able to process replicate information and controls, like *PureCLIP* [12], can be used.

Acknowledgements

The authors are all members of the Hentze group. We thank Matthias Hentze for his mentorship, scientific advice and support. We also thank Wolfgang Huber for his feedback, scientific advice and support.

References

1. Gebauer F, Schwarzl T, Valcárcel J, Hentze MW (2020) RNA-binding proteins in human genetic disease. Nat Rev Genet 22:185–198
2. König J, Zarnack K, Rot G et al (2010) ICLIP reveals the function of hnRNP particles in splicing at individual nucleotide resolution. Nat Struct Mol Biol 17:909–915
3. Van Nostrand EL, Pratt GA, Shishkin AA et al (2016) Robust transcriptome-wide discovery of RNA-binding protein binding sites with enhanced CLIP (eCLIP). Nat Methods 13:508–514
4. Hafner M, Landthaler M, Burger L et al (2010) Transcriptome-wide identification of RNA-binding protein and MicroRNA target sites by PAR-CLIP. Cell 141:129–141
5. Licatalosi DD, Mele A, Fak JJ et al (2008) HITS-CLIP yields genome-wide insights into brain alternative RNA processing. Nature 456:464–469
6. Love MI, Huber W, Anders S (2014) Moderated estimation of fold change and dispersion for RNA-seq data with DESeq2. Genome Biol 15:550
7. Girardot C, Scholtalbers J, Sauer S et al (2016) Je, a versatile suite to handle multiplexed NGS libraries with unique molecular identifiers. BMC Bioinformatics 17:419
8. Martin M (2011) Cutadapt removes adapter sequences from high-throughput sequencing reads. EMBnet J 17:10–12
9. Dobin A, Davis CA, Schlesinger F et al (2013) STAR: ultrafast universal RNA-seq aligner. Bioinformatics 29:15–21
10. Ignatiadis N, Klaus B, Zaugg JB, Huber W (2016) Data-driven hypothesis weighting increases detection power in genome-scale multiple testing. Nat Methods 13:577–580
11. Robinson JT, Thorvaldsdóttir H, Wenger AM et al (2017) Variant review with the integrative genomics viewer. Cancer Res 77:e31–e34
12. Krakau S, Richard H, Marsico A (2017) Pure-CLIP: capturing target-specific protein-RNA interaction footprints from single-nucleotide CLIP-seq data. Genome Biol 18:240

Chapter 11

Identification of miRNAs Bound to an RNA of Interest by MicroRNA Capture Affinity Technology (miR-CATCH)

Andrea Zeni, Margherita Grasso, and Michela A. Denti

Abstract

microRNA capture affinity technology (miR-CATCH) uses affinity capture biotinylated antisense oligonucleotides to co-purify a target transcript together with all its endogenously bound miRNAs. The miR-CATCH assay is performed to investigate miRNAs bound to a specific mRNA. This method allows to have a total vision of miRNAs bound not only to the 3′UTR but also to the 5′UTR and Coding Region of target messenger RNAs (mRNAs).

Key words miR-CATCH, miRNAs, Post-transcriptional regulation, Pull-down, miRNA target analysis

1 Introduction

In the last years, different methods to identify and validate miRNAs regulating mRNAs of interest have been developed. We tested luciferase assays, western blot analyses, and the miR-CATCH method.

The luciferase reporter assay is commonly performed as a tool to study gene expression at the transcriptional level. It is widely used because it has high sensitivity and is convenient, quick, and relatively inexpensive. The luciferase reporter gene is expressed downstream of the target gene and detected after exogenous addition of luciferin substrate. Specifically, to test the effect of miRNAs on their target, the experimental procedure consists firstly in cloning the 3′UTR of the miRNA target downstream of a firefly luciferase reporter. Then, each construct is co-transfected with miRNA-overexpressing plasmids or miRNA mimics into a suitable cell line. With this technique it is possible to analyze whether or not a particular miRNA binds to the 3′UTR of the mRNA under

Andrea Zeni and Margherita Grasso contributed equally to this work.

Erik Dassi (ed.), *Post-Transcriptional Gene Regulation*, Methods in Molecular Biology, vol. 2404,
https://doi.org/10.1007/978-1-0716-1851-6_11,

investigation, by monitoring changes in luciferase levels due to the translational repression of the luciferase reporter by the overexpressed miRNA [1].

The western blot is an analytical technique used to detect specific proteins in a sample of tissue homogenate or extract. It uses gel electrophoresis to separate denatured proteins by the length of the polypeptide. The proteins are then transferred to a nitrocellulose membrane, where they are stained with antibodies specific to the target protein. Regarding miRNAs, with this technique it is possible to analyze down- or upregulation of a protein translated from a mRNA of interest, upon miRNAs overexpression or inhibition [2–4].

Unlike the techniques previously described, the miR-CATCH method focuses on the isolation of those miRNAs that, in conjunction with the RISC complex, physically interact with the RNA of interest, in a physiological context. In fact, the method does not rely on the artificial upregulation of miRNA levels, as the other two methods do. Additionally, the miR-CATCH method has the advantage to identify miRNAs that bind to the entire full-length mRNA, including the 5′UTR, the coding region and the 3′ UTR. The miR-CATCH technique, first described by Catherine M. Greene and collaborators [5, 6], uses biotinylated antisense oligonucleotides complementary to predicted single-stranded regions of the target mRNA, to capture a target mRNA together with all its endogenously bound miRNAs. Oligonucleotides are immobilized to streptavidin magnetic beads. Following a formaldehyde cross-linking step, cells are lysed. Oligo-prepared beads are mixed with the cell lysate and pulled down with a magnet. A final step is used to fully reverse the cross-linking, leading to release of all components of the complex for further analyses, such as mRNA quantification and miRNA expression analysis [5, 6] (Fig. 1).

The miR-CATCH method has been used in combination with several methods either for the quantitation of specific miRNAs or for a more comprehensive miRNome profiling. It was first designed to identify miRNAs binding alpha-1-antitrypsin mRNA, interleukin-8 mRNA, and secretory leucoprotease inhibitor mRNA in 2013 by Hassan et al. [5] who profiled the enriched miRNAs with an nCounter miRNA Expression Assay (Nanostring). It was then used in 2016 in combination with Taqman RT-qPCR single miRNA assays, to confirm the interaction of miR-659-3p with GRN mRNA in neuroblastoma cell lines [7] and of miR-96 and miR-182 with Rac1 mRNA in mice retinae [8].It has been coupled with next-generation sequencing to identify miRNAs that interact with MSLN mRNA in a mesothelial cell line [9]. Followed by a Taqman RT-qPCR low-density array analysis of the miRNome, the method has been used to identify miRNAs binding to XIAP mRNA in Chinese Hamster Ovary cells [10] and oocyte miRNAs bound by a circular RNA present in spermatozoa [11].

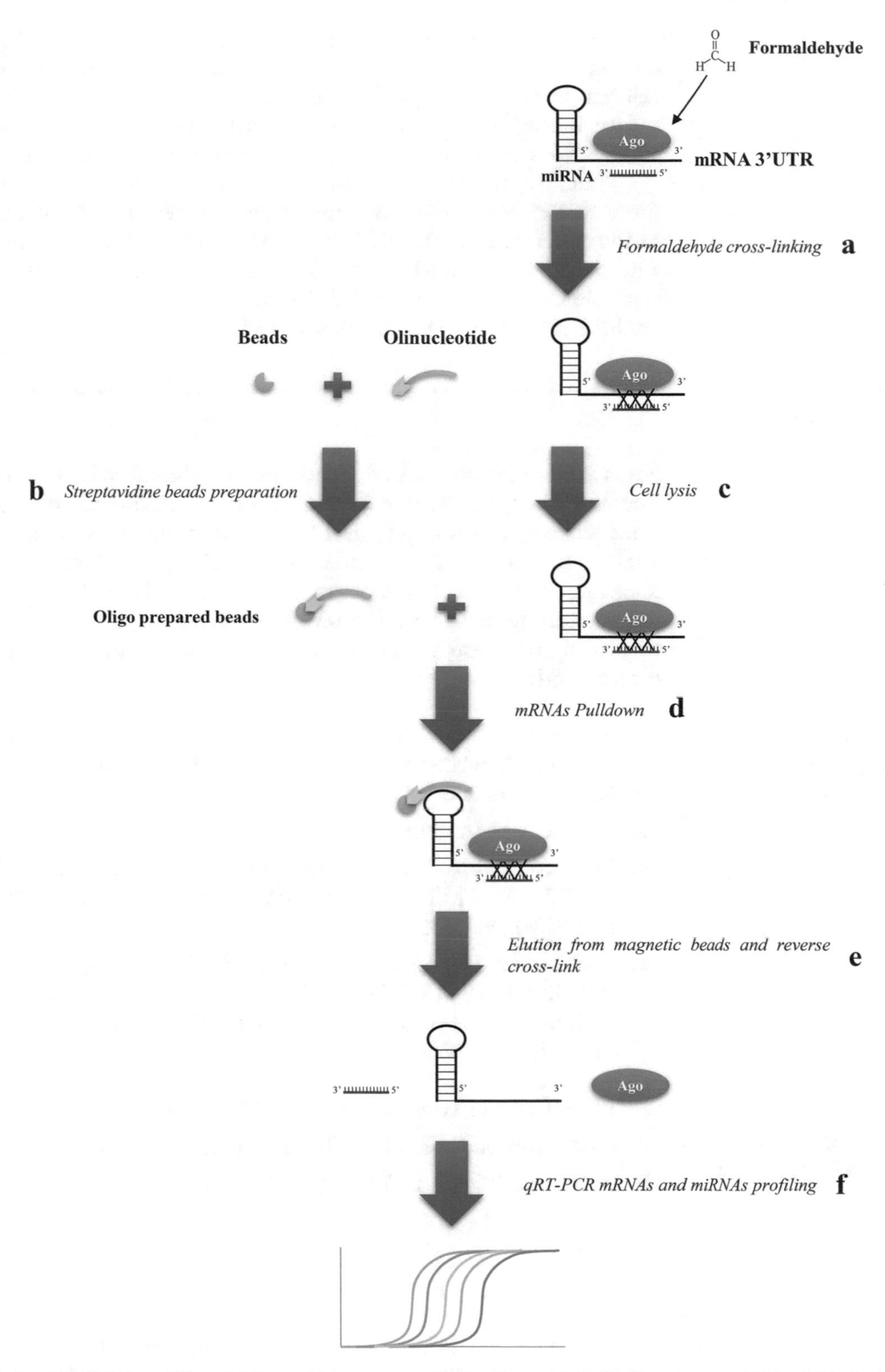

Fig. 1 miR-CATCH workflow. (**a**) Formaldehyde cross-linking:Formaldehyde allows to cross link all the factors, such as miRNAs and proteins, to mRNAs; (**b**) Streptavidin beads preparation and oligonucleotide immobilization: oligonucleotides are immobilized to magnetic streptavidin beads; (**c**) Cell lysis; (**d**) mRNAs pull-down: the

Recently, an implemented method (miRCATCHv2.0) has been described, which improves pull-down's efficiency by introducing cell lysis by sonication (which also fragments the RNA), the use of twelve biotinylated tiling probes, in two independent sets, and small RNA sequencing for the analysis of the miRNome [12].

Here, we describe the original miR-CATCH method, as it is a more cost-effective and less labor-intensive means for the identification or validation of miRNAs interacting with an RNA of interest. The pulled-down miRNAs can be identified or analyzed by any available method. For a detailed description of these methods, the reader is referred to a recent review [13].

2 Materials

RNase-free conditions are critical: all consumables should be RNase free and filter pipette tips should be used. Prepare all solutions using ultrapure water (prepared by purifying deionized water, to attain a sensitivity of 18 MΩ-cm at 25 °C) and analytical grade reagents. All buffers should be stored at 4 °C and brought to room temperature prior to use. The protocol can be adapted to alternative equipment of similar purpose, although optimization is recommended.

2.1 Formaldehyde Cross-Linking and Cell Lysis

1. Formaldehyde 2%: From formaldehyde, prepare a fresh 10 ml stock of 2% formaldehyde in DPBS at room temperature (*see* **Notes 1** and **2**).
2. Glycine 3 M (*see* **Note 3**).
3. Dulbecco's phosphate-buffered saline (DBPS): 0.137 M NaCl, 0.0027 M KCl, 0.0119 M phosphates (Na_2HPO_4, KH_2PO_4) (*see* **Note 3**).
4. Lysis buffer: 140 mM NaCl, 50 mM HEPES (adjust to pH 7.5 with KOH), 1 mM EDTA, 1% v/v Triton-X 100, 0.1% w/v sodium deoxycholate. Always add PMSF, protease inhibitor cocktail, and RNase inhibitor fresh before use (*see* **Note 3**).

2.2 Streptavidin Bead Preparation and mRNAs Pull-Down

1. TE buffer 1×: 10 mM Tris (pH 7.5), 1 mM EDTA.
2. Hybridization buffer: 2× TE buffer (pH 7.5), 1 M LiCl.
3. Washing buffer: 0.1 M NaCl, 0.5% SDS.

Fig. 1 (continued) oligonucleotide-bead complexes capture mRNAs target; (**e**) Elution and reverse cross-link: the formaldehyde cross-links are reversed and the target of interest and its associated miRNAs are released; (**f**) Selected mRNAs and associated miRNAs are profiled and quantified

2.3 miR-CATCH: Total RNA and Capture Validations

1. Trizol for Total RNA extraction (*see* Subheading 3.6).
2. Quantitative Real Time PCR Kit use according to the manufacturer's instruction (*see* **Note 7**).

2.4 Analysis of Pulled-Down miRNAs

1. miRNome Analysis with microRNA Ready-to-Use PCR Human Panels based on LNA technology: according to manufacturer's instruction.
2. TaqMan Real Time PCR for miRNAs quantification: according to manufacturer's instruction.

2.5 Laboratory Equipment

1. Humidified 37 °C, 5% CO_2 incubator.
2. Cell culture hood.
3. Magnetic rack for 15 ml conical tubes.
4. Magnetic rack for microcentrifuge tubes.
5. Rotating wheel for microcentrifuge tubes.
6. Thermoblock with shaking.
7. Refrigerated centrifuge for conical 15 ml tubes.
8. Refrigerated benchtop centrifuge.
9. Vortex.
10. Magnet optimized for efficient magnetic separation of all types of Dynabeads (range 1–4.5 μm in diameter) in small (<2 ml) sample volume (Thermo Fisher Scientific).
11. Homogenizer specific for any tissues and cells allowing easy and reproducible isolation of stable RNA, active proteins, and full-length genomic DNA (MP Biomedicals).
12. Thermocycler for PCR amplification.

2.6 Other Labware and Consumables

1. 0.5 mm RNase-free glass beads.
2. Dynabeads MyOne Streptavidin C1.
3. RNase Inhibitor.
4. Protease Inhibitor Cocktail.
5. Phenylmethylsulfonyl fluoride (PMSF).
6. Biotin-triethylene glycol (TEG) 5′-modified single-stranded DNA oligonucleotides.

3 Methods

3.1 Design of Biotinylated Capture and Control DNA Oligonucleotides

1. Model the secondary structures of the transcript of interest using an RNA secondary structure prediction software, such as M-FOLD ([14]*http://www.unafold.org/*) or similar (Fig. 2).

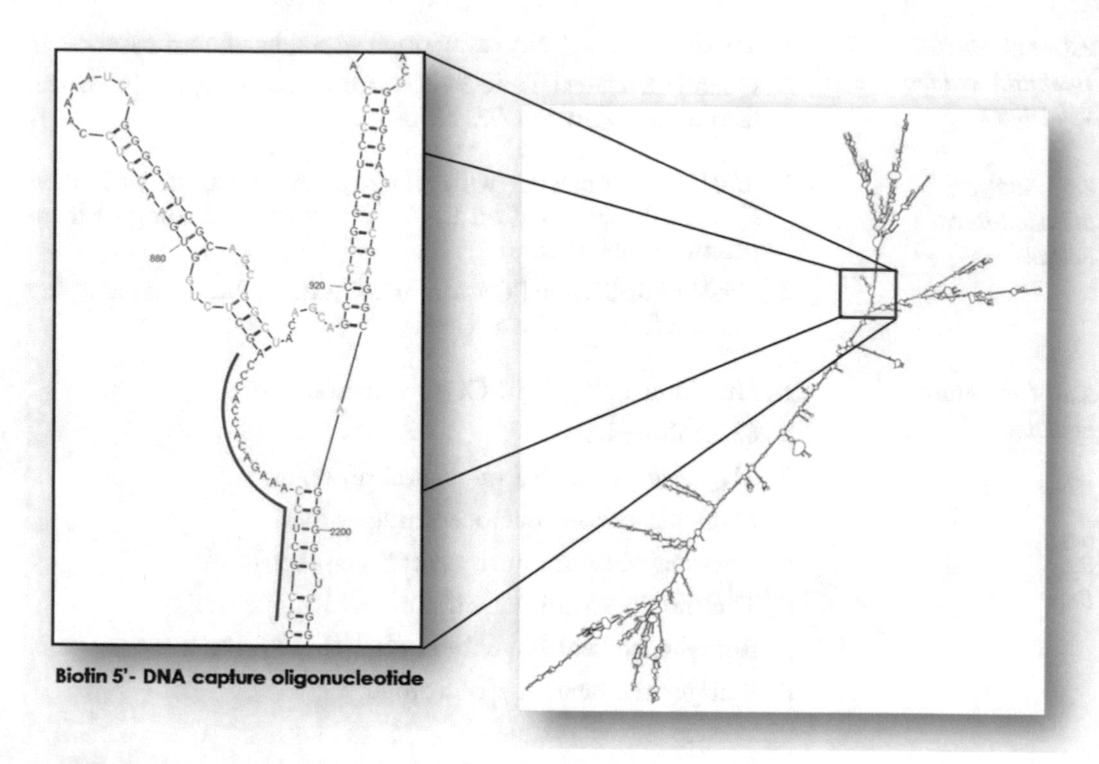

Fig. 2 Design of mRNA capture oligonucleotide. Secondary structure of selected mRNA was elaborated through M-Fold and the exposed single strand region of the transcript was used as a target

2. In order to identify suitable sequences for oligo design, analyze the most thermodynamically stable structures with UGENE program ([15]*http://ugene.net/*), a software that allows the identification of the most probable single-stranded regions present on the transcript.
3. Use accessible single-stranded region of the transcript as target sequences to design one to three specific 20- to 22-nt-long DNA capture oligonucleotides. An optimal capture oligonucleotide should have a low hairpin melting temperature (below 37 °C at 1 M NaCl) and a high hybrid melting temperature (over 50 °C at 1 M NaCl) to hybridize with the single-stranded region. We design up to three different capture oligonucleotides per target RNA and chose the one which best performs in the capture validation steps (*see* Subheading 3.4).
4. Check the specificity of the designed capture oligonucleotides by performing a BlastNsearch with Basic Local Alignment Search Tool (BLAST) ([16]*https://blast.ncbi.nlm.nih.gov/Blast.cgi*), over the entire transcriptome of the organism under investigation (human, mouse, etc.). To predict specificity we apply an arbitrary rule of thumb: this control oligonucleotide should not recognize any transcript other than the intended one, over a threshold of 14 consecutive nucleotides

and should have at least 6 mismatches (either concentrated or distributed over the length of the oligonucleotide) if the oligonucleotide's length is 20 nucleotides (7 if a 21-mer, 8 if a 22-mer). Since a G:T mismatch has a thermodynamically relevant contribution to the binding of the oligonucleotide to the target RNA when the G belongs to the RNA (rG) and the T to the DNA (dT) [17], we tend to consider rG:dT as paired bases.

5. Design a control oligonucleotide by scrambling the DNA sequence of the capture oligonucleotide and performing a BlastN search (see above) against the relevant transcriptome: this control oligonucleotide should not recognize any transcript, based on the rules given above.
6. Buy capture and control oligonucleotides with a biotin coupled at the 5′ (*see* **Note 4**).

3.2 Formaldehyde Cross-Linking

1. Grow cell lines in T75 culture flasks, using them as a starting material for each miR-CATCH experiment.
2. Fix cells with 2% formaldehyde to a final concentration of 1% and mix on a plate rocker for 10 min. Formaldehyde allows to cross link all the factors, such as miRNAs and proteins, to selected mRNAs.
3. Stop the formaldehyde reaction by adding Glycine 3 M to a final concentration of 0.2 M and mix on the plate rocker for 5 min.
4. Centrifuge at 2000 × *g* for 5 min at 4 °C.
5. Discard the supernatant, resuspend in 50 ml ice-cold DPBS, and centrifuge at 2000 × *g* for 5 min at 4 °C.
6. Repeat **step 5** once.
7. Resuspend the pellet in 0.6 ml ice-cold DPBS. Transfer the suspension to a 1.5 ml eppendorf tube and resuspend any remaining cells in the 50 ml tube with another 0.6 ml ice-cold DPBS.
8. Centrifuge the cells at 2000 × *g* for 5 min at 4 °C and discard the supernatant. Remove the last drops of supernatant with a pipette.
9. Flash-freeze the cell pellet in liquid nitrogen. Flash-frozen cell pellet can be stored at −80 °C until use.

3.3 Cell Lysis

1. Prepare 1.4 ml of Lysis buffer.
2. Homogenize cells in 0.4 ml of Lysis buffer, 0.4 ml of glass beads using the homogenizer four times for 30 s at speed 5.5, ensuring that the lysate remains on ice during the 5-min cool-down periods.

3. Separate the lysate from the glass beads by puncturing a hole in the bottom of the eppendorf tube and centrifuging it into another eppendorf tube at 3500 × *g* for 5 min at 4 °C.
4. Wash the glass beads by adding 1 ml of Lysis buffer and centrifuging at 3500 × *g* for 5 min at 4 °C.
5. Finally, pellet cells by centrifuging at 18,000 × *g* for 5 min at 4 °C, transfer the supernatant to a new tube and keep on ice until use (*see* **Note 5**).

3.4 Streptavidin Beads Preparation, Oligonucleotide Immobilization, and mRNAs Pull-Down

1. Wash MyOne streptavidin magnetic beads according to the manufacturer's instruction.
2. Add biotinylated capture oligonucleotide to the beads in 1× TE buffer, rotating the tubes for 15 min at room temperature.
3. Briefly wash the beads twice in 1 ml hybridization buffer.
4. Resuspend the beads with 0.7 ml hybridization buffer, 0.5 ml 1× TE buffer and mixed with 0.2 ml of formaldehyde-treated cell lysate for 90 min at 37 °C for adequate annealing.
5. Wash the beads four times using washing buffers (capturing mRNA:miRNA complexes using a magnet).

3.5 Elution and Cross-Link Reversal

1. Resuspend the beads in TE buffer.
2. Incubate the beads for 5 min at 60 °C to reverse the interaction between the biotin-labeled DNA:mRNA:miRNA complexes and the magnetic beads.
3. Transfer the supernatant to a new tube ensuring that no beads are transferred. Also incubate a volume of original lysate as a reference sample to extract total RNA.
4. Incubate the eluted samples for 45 min at 70 °C to reverse the cross-linkages. Also incubate the volume of lysate from **step 3**.
5. Subsequently used samples for the following validations.

3.6 miR-CATCH: Total RNA Validation

Analyze the total amount of selected mRNAs in cell lysates that did not undergo the capture procedure (*see* **Note 6**) with a Quantitative Real Time PCR kit and specific primers which amplify a product of about 150 bp.

1. Add Trizol and resuspend the lysate (up and down by pipetting).
2. Incubate for 5 min at room temperature.
3. Add chloroform to obtain phase separation and vortex for 15 s, then incubate 2–3 min at room temperature.
4. Centrifuge at 12,000 × *g* for 15 min at 4 °C.
5. Take carefully only upper aqueous phase containing RNA and put it in a new eppendorf.

6. Add isopropanol to precipitate RNA and incubate for 15 min at room temperature.
7. Centrifuge at 12,000 × *g* for 10 min at 4 °C.
8. Remove carefully isopropanol and add 75% ethanol to wash RNA.
9. Centrifuge at 7500 × *g* for 5 min at 4 °C.
10. Remove ethanol and let dry the pellet.
11. Resuspend in ddH_2O and determine RNA concentration.
12. Perform cDNA retrotranscription according to the manufacturer's instruction and then analyze the expression of selected RNAs using specific primers and the housekeeping gene expression to normalize the results.

3.7 miR-CATCH: Capture Validation

After the miR-CATCH protocol, in order to validate the efficacy of transcripts isolation using the capture oligonucleotides, analyze the eluted samples with a Quantitative Real Time PCR kit and specific primers, to quantify the amount of the target transcript in each sample.

1. Perform cDNA retrotranscription and Quantitative Real Time PCR starting directly from eluted samples (*see* **Note 7**).
2. Then, evaluate the levels of mRNAs measured in Capture samples compared to the levels obtained in total RNA and in Scramble samples as represented in Fig. 3.

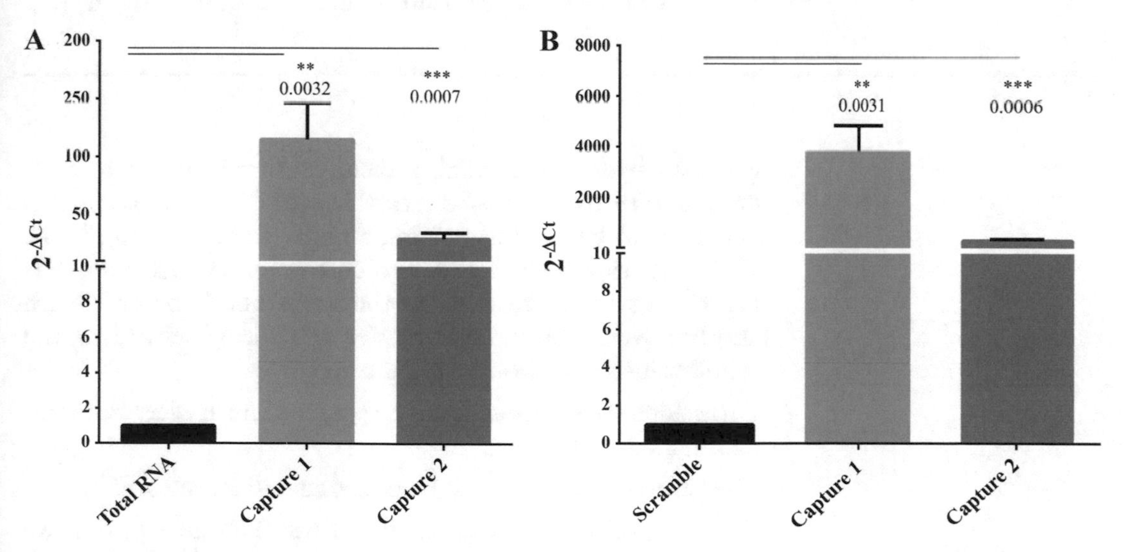

Fig. 3 Captures enrichment in the selected cell line. Representation of the fold enrichment ($2^{-\Delta Ct}$) of two different mRNAs (Capture 1 and Capture 2, obtained using two different oligonucleotides) compared to total RNA (**a**) and specific Scramble (**b**)

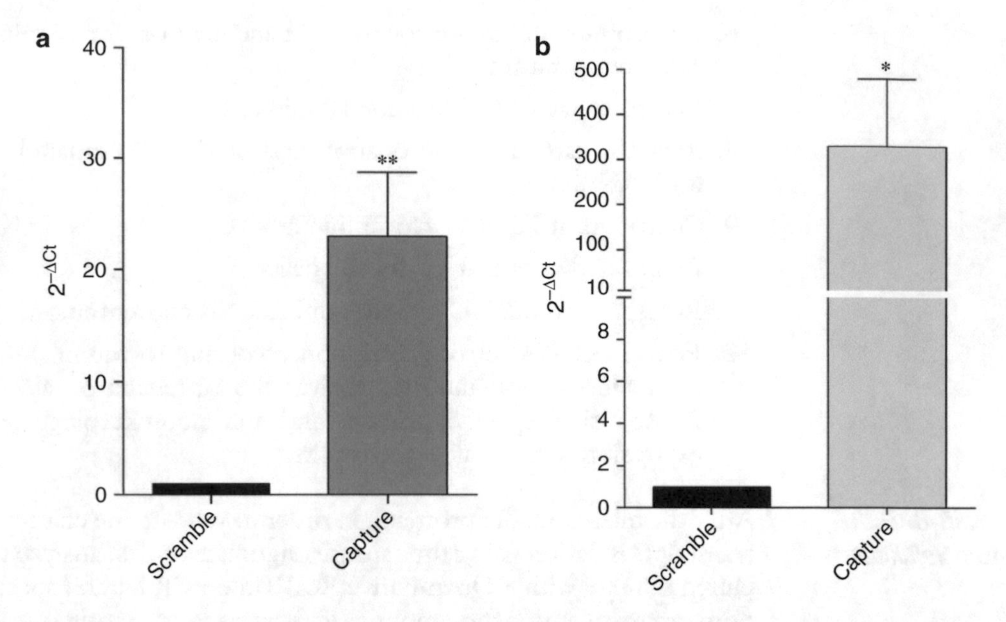

Fig. 4 mRNA:miRNA isolation technique. A capture antisense DNA oligonucleotide with a biotin modification at the 5′ end was used to pull-down mRNA of interest. RT-qPCR showed enrichment of mRNA (**a**) and miRNA (**b**) in the selected cell line compared to a Scramble oligonucleotide used as negative control (adapted from [7])

3.8 Analysis of Captured miRNAs

In this step, the enrichment of single miRNAs in the captured sample (versus the samples treated with scramble oligonucleotide) can be analyzed via single-miRNA RT-qPCR assays. Alternatively, high-throughput methodologies such as TaqMan Arrays, Exiqon RT-PCR panels, and Nanostring nCounter can be used to compare the captured miRNome to the reference miRNome (in the RNA pulled down by the scramble control oligonucleotide) (Fig. 4) [7].

4 Notes

1. Formaldehyde is a sensitizing agent causing irritation to eyes, nose, and throat and also a cancer hazard. For this reason, it is necessary to be careful, wearing a mask when weighing it. To avoid exposing formaldehyde to co-workers, when transporting the cover weigh boat to the fume hood, cover it with another weigh boat, and transfer the formaldehyde to the cylinder inside the hood mixing on a stirrer.
2. Formaldehyde solution has to be prepared fresh every time and maintained in the dark.
3. All the other solutions can be prepared and stored at 4 °C.
4. To test specificity and selectivity of the oligonucleotides, we recommend: to perform a capture with non-targeting oligonucleotides with a scrambled sequence (specificity) and to analyze

the expression of non-targeted genes with the highest degree of complementarity with the oligo (selectivity).

5. Different captures can be performed from the same lysate, although freezing and thawing lysate reduce the efficiency.
6. RNA extraction with Trizol is the best option to extract RNA from cell lysates. However, extraction can be optimized based on starting material (cells, lysates, tissues...).
7. Before Quantitative Real Time PCR analysis, the eluted samples could undergo an additional RNA extraction and purification step. We tried to perform RNA extraction step with Trizol and RNA extraction kit but we observed a loss of total RNA and, consequently, low-expressed miRNAs became undetectable. For this reason, in the present protocol we directly use eluted samples to perform Quantitative Real Time PCR experiments.

Acknowledgments

We wish to thank Catherine M. Greene (Royal College of Surgeons in Ireland, Dublin) for having hosted Francesca Fontana in her laboratory in the Fall of 2013, and Sebastian Vencken (Royal College of Surgeons in Ireland, Dublin) for having taught Francesca the miR-CATCH protocol. We also wish to thank Francesca Fontana for having shared with the rest of the Denti Lab her knowledge of the miR-CATCH methodology.

References

1. Jin Y, Chen Z, Liu X, Zhou X (2013) Evaluating the MicroRNA targeting sites by luciferase reporter gene assay. Methods Mol Biol 936:117–127
2. Denti MA, Rosa A, Sthandier O, De Angelis FG, Bozzoni I (2004) A new vector, based on the polII promoter of the U1 snRNA gene, for the expression of siRNAs in mammalian cells. Molecular Ther 10:191–199
3. Obad S, dos Santos CO, Petri A, Heidenblad M, Broom O, Ruse C, Fu C, Lindow M, Stenvang J, Straarup EM, Hansen HF, Koch T, Pappin D, Hannon GJ, Kauppinen S (2011) Silencing of microRNAs families by seed-targeting tiny LNAs. Nat Genet 43:371–378
4. Li P, Chen Y, Juma CA, Yang C, Huang J, Zhang X, Zeng Y (2019) Differential inhibition of target gene expression by human microRNAs. Cell 8:791
5. Hassan T, Smith SGJ, Gaughan K, Oglesby IK, O'Neill S, McElvaney NG, Greene CM (2013) Isolation and identification of cell-specific microRNAs targeting a messenger RNA using a biotinylated antisense oligonucleotide capture affinity technique. Nucleic Acids Res 41: e71
6. Vencken S, Hassan T, McElvaney NG, Smith SGJ, Greene CM (2015) miR-CATCH: microRNA capture affinity technology. In: Sioud M (ed) RNA interference: challenges and therapeutic opportunities, methods in molecular biology, vol 1218. Springer, Heidelberg, pp 365–373
7. Piscopo P, Grasso M, Fontana F, Crestini A, Puopolo M, Del Vescovo V, Venerosi A, Calamandrei G, Vencken SF, Greene CM, Confaloni A, Denti MA (2016) Reduced miR-659-3p levels correlate with progranulin increase in hypoxic conditions: implications for frontotemporal dementia. Front Mol Neurosci

9:31. https://doi.org/10.3389/fnmol.2016.00031

8. Palfi A, Hokamp K, Hauck SM, Vencken S, Millington-Ward S, Chadderton N, Carrigan M, Kortvely E, Greene CM, Kenna PF, Farrar GJ (2016) microRNA regulatory circuits in a mouse model of inherited retinal degeneration. Sci Rep 6:31431. https://doi.org/10.1038/srep31431
9. De Santi C, Vencken S, Blake J, Haase B, Benes V, Gemignani F, Landi S, Greene CM (2017) Identification ofMiR-21-5p as a functional regulator of MesothelinExpression using MicroRNA capture AffinityCoupled with next generation sequencing. PLoS One 12: e0170999. https://doi.org/10.1371/journal.pone.0170999
10. Griffith A, Kelly PS, Vencken S, Lao NT, Greene CM, Clynes M, Barron N (2018) miR-CATCH identifies biologically active miRNA regulators of the pro-survival gene XIAP, in Chinese hamster ovary cells. Biotechnol J 13:e1700299. https://doi.org/10.1002/biot.201700299
11. Ragusa M, Barbagallo D, Chioccarelli T, Manfrevola F, Cobellis G, Di Pietro C, Brex D, Battaglia R, Fasano S, Ferraro B, Sellitto C, Ambrosino C, Roberto L, Purrello M, Pierantoni R, Chianese R (2019) CircNAPEPLD is expressed in human and murine spermatozoa and physically interacts with oocyte miRNAs. RNA Biol 16:1237–1248. https://doi.org/10.1080/15476286.2019.1624469
12. Marranci A, D'Aurizio R, Vencken S, Mero S, Guzzolino E, Rizzo M, Pitto L, Pellegrini M, Chiorino G, Greene CM, PolisenoL (2019) Systematic evaluation of the microRNAome through miR-CATCHv2.0identifies positive and negative regulators of BRAF-X1 mRNA. RNA Biol 16:865–878. https://doi.org/10.1080/15476286.2019.1600934
13. Precazzini F, Detassis S, Imperatori AS, Denti MA, Campomenosi P (2021) Measurement methods for the development of microRNA-based tests for cancer diagnosis. Int J Mol Sci 21:1176
14. Zuker M (2003) Mfold web server for nucleic acid folding and hybridization prediction. Nucleic Acids Res 31:3406–3415. https://doi.org/10.1093/nar/gkg595
15. Okonechnikov K, Golosova O, Fursov M, the UGENE team (2012) Unipro UGENE: a unified bioinformatics toolkit. Bioinformatics 28:1166–1167. https://doi.org/10.1093/bioinformatics/bts091
16. Altschul SF, GishW MW, Myers EW, Lipman DJ (1990) Basic local alignment search tool. J Mol Biol 215:403–410. https://doi.org/10.1016/S0022-2836(05)80360-2
17. Sugimoto N, Nakano M, Nakano S (2000) Thermodynamics-structure relationship of single mismatches in RNA/DNA duplexes. Biochemistry 39:11270–11281. https://doi.org/10.1021/bi000819p

Chapter 12

Identifying Protein Interactomes of Target RNAs Using HyPR-MS

Katherine B. Henke, Rachel M. Miller, Rachel A. Knoener, Mark Scalf, Michele Spiniello, and Lloyd M. Smith

Abstract

RNA–protein interactions are integral to maintaining proper cellular function and homeostasis, and the disruption of key RNA–protein interactions is central to many disease states. HyPR-MS (hybridization purification of RNA–protein complexes followed by mass spectrometry) is a highly versatile and efficient technology which enables multiplexed discovery of specific RNA–protein interactomes. This chapter provides extensive guidance for successful application of HyPR-MS to the system and target RNA(s) of interest, as well as a detailed description of the fundamental HyPR-MS procedure, including: (1) experimental design of controls, capture oligonucleotides, and qPCR assays; (2) formaldehyde cross-linking of cell culture; (3) cell lysis and RNA solubilization; (4) isolation of target RNA(s); (5) RNA purification and RT-qPCR analysis; (6) protein preparation and mass spectrometric analysis; and (7) mass spectrometric data analysis.

Key words HyPR-MS, RNA–protein interactomes, RNA-binding proteins, Mass spectrometry, Hybridization capture, Proteomics, RNA, Interactomics

1 Introduction

RNA–protein interactions are crucial to multiple aspects of cellular function and homeostasis. Proteins bind to both coding and non-coding RNA sequences to mediate processes including RNA transcription, splicing, localization, translation, and degradation [1–7], and disruptions in RNA–protein interactions are central to many different disease states [8–10]. Characterizing the protein interactome of a specific RNA is therefore essential to understanding its biology both under normal conditions and in pathological states, and may aid in the discovery of therapeutic targets.

The updated online version of this chapter was revised. The correction to this chapter is available at https://doi.org/10.1007/978-1-0716-1851-6_23

Erik Dassi (ed.), *Post-Transcriptional Gene Regulation*, Methods in Molecular Biology, vol. 2404,
https://doi.org/10.1007/978-1-0716-1851-6_12,

Strategies for the interrogation of RNA–protein interactions can be broadly classified as either protein-centric or RNA-centric [11]. Protein-centric approaches isolate a protein of interest using immunoprecipitation then identify its associated RNAs, generally through high-throughput RNA sequencing. Such approaches include CLIP (cross-linking immunoprecipitation) [12] and variants like HITS-CLIP [13], PAR-CLIP [14], iCLIP [15], eCLIP [16], and fCLIP [17], among others [11]. Conversely, RNA-centric approaches utilize sequence-specific hybridization probes to isolate a specific RNA, then identify the associated proteins by mass spectrometry. These techniques include CHART-MS [18], ChIRP-MS [19], RAP-MS [20], and HyPR-MS [21–24], described here.

HyPR-MS (hybridization purification of RNA–protein complexes followed by mass spectrometry) is a versatile strategy for probing the in vivo protein interactomes of one or more target RNAs (Fig. 1). Briefly, cell culture is subjected to in vivo formaldehyde cross-linking to covalently stabilize RNA–protein and protein–protein interactions. After cell lysis, biotinylated capture oligonucleotides, designed to be complementary to the target RNA(s), facilitate the capture of the target RNA–protein

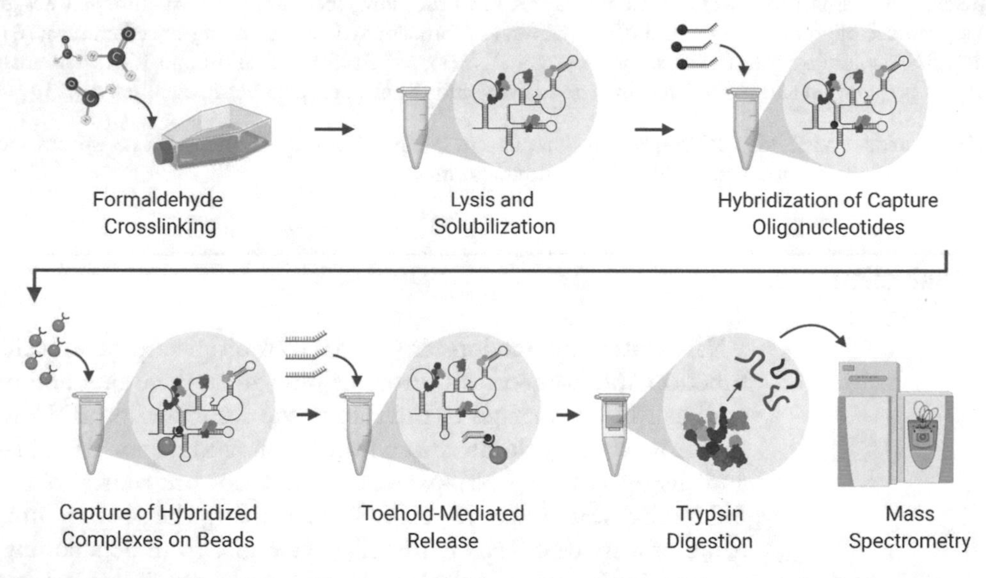

Fig. 1 Overview of the HyPR-MS technology. HyPR-MS begins with formaldehyde cross-linking of cell culture to covalently fix protein–RNA and protein–protein interactions. Cells are then lysed and RNA–protein complexes are solubilized. Biotinylated capture oligonucleotides (magenta), which include a sequence that is specifically complementary to the target RNA, are then added to the lysate and hybridize to the target RNA. Hybridized complexes are captured on streptavidin-coated magnetic beads, and the beads are washed to remove nonspecific interactors. Release oligonucleotides (green), which are complementary to the entire sequence of the capture oligonucleotide, are then added, and the target RNA–protein complexes are released from the beads via toehold-mediated strand displacement. The proteins are then digested with trypsin prior to mass spectrometric analysis to identify the protein interactome of the target RNA. (This figure was adapted from Knoener et al. [24] under CC BY 4.0 (https:/creativecommons.org/licenses/by/4.0/))

complexes. The oligonucleotide–target hybrids are isolated using streptavidin-coated magnetic beads and then released using a toehold-mediated release strategy. The proteins associated with the RNA target(s) are then purified, digested with a protease, and analyzed via mass spectrometry. Strengths of HyPR-MS include its high efficiency and specificity, its versatility, and its capacity for multiplexed discovery of specific RNA–protein interactomes.

The protocol presented here describes the general design and execution of a HyPR-MS experiment, and offers guidance for successful application to the system and target RNA(s) of interest. The Methods section (Subheading 3) describes the HyPR-MS workflow, including (1) experimental design of controls, capture oligonucleotides, and qPCR assays; (2) formaldehyde cross-linking of cells in culture; (3) cell lysis and RNA solubilization; (4) isolation of target RNA(s); (5) RNA purification and RT-qPCR analysis; (6) protein preparation and mass spectrometric analysis; and (7) mass spectrometric data analysis (Fig. 2). Importantly, some HyPR-MS parameters are target RNA-specific and may require optimization through empirical testing. Before performing full-scale HyPR-MS experiments for protein identification, we recommend first performing small-scale experiments on relatively few cells to evaluate capture oligonucleotide performance and to establish appropriate capture parameters. These small-scale experiments will provide enough RNA to monitor capture efficiency and specificity via RT-qPCR, but will not provide enough protein for mass spectrometric analysis. Once appropriate capture parameters have been established, one can scale-up to perform the entire HyPR-MS experiment, including mass spectrometric analysis, on a larger number of cells.

2 Materials

2.1 Formaldehyde Cross-Linking

1. Cell line of interest growing in culture.
2. Formaldehyde solution.
3. Tris–HCl solution pH 8.0.
4. Phosphate-buffered saline (PBS).
5. Cell scraper or trypsin (for adherent cell lines).
6. Liquid nitrogen.
7. Orbital shaker.
8. Centrifuge with a swinging-bucket rotor.

2.2 Cell Lysis, Target RNA Isolation, and RT-qPCR

All solutions used in this portion of the experiment should be prepared using certified RNase-free components. Similarly, all pipette tips and tubes should be certified RNase-free. The use of RNase decontamination wipes to wipe down pipettes, lab benches, and other lab surfaces prior to beginning this portion of the experiment is highly recommended.

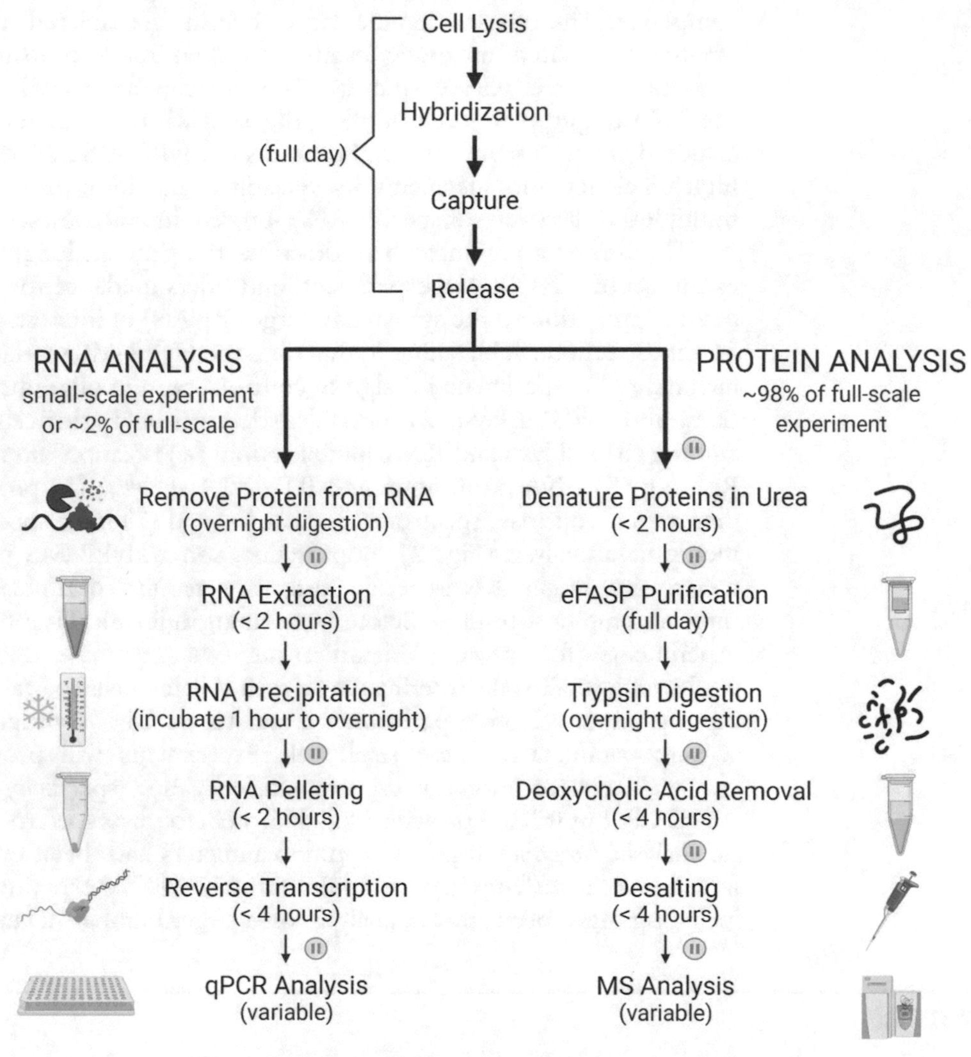

Fig. 2 RNA and protein purification/analysis steps in HyPR-MS. After isolating the target RNA–protein complexes from cell lysate, the sample should be divided into two aliquots for RNA and protein analysis. For a full-scale experiment, the majority of the sample (~98%) should be used for protein analysis, while ~2% should be used for RNA analysis. For a small-scale experiment, the entire sample can be used for RNA analysis. The major steps for both RNA and protein purification/analysis are indicated, with the approximate duration of each step noted. Optional places to pause the experiment and store the samples at −20 °C are indicated. In general, we recommend completing the entire experiment (from cell lysis through qPCR and mass spectrometric analysis) in the span of approximately 1 week and minimizing the number of freeze–thaw cycles

1. Lysis buffer (prepare fresh): 469 mM LiCl, 62.5 mM Tris–HCl pH = 7.5, 1.25% lithium dodecyl sulfate (LiDS), 1.25% Triton X-100, 12.5 mM ribonucleoside vanadyl complex, 12.5 mM dithiothreitol (DTT), 125 U/mL RNasin Plus (Promega), 1.25× protease/phosphatase inhibitor cocktail.

2. Nuclease-free water.
3. Target RNA and control capture oligonucleotides (*see* Subheading 3.1.2 for discussion of capture oligonucleotide design).
4. Streptavidin-coated magnetic Sera-Mag SpeedBeads (Thermo Fisher Scientific).
5. Wash buffer: 375 mM LiCl, 50 mM Tris–HCl pH = 7.5, 0.2% LiDS, 0.2% Triton X-100.
6. Release buffer: 375 mM LiCl, 50 mM Tris–HCl pH = 7.5, 0.1% LiDS, 0.1% Triton X-100.
7. Target RNA and control release oligonucleotides (complementary to respective capture oligonucleotides).
8. Proteinase K.
9. $CaCl_2$ solution.
10. TRI Reagent.
11. Chloroform.
12. Ethanol.
13. Reverse transcription kit.
14. qPCR master mix.
15. Target-specific hydrolysis probe qPCR assays (*see* Subheading 3.1.3 for discussion of qPCR assay design).
16. Vortex mixer.
17. Probe sonicator.
18. Benchtop centrifuge.
19. Low-protein-binding tubes.
20. Nutating mixer.
21. Laboratory incubator.
22. Magnetic tube rack.
23. Low-retention PCR tubes.
24. Thermal cycler.
25. qPCR plates and sealing film.
26. Real-time PCR instrument.

2.3 Protein Preparation and Mass Spectrometric Analysis

1. 1% 3-[(3-cholamidopropyl)dimethylammonio]-1-propanesulfonate (CHAPS) solution.
2. LC-MS grade water.
3. Urea.
4. Deoxycholic acid.
5. DTT solution.

6. eFASP exchange buffer (prepare fresh): 8 M urea, 50 mM ammonium bicarbonate, 0.1% deoxycholic acid.
7. eFASP reducing buffer (prepare fresh): 8 M urea, 50 mM ammonium bicarbonate, 20 mM DTT.
8. eFASP alkylation buffer (prepare fresh and protect from light): 8 M urea, 50 mM ammonium bicarbonate, 50 mM iodoacetamide.
9. eFASP digestion buffer (prepare fresh): 1 M urea, 50 mM ammonium bicarbonate, 0.1% deoxycholic acid.
10. Sequencing- or mass spectrometry-grade trypsin.
11. 50 mM ammonium bicarbonate.
12. Trifluoroacetic acid (TFA).
13. Ethyl acetate.
14. LC-MS grade acetonitrile.
15. Formic acid (FA).
16. 50 kDa molecular weight cutoff filters (0.5 mL) and collection tubes.
17. Benchtop centrifuge.
18. Laboratory incubator.
19. Low-protein-binding tubes.
20. Vortex mixer.
21. Vacuum centrifuge concentrator.
22. C18 solid-phase extraction pipette tips (100 μL).
23. nanoAcquity high performance liquid chromatography system (Waters) coupled on-line to a Q Exactive HF mass spectrometer (Thermo Fisher Scientific) (or similar LC-MS setup).
24. LC-MS column: 100 μm id × 365 μm od fused silica capillary microcolumn packed with 20 cm of 1.7 μm diameter, 130 Å pore size C18 beads with an emitter tip pulled to ~1 μm using a laser puller (or similar column).
25. Column oven.

3 Methods

3.1 Experimental Design

3.1.1 Design of Control Experiment(s)

A successful HyPR-MS experimental design must include a control(s) to help determine which proteins identified in a target RNA pulldown sample are target-specific interactors (*see* **Note 1**). Three potential controls are listed below, with descriptions provided in the corresponding Notes. Alternative controls may be appropriate depending on the goal(s) of the particular HyPR-MS experiment.

1. Scrambled oligonucleotide pulldown control (*see* **Note 2**).
2. Poly(dT) oligonucleotide pulldown control (*see* **Note 3**).
3. Lysate control (*see* **Note 4**).

3.1.2 Design of Capture Oligonucleotides

In HyPR-MS, biotinylated DNA oligonucleotides are used to target and capture RNA molecules of interest in a sequence-specific manner. Designing quality capture oligonucleotides is therefore critical to the success of a HyPR-MS experiment. Several factors to consider when designing capture oligonucleotides are outlined below, with in-depth discussion provided in the associated Notes.

1. Number of capture oligonucleotides required for a target RNA (*see* **Note 5**).
2. Secondary structure of the target RNA (*see* **Note 6**).
3. Capture oligonucleotide specificity and potential for off-target hybridization (*see* **Note 7**).
4. Capture oligonucleotide–target RNA melting temperature (T_m) under HyPR-MS experimental conditions (*see* **Note 8**).
5. Secondary structure of the capture oligonucleotide (*see* **Note 9**).
6. Potential for oligonucleotide–oligonucleotide hybridization (*see* **Note 10**).
7. Toehold-mediated release of the target RNA–protein complexes from the beads (*see* **Note 11**).

3.1.3 Design of qPCR Assay(s)

It is important to monitor RNA capture efficiency and specificity in any HyPR-MS experiment, and RT-qPCR is a useful technique for making these measurements. "Capture efficiency" is the percentage of target RNA present in the lysate at the beginning of the experiment that is captured on and subsequently released from the streptavidin-coated beads. "Capture specificity" is the ratio of target RNA captured using the target RNA capture oligonucleotide(s) relative to that captured by a scrambled oligonucleotide or other negative control (*see* **Note 12**). Typically, one qPCR assay should be designed within ~500 nt of each capture oligonucleotide to monitor the capture efficiency of that region of the target RNA (*see* **Notes 13** and **14**). Designing at least one qPCR assay to monitor the presence of a nontarget, housekeeping RNA, such as GAPDH, is also recommended.

3.2 Formaldehyde Cross-Linking

All mammalian cell lines investigated thus far have been amenable to HyPR-MS, including both suspension and adherent cell lines [21–24]. Cells have been cultured using standard medium (e.g., DMEM with 10% fetal bovine serum and 1% penicillin–streptomycin). Appropriate cell culture conditions should be determined for the specific cell line of interest. In general, we have found that small-scale HyPR-MS experiments require on the order of 10^5 to 10^6 cells, while full-scale experiments require on the order of 10^7 to 10^8 cells (*see* **Note 15**).

1. Add formaldehyde to the cell culture medium to a final concentration of 1% and gently shake for 10 min at room temperature (*see* **Note 16**).
2. Add Tris–HCl pH = 8.0 to 250 mM and gently shake for 10 min at room temperature to quench excess formaldehyde.
3. Remove the culture medium and wash the cells twice with cold 1× PBS.

   For suspension cell culture, collect the cells via centrifugation (~125 × *g* for 10 min at 4 °C) after each wash and pipet off and discard the supernatant.

   For adherent cell culture, washing can be done in the culture plate. After the second wash, use trypsin or a cell scraper to detach the cells from the plate and transfer the cell suspension to a centrifuge tube. Collect the cells via centrifugation (~125 × *g* for 10 min at 4 °C) and pipet off and discard the supernatant.
4. Flash-freeze the cell pellet with liquid nitrogen and store at −80 °C (*see* **Note 17**).

3.3 Cell Lysis and RNA Solubilization

1. Thaw the cell pellet on ice and resuspend the cells in freshly prepared, cold lysis buffer to a concentration of 5 × 10^6 cells/mL (*see* **Notes 15** and **18**).
2. Lyse the cells on ice for 10 min, vortexing periodically to help break up the cell pellet.
3. Sonicate the lysate with a probe sonicator to break up chromatin and solubilize RNA–protein complexes (*see* **Note 19**).
4. Centrifuge the lysate at 1,000 × *g* for 2 min at 4 °C to pellet any insoluble material. Transfer the supernatant to a new low-protein-binding tube.
5. Reserve two aliquots of the clarified cell lysate for downstream RT-qPCR (~2% of the total lysate volume) and mass spectrometric (~20 μL aliquot) analyses. Store these aliquots in low-protein-binding tubes at 4 °C until ready for use.

3.4 Hybridization Capture and Elution

1. Add the appropriate amount of capture oligonucleotide(s) to the lysate (*see* **Note 20**), then add RNase-free water to increase the lysate volume by 25% (new buffer component concentrations for hybridization: 375 mM LiCl, 50 mM Tris–HCl, 1% LiDS, 1% Triton X-100, 10 mM ribonucleoside vanadyl complex, 10 mM DTT, 100 U/mL RNasin Plus, 1× protease/phosphatase inhibitors).

2. Gently rock the lysate at 37 °C for 3 h to allow for hybridization (*see* **Note 21**).
3. Toward the end of the hybridization period (~20 min remaining), transfer an appropriate volume of streptavidin-coated magnetic beads to a low-protein-binding tube (henceforth, this volume will be referred to as the "bead volume") (*see* **Note 22**). Place the tube on a magnet stand and wait 3–4 min for the beads to be drawn to the magnet. Remove the supernatant with a pipette and discard.
4. Remove the tube of beads from the magnet stand and resuspend the beads in one bead volume of wash buffer. Place the tube on a magnet stand and wait 3–4 min for the beads to be drawn to the magnet. Remove the supernatant with a pipette and discard.
5. Repeat **step 4** twice more, for a total of three washes.
6. Remove the tube of beads from the magnet stand and resuspend the beads in one bead volume of 37 °C wash buffer.
7. Once the 3 h hybridization period is complete, add the bead slurry to the lysate and gently rock the lysate–bead mixture at 37 °C for 1 h to capture the hybridized RNA–protein complexes (*see* **Note 23**). Ensure that the rocking is sufficient to prevent the beads from aggregating at the bottom of the tube during capture.
8. After the 1 h incubation period, place the tube containing the lysate–bead mixture on a magnet stand and wait 3–4 min for the beads to be drawn to the magnet. Remove the supernatant with a pipette and store at 4 °C in a low-protein-binding tube for eventual RT-qPCR analysis.
9. Resuspend the beads in one bead volume of 37 °C wash buffer and gently rock at 37 °C for 15 min.
10. Place the tube on a magnet stand and wait 3–4 min for the beads to be drawn to the magnet. Remove the supernatant with a pipette and discard.
11. Repeat **steps 9** and **10** for a second wash.
12. Resuspend the beads in one bead volume of release buffer and gently rock at room temperature for 5 min.
13. Place the tube on a magnet stand and wait 3–4 min for the beads to be drawn to the magnet. Remove the supernatant with a pipette and discard.
14. Resuspend the beads in one bead volume of release buffer and add release oligonucleotide(s) (*see* **Note 24**).
15. Gently rock the bead slurry for 30 min at room temperature to release the target RNA–protein complexes from the beads.

16. Place the tube on a magnet stand and wait 3–4 min for the beads to be drawn to the magnet. Remove the supernatant with a pipette and transfer to a new low-protein-binding tube. Store the supernatant at 4 °C for eventual RT-qPCR and mass spectrometric analyses (*see* **Note 25**).
17. Resuspend the beads in one bead volume of release buffer and store the bead slurry at 4 °C for eventual RT-qPCR analysis (*see* **Note 26**).

3.5 RNA Purification and RT-qPCR Analysis

Small aliquots of the precapture lysate sample (from Subheading 3.3, **step 5**), the postcapture lysate sample (from Subheading 3.4, **step 8**), the RNA capture sample(s) (from Subheading 3.4, **step 16**), and the bead sample (from Subheading 3.4, **step 17**) should be prepared for RT-qPCR analysis. It is important to note what proportion (percentage) of each sample is used for RT-qPCR, as this information is critical for downstream calculations. In general, aliquoting ~2% of the volume of each sample for RT-qPCR analysis is sufficient (*see* **Note 27**) (Fig. 2).

3.5.1 RNA Purification

1. Bring all RT-qPCR aliquots to the same final volume (300 μL) and same buffer component concentrations as the postcapture lysate sample (from Subheading 3.4, **step 8**). Add $CaCl_2$ to 4 mM and proteinase K to 1 mg/mL.
2. Gently rock the samples at 37 °C overnight to digest proteins.
3. After allowing the samples to cool to room temperature, add 500 μL of TRI Reagent to each sample and vortex.
4. Allow the samples to sit at room temperature for 5 min, vortexing periodically.
5. Add 100 μL of chloroform to each sample, shake the samples vigorously for 15 s, and allow to sit at room temperature for 10 min.
6. Centrifuge the samples at 12,000 × *g* for 15 min at 4 °C and quantitatively transfer the top, aqueous layer to a clean tube (*see* **Note 28**).
7. Add ethanol to each sample to a final concentration of 75% (*see* **Note 29**) and incubate at −20 °C for at least 1 h or overnight.
8. Centrifuge the samples at 20,800 × *g* for 15 min at 4 °C to pellet the RNA.
9. Carefully remove the supernatant with a pipette and discard (*see* **Note 30**).
10. Wash each RNA pellet with 750 μL of room temperature 75% ethanol.
11. Centrifuge the samples at 20,800 × *g* for 15 min at room temperature to pellet the RNA.

12. Carefully remove the supernatant with a pipette and discard (*see* **Note 30**).
13. Allow the RNA pellets to air-dry for ~5 min, then resuspend each pellet in 15 μL of RNase-free water. Store samples at 4 °C prior to performing reverse transcription.

3.5.2 Reverse Transcription

1. Reverse transcription should be performed according to the kit manufacturer's protocol (*see* **Note 31**). We advise including several control reactions (*see* **Note 32**).
2. After reverse transcription, store the cDNA samples at 4 °C if qPCR will be performed on the same day, or at −20 °C if performed on a different day.

3.5.3 qPCR

1. Prepare qPCR plate(s) according to the qPCR master mix manufacturer's protocol using the cDNA samples and controls from Subheading 3.5.2 (*see* **Note 33**).
2. Perform qPCR using cycling parameters appropriate for each qPCR assay of interest (*see* **Note 34**).
3. Calculate capture efficiency (*see* **Note 35**), release efficiency (*see* **Note 36**), capture specificity (*see* **Note 37**), and target RNA enrichment (*see* **Note 38**) based on the qPCR results. Details about these calculations are discussed in the corresponding Notes.

3.6 Protein Preparation and Mass Spectrometric Analysis

Samples to prepare for mass spectrometric analysis include the precapture lysate sample (from Subheading 3.3, **step 5**) and all target RNA/control capture samples (from Subheading 3.4, **step 16**). The eFASP procedure described below has been adapted from the method described by Erde et al. [25] (Fig. 2).

3.6.1 eFASP

1. Prepare one CHAPS-passivated molecular weight cutoff filter and collection tube per sample 1 day prior to performing eFASP. Passivation is achieved by filling the collection tube with ~1 mL of 1% CHAPS, inserting the filter, filling the filter with ~0.5 mL of 1% CHAPS, and letting the tube/filter sit overnight at room temperature. Prior to use, the CHAPS solution should be discarded, and the tube and filter rinsed at least five times to remove excess CHAPS. For each rinse, the tube/filter should be placed in a clean beaker containing a large volume of LC-MS grade water and gently stirred for 30 min.
2. Add solid urea and deoxycholic acid to bring each protein sample to 8 M urea, 0.1% deoxycholic acid.
3. For each sample, place a passivated filter inside a nonpassivated collection tube and add 450 μL of sample to the filter. Centrifuge the sample at 14,000 × *g* for 10 min at room temperature and discard the flow-through. Continue passing the sample

through the filter in this manner until the entire sample volume has passed through.

4. Add 400 μL of eFASP exchange buffer to the filter and centrifuge at 14,000 × *g* for 10 min at room temperature. Discard the flow-through. Repeat twice more, for a total of three washes.
5. Add 200 μL of eFASP reducing buffer to the filter and incubate at room temperature for 30 min.
6. Centrifuge the sample at 14,000 × *g* for 10 min at room temperature and discard the flow-through.
7. Add 200 μL of eFASP alkylation buffer to the filter and incubate at room temperature in the dark for 1 h.
8. After the 1 h incubation period, add DTT to 75 mM and incubate at room temperature for an additional 10 min.
9. Centrifuge the sample at 14,000 × *g* for 10 min at room temperature and discard the flow-through.
10. Add 400 μL of eFASP digestion buffer to the filter and centrifuge at 14,000 × *g* for 10 min at room temperature. Discard the flow-through. Repeat twice more, for a total of three washes.
11. Transfer the filter to a clean, passivated collection tube and add 100 μL of eFASP digestion buffer containing an appropriate amount of trypsin to the filter (*see* **Note 39**).
12. Close the tube cap and seal with parafilm. Incubate the tube/filter apparatus overnight at 37 °C without rocking.
13. Remove the parafilm and centrifuge the sample at 14,000 × *g* for 10 min at room temperature.
14. Leaving the flow-through in the bottom of the tube, add 50 μL of 50 mM ammonium bicarbonate to the filter and centrifuge at 14,000 × *g* for 10 min at room temperature. Repeat once more for a total of two washes, each time allowing the wash volume to accumulate in the bottom of the tube.
15. Transfer the complete flow-through volume (200 μL) to a clean, low-protein-binding tube and add 200 μL of ethyl acetate.
16. Add trifluoroacetic acid (TFA) to 0.5% and vortex the sample for 1 min.
17. Centrifuge the sample at 15,800 × *g* for 2 min at room temperature.
18. Remove the top (ethyl acetate) layer with a pipette and discard.

19. Repeat the ethyl acetate extraction twice more, for a total of three extractions. Each time, add 200 μL of ethyl acetate and vortex the sample for 1 min prior to repeating **steps 17** and **18**.
20. Dry the peptide sample in a vacuum centrifuge concentrator.

3.6.2 C18 Solid-Phase Extraction

1. Reconstitute the dried peptide sample in 150 μL of 0.1% TFA.
2. Condition a C18 solid-phase extraction pipette tip by washing it at least three times with 150 μL aliquots of 70% acetonitrile (ACN). All pipetting with the C18 tip should be performed slowly.
3. Equilibrate the tip by washing it at least three times with 150 μL aliquots of 0.1% TFA.
4. Load the peptides onto the tip by pipetting the complete peptide sample up and down at least five times.
5. Wash the tip at least ten times with 150 μL aliquots of 0.1% TFA.
6. Elute the peptides from the tip by pipetting a 150 μL aliquot of 70% ACN/0.1% TFA up and down at least five times.
7. Dry the desalted peptides in a vacuum centrifuge concentrator and reconstitute in 95:5 H_2O:ACN with 0.2% formic acid (FA). Store the sample at −20 °C prior to mass spectrometric analysis.

3.6.3 Mass Spectrometry

We describe here the analysis of HyPR-MS samples using a high-performance liquid chromatography system (nanoAcquity, Waters) coupled to an electrospray ionization (ESI) orbitrap mass spectrometer (Q Exactive HF, Thermo Fisher Scientific). The column (described in Subheading 2.3) should be operated at 60 °C using a column oven. Comparable LC-MS setups could also be used. Similarly, the LC-MS/MS method described below has worked well in our hands for the analysis of HyPR-MS samples, but a variety of routine bottom-up LC-MS/MS methods may be acceptable. Provided that the samples contain enough peptides, we recommend performing two technical replicate LC-MS/MS injections of each sample. We also recommend running ACN and water blank injections between samples to minimize carryover from previous injections.

1. LC method: Load peptides on-column with 2% ACN in 0.2% FA at a flow rate of 400 nL/min for 30 min. Elute peptides over 120 min at a flow rate of 300 nL/min with the following gradient (all in 0.2% FA): 8% ACN at time 1 min; 34% ACN at time 81 min; 44% ACN at time 91 min; 64% ACN from 92–99 min; equilibrate to 2% ACN from 103–120 min.
2. MS/MS method: Perform full-mass profile scans (375–1500 *m/z*) in the orbitrap at a resolution of 120,000,

automatic gain control (AGC) target of 1×10^6, and maximum injection time of 100 ms. Follow each full-mass profile scan by MS/MS HCD scans of the 10 highest intensity parent ions with $z > 1$ at 30% relative collision energy and 15,000 resolution with a mass range starting at 100 *m/z*, AGC target of 1×10^5, and a maximum injection time of 50 ms. Enable dynamic exclusion with a repeat count of one over a duration of 15 s.

3.7 Mass Spectrometric Data Analysis

In general, the goal of mass spectrometric data analysis in HyPR-MS is to identify which proteins are significantly enriched in the target RNA capture sample as compared to control samples, indicating their interaction with the target RNA. Because the determination of enriched proteins is a statistical process, consideration should be given to how many biological replicates will be necessary for drawing meaningful conclusions from the experiment. Typically, at least three replicates are necessary, though more replicates are helpful for obtaining greater confidence in the results.

1. Obtain a protein database for the relevant organism (*see* **Note 40**).
2. Load the protein database and all spectral files from the target and control samples into a suitable proteomic search software program and perform the search (*see* **Note 41**). The default search parameters for many search software programs will be sufficient. Ensure that oxidation of methionine is set as a variable modification and that carbamidomethylation of cysteine is set as a fixed modification. Also ensure that trypsin is selected as the protease to be used for in silico digestion and set the maximum number of missed cleavages to two and the minimum peptide length to seven amino acids.
3. Filter the search results to remove low-confidence identifications (e.g., apply a 1% false discovery rate (FDR) for both peptides and proteins) (*see* **Note 42**).
4. Perform label-free peptide and protein quantification using the results obtained from the search software (*see* **Note 43**).
5. Using the protein abundances calculated by the quantification software, perform statistical analyses to determine which proteins are significantly enriched in the target RNA capture sample as compared to the control. Large-scale statistical analyses can be performed using a software platform such as Perseus [26]. The best approach for normalizing the quantification data (*see* **Note 44**) and determining proper statistical thresholds will be specific to each application of HyPR-MS and experimental design. There are multiple acceptable ways to analyze the data, and there are no universal parameters that guarantee meaningful results. The most appropriate method,

and the confidence one can have in the results, will depend on the data. One standard approach is to begin by $\log_2$-transforming the protein intensity values for each sample to obtain normal distributions. The protein intensity values can then be grouped by condition (i.e., target RNA capture samples and control samples), and the proteins can be filtered to remove those with an insufficient number of observations (e.g., proteins observed in fewer than two-thirds of the biological replicates for a particular sample type). Any remaining missing values can be imputed (*see* **Note 45**). Next, two-sample T-tests can be used to compare protein abundances between the target and control samples to determine which proteins are differentially abundant at a specified p-value (e.g., $p = 0.05$). A permutation-based FDR cutoff (e.g., 1–10%) can be used to correct for multiple hypothesis testing. Proteins within this FDR cutoff are statistically differentially abundant between the target RNA capture and control samples, subject to the criteria applied (*see* **Note 46**). Additional approaches for analyzing quantitative proteomics data from a HyPR-MS experiment can be found in our previous publications [21–24].

4 Notes

1. Carefully designed, target-specific capture oligonucleotides and stringent wash steps in the HyPR-MS protocol reduce the presence of nonspecific protein binders in the target RNA pulldown samples; however, nonspecific binders cannot be completely eliminated.
2. A scrambled pulldown control utilizes a scrambled sequence oligonucleotide which is not significantly complementary to any region of the genome or transcriptome (determined using BLAST [27] (https://blast.ncbi.nlm.nih.gov/Blast.cgi)), and therefore does not target any specific RNA. Generally, this oligonucleotide is designed to have approximately the same G/C content and T_m as the target capture oligonucleotide(s). The proteins identified in this pulldown are likely nonspecific binders inherent to the HyPR-MS procedure. Therefore, proteins that overlap between this control and the target RNA pulldown are probably not specific binders of the target RNA.
3. A poly(dT) pulldown control captures RNA molecules with poly(A) tails, such as mRNAs. We have found that a poly (dT) oligonucleotide containing ~20 Ts, in addition to the 8 nt toehold sequence, is sufficient for capture. This control experiment enables one to determine which of the proteins identified in the target RNA pulldown experiment are distinct

to that particular RNA, and which are general RNA-binding proteins.

4. A lysate control involves the proteomic analysis of whole cell lysate to determine the most abundant proteins present in the sample. Proteins which are highly abundant in cell lysate have the potential to be carried through to the final RNA pulldown sample simply due to their abundance and not due to specific interaction with the target RNA. If many of the proteins identified in the target RNA pulldown sample are the same as the most abundant proteins in whole cell lysate, this may be a sign that the pulldown parameters require further optimization and/or that more stringent washing steps should be included.

5. It is advisable to design capture oligonucleotides complementary to multiple regions of the target RNA. The sonication step of HyPR-MS may cause RNA fragmentation, thus the use of multiple capture oligonucleotides spanning the length of the target RNA provides a more complete characterization of the proteins bound along the entire length of the RNA than would the use of a single capture oligonucleotide. The appropriate number of capture oligonucleotides depends greatly on the target RNA length and sonication intensity. Begin by designing several (~3–8) capture oligonucleotides and qPCR assays to span the length of the target RNA. Then, monitor the capture efficiency of different regions of the target transcript when different combinations of capture oligonucleotides are used. Ideally, one should use the minimum number of capture oligonucleotides that are necessary to ensure sufficient capture along the entire length of the transcript. Additional capture oligonucleotides may marginally increase capture efficiency, but they also increase the potential for off-target hybridization and undesirable oligonucleotide–oligonucleotide interactions. If it proves difficult to obtain acceptable levels of capture along the entire length of the target transcript, it may be helpful to investigate the integrity of the RNA in the sample by purifying the RNA and analyzing it on an agarose gel or Bioanalyzer. If the RNA looks very degraded (determined by analyzing the ratio of the 28S:18S rRNA band intensities or the RNA integrity number [28]), it may be necessary to use more RNase inhibitors in the experiment and/or to decrease sonication intensity.

6. It is important to consider the secondary structure of the target RNA when designing capture oligonucleotides. Capture oligonucleotides provide better capture efficiency when they are designed to complement regions of the target RNA that are single-stranded. If the secondary structure of the target RNA is uncharacterized, software tools like Mfold [29] (www.unafold.org) can be useful for making predictions.

7. Capture oligonucleotides should be designed so that they have minimal potential for hybridization to off-target RNAs. Capture of off-target RNAs and their associated protein interactors cause false positives in the HyPR-MS protein data analysis. BLAST should be used to assess the specificity of all proposed capture oligonucleotides in the context of the relevant complete genome/transcriptome.

8. Hybridization parameters (temperature, salt concentration, incubation time, etc.) for HyPR-MS will need to be optimized on a case-by-case basis to allow for stable oligonucleotide–target RNA hybridization whilst minimizing the formation of off-target hybrids. In general, capture oligonucleotides ~30 nt in length, a hybridization buffer containing 375 mM LiCl, a hybridization temperature of 37 °C, and a hybridization time of 3 h have provided satisfactory stability, specificity, and efficiency results. The T_m of a given capture oligonucleotide–target RNA hybrid under various hybridization conditions can be assessed using freely available webtools such as the IDT OligoAnalyzer (https://www.idtdna.com/calc/analyzer) or OligoCalc [30] (http://biotools.nubic.northwestern.edu/OligoCalc.html).

9. It is important to design capture oligonucleotides that do not form stable secondary structures, such as hairpins, under HyPR-MS hybridization conditions. It is more favorable for an oligonucleotide which does not have stable intramolecular interactions to hybridize to the target RNA than it is for an oligonucleotide which adopts a stable secondary structure. There are multiple freely available webtools that can be used to determine the propensity of any given capture oligonucleotide to form such structures, such as the IDT OligoAnalyzer (https://www.idtdna.com/calc/analyzer) or OligoCalc [30] (http://biotools.nubic.northwestern.edu/OligoCalc.html).

10. Capture oligonucleotides should be designed so that stable homodimers and heterodimers do not form under HyPR-MS experimental conditions. These dimers compete with the desired capture oligonucleotide–target RNA hybrid. There are multiple freely available webtools that can be used to determine the propensity of the capture oligonucleotides used in any given experiment to form such hybrids, such as the IDT OligoAnalyzer (https://www.idtdna.com/calc/analyzer) or the Thermo Fisher Scientific Multiple Primer Analyzer (https://www.thermofisher.com/us/en/home/brands/thermo-scientific/molecular-biology/molecular-biology-learning-center/molecular-biology-resource-library/thermo-scientific-web-tools/multiple-primer-analyzer.html).

11. HyPR-MS utilizes a "toehold-mediated" strategy to release the purified target RNA–protein complexes from the streptavidin-coated magnetic beads [21–24]. This "toehold" must be incorporated into the design of the capture oligonucleotide. Typically, a 30 nt sequence complementary to the target RNA is designed first. Then, on the 5′ or 3′ end of that sequence, an 8 nt sequence which is not complementary to the target RNA is added, making the entire capture oligonucleotide 38 nt in length. With this design, a 30 nt stretch of the capture oligonucleotide forms a hybrid with the target RNA, while the 8 nt toehold remains single-stranded (Fig. 1). During the release step of HyPR-MS, a release oligonucleotide which is completely complementary to the capture oligonucleotide (38 nt in length, in this example) is added to the bead slurry. The 8 nt, single-stranded region of the capture oligonucleotide serves as a toehold for hybridization of the entire 38 nt capture oligonucleotide with the 38 nt release oligonucleotide. Because the capture oligonucleotide–release oligonucleotide hybrid is more thermodynamically stable than the capture oligonucleotide–target RNA hybrid (a 38 nt hybrid is more stable than a 30 nt hybrid), the release oligonucleotide displaces the target RNA from the capture oligonucleotide, thereby releasing the target RNA–protein complexes into solution (Fig. 1). A notable feature of toehold-mediated release is that it enables multiplexing of HyPR-MS experiments, allowing for the analysis of multiple RNA targets from the same cell lysate preparation [21–24]. When designing capture oligonucleotides, the entire 38 nt sequence (complementary region plus toehold region) should be assessed for off-target hybridization via BLAST.
12. An alternative measure of capture specificity is the ratio of target to nontarget RNA in the target RNA capture sample. RNA-seq is better suited to making these measurements than is RT-qPCR.
13. There are benefits to designing qPCR assays to amplify regions both adjacent to and further away from the locations targeted by capture oligonucleotides. A qPCR assay designed to amplify a region directly adjacent to a capture oligonucleotide gives the best measurement of the capture of that region of the transcript. However, that qPCR assay gives very little information about the length of the captured RNA molecules. For this information, a qPCR assay which amplifies a region further away from the capture oligonucleotide should be designed. If the capture efficiency measured by this qPCR assay is lower than desired, adjusting sonication parameters to decrease RNA fragmentation may be helpful.

14. The use of hydrolysis probe qPCR assays (rather than SYBR Green) is recommended to maximize the specificity and sensitivity of qPCR measurements. Information on how to design qPCR assays is available elsewhere [31].

15. The number of cells required per HyPR-MS experiment will vary and should be determined empirically for each distinct RNA target and cell line. The goal of a HyPR-MS experiment is to identify the ensemble of proteins that are associated with a target RNA, therefore a successful HyPR-MS experiment must yield enough protein in the target RNA capture sample for analysis via bottom-up mass spectrometry. The amount of any given protein in the capture sample is a function of the stoichiometry with which the protein binds the target RNA, the abundance of the target RNA in the cell line of interest, and target RNA capture efficiency. Higher protein:RNA stoichiometry, target RNA abundance, and/or capture efficiency will decrease the number of cells required per HyPR-MS experiment. To determine the number of cells required for any given experiment, the abundance of the target RNA in the cell line of interest should first be estimated via RNA-seq and/or RT-qPCR. Then, "small-scale" RNA capture experiments should be performed using relatively few cells (on the order of 10^5 to 10^6) to optimize capture efficiency and specificity. Note that these small-scale experiments will provide enough RNA for RT-qPCR analyses, but will not provide enough protein for mass spectrometric analysis. Once capture parameters have been optimized in small-scale experiments, one can estimate the number of cells required for full-scale HyPR-MS experiments by taking into account the measured target RNA abundance and capture efficiency, and by making a few approximations regarding protein:target RNA stoichiometry and mass spectrometric detection limit (e.g., a 1:1 protein: target RNA stoichiometry and a mass spectrometric detection limit of ~1 fmol of peptide). Previous studies have successfully used on the order of 10^7 to 10^8 cells for full-scale HyPR-MS experiments [21–24].

16. Formaldehyde concentration and cross-linking time may need to be optimized based on the cell line and RNA target of interest.

17. Cells should be pelleted in quantities appropriate for individual HyPR-MS experiments because cross-linked cells are difficult to resuspend without lysing. To preserve the integrity of the RNA, cross-linked cells should not be thawed and then refrozen.

18. We have found that it is important to lyse cells at a consistent concentration of 5×10^6 cells/mL. In our hands, this

concentration has provided a sufficient protease/RNase inhibitor:cell ratio to maintain protein and RNA integrity.

19. Appropriate sonication parameters will need to be determined for each system and sonicator. Sonication aids in breaking up chromatin and solubilizing RNA–protein complexes, but it also causes RNA fragmentation. Ideal sonication parameters will effectively solubilize the target RNA whilst minimizing RNA fragmentation. Optimal parameters should be determined empirically by monitoring (a) RNA solubilization via RT-qPCR and/or absorbance at 260 nm and (b) the general degree of RNA fragmentation via analysis of sonicated RNA on an agarose gel or Bioanalyzer (to investigate RNA integrity via the ratio of the 28S:18S rRNA band intensities or the RNA integrity number [28]). In general, ~4–12 s of light sonication is appropriate for a ~1 mL aliquot of lysate. Because sonication heats the lysate and heat can reverse formaldehyde cross-links [32, 33], sonicating the lysate on ice and performing the sonication in ~4 s bursts with ~4 s of rest between each burst is recommended.

20. Appropriate capture oligonucleotide concentrations will depend on target RNA abundance and will need to be determined empirically. To start, try small-scale experiments to test oligonucleotide concentrations ranging from ~1 to 15 nM. Following RT-qPCR analysis, determine which concentration provides the optimal capture efficiency and specificity. Additional experiments to titrate up or down may be necessary.

21. Hybridization time and temperature may be optimized for the specific melting temperature(s) of the capture oligonucleotide–target RNA hybrids and the abundance of the target RNA.

22. The appropriate volume of streptavidin-coated magnetic beads should be determined empirically. Typically, 3 μL of beads for every picomole of capture oligonucleotide is sufficient.

23. If the bead volume exceeds ~20% of the hybridization volume, it is best to first transfer the bead slurry to clean tube(s), remove the supernatant, then transfer the lysate to the tube(s) containing the beads. This way, the RNase and protease inhibitors present in the lysate do not get diluted by a large bead volume.

24. Appropriate release oligonucleotide concentrations will depend on capture oligonucleotide concentrations and will need to be determined empirically. To start, try using 100–1,000× more release oligonucleotide than the corresponding capture oligonucleotide and adjust based on measurements of capture and release efficiency (*see* **Notes 35** and **36**).

25. If multiplexing the HyPR-MS experiment to analyze multiple RNA targets from the same cell lysate preparation, or if including a scrambled or poly(dT) oligonucleotide control, one should perform any additional RNA release steps prior to proceeding to **step 17**. For each sequential release step, the beads should first be washed by resuspending in one bead volume of release buffer and gently rocking at room temperature for 5 min prior to repeating **steps 13–16**.
26. The RNA from an aliquot of these resuspended beads will be analyzed via RT-qPCR to measure the amount of target RNA that remains on the beads after all release steps. If a significant amount of target RNA remains on the beads, release conditions may require further optimization (e.g., increasing release oligonucleotide concentration, increasing release time, or decreasing release temperature).
27. For small-scale experiments, the entire sample volume can be used.
28. It is important to avoid transferring any of the interphase (containing DNA) or organic phase (containing protein). Additionally, transferring a consistent volume for each sample (e.g., 500 μL for the parameters described here) will ensure accurate downstream calculations of capture efficiency, release efficiency, etc.
29. The addition of a coprecipitant, such as glycogen, is recommended to facilitate RNA precipitation and to make the RNA pellet easier to observe in subsequent steps.
30. It is important not to disturb the RNA pellet while removing the supernatant. To minimize disruption of the pellet, first use a large (~1,000 μL) pipette tip to remove the majority of the supernatant, then centrifuge the sample briefly and remove the remainder of the supernatant using a smaller (~10 μL) pipette tip.
31. It is important that each reverse transcription reaction contain an appropriate amount of RNA, which will be dictated by the kit manufacturer. We recommend using a spectrophotometer to measure the concentration of RNA in each sample following ethanol precipitation.
32. The following controls should be considered for reverse transcription experiments: (1) A reaction without any RNA to verify that the reverse transcription reactions are not contaminated with exogenous RNA/DNA. For this control, omit RNA from the reaction and use water to bring the reaction to the appropriate final volume; (2) A no–reverse transcriptase control to monitor for genomic DNA contamination in the RNA samples. Most qPCR polymerases will not amplify RNA, therefore any qPCR signal from a reaction without reverse

transcriptase may be attributed to genomic DNA contamination. For this control, omit the reverse transcriptase enzyme from the reaction and use water to bring the reaction to the appropriate final volume; (3) Serial dilutions of the RNA samples of interest. Reverse transcription efficiency is sensitive to the complexity of the RNA sample being reverse transcribed. If the qPCR signal from pre–reverse transcription serial dilutions of an RNA sample is not dropping linearly, it can indicate that the initial RNA sample may be too complex and qPCR results from that sample may be unreliable. For these controls, prepare serial dilutions (e.g., three 10-fold dilutions) of the RNA sample in water and perform reverse transcription on each of the dilutions.

33. Technical duplicate or triplicate qPCR reactions should be performed for each cDNA sample/qPCR assay combination. Additionally, a control qPCR reaction without any cDNA should be included to verify that the qPCR reactions are not contaminated with exogenous DNA. For this control, omit cDNA from the reaction and use water to bring the reaction to the appropriate final volume.
34. Appropriate qPCR cycling parameters should be determined empirically for each qPCR assay. qPCR master mixes often come with a recommended protocol, which can be a helpful place to start. Online resources can also be useful for approximating appropriate annealing temperatures for each qPCR assay.
35. Target RNA capture efficiency can be determined using a target-specific qPCR assay and a calibration curve made up of serial dilutions of the precapture lysate sample (from Subheading 3.3, **step 5**). These dilutions can be made prior to performing reverse transcription. By comparing the quantification cycle (C_q) of the target RNA capture sample (from Subheading 3.4, **step 16**) to the calibration curve, one can quantify the relative amount of target RNA present in that sample and calculate capture efficiency. The same approach can be used to quantify the percentage of the target RNA present in the postcapture lysate sample (from Subheading 3.4, **step 8**), any other RNA capture samples (from Subheading 3.4, **step 16**), and the bead sample (from Subheading 3.4, **step 17**). Capture efficiencies can vary substantially depending on the RNA target and capture oligonucleotide(s) used, and capture efficiencies ranging from ~10 to 60% are commonly observed. If the observed capture efficiency is lower than desired, increasing capture oligonucleotide concentration(s), adjusting hybridization time and temperature, and/or designing new capture oligonucleotide(s) to target more open regions of the target RNA may be helpful.

36. Release efficiency can be calculated by dividing the percentage of the target RNA in the target RNA capture sample by the summed percentage in the target RNA capture sample, any other capture sample(s), and the bead sample. Release efficiencies ranging from ~50% to 90% are commonly observed. If the observed release efficiency is lower than desired, adjusting release time or temperature or increasing the concentration of release oligonucleotides used may be helpful.
37. Capture specificity can be calculated by dividing the capture efficiency of the target RNA in the target RNA capture sample by the capture efficiency of the target RNA in a nontarget RNA capture sample. The nontarget RNA capture sample could be a scrambled oligonucleotide control capture sample or a capture sample for a different RNA target.
38. Target RNA enrichment can be calculated by comparing the ratio of target RNA:GAPDH (or some other housekeeping RNA) in the lysate prior to hybridization capture to the same ratio in the final RNA capture sample. Enrichment can be calculated either by using the $2^{-\Delta\Delta Ct}$ method [34] or by comparing to a genomic DNA standard curve (for absolute transcript quantification).
39. Trypsin should be added so that the final trypsin:protein ratio is ~1:20 to 1:100.
40. Reference protein databases are commonly used for mass spectrometric data analysis. Reference databases can be obtained from UniProt [35] (https://www.uniprot.org/proteomes/), GENCODE [36] (https://www.gencodegenes.org/), and RefSeq [37] (https://www.ncbi.nlm.nih.gov/refseq/). In the proteomics community, UniProt databases are often used. We also recommend including a database of common contaminants in the search, so that exogenous proteins present in the samples can be identified (e.g., trypsin, streptavidin, keratin, etc.).
41. Search software programs such as MetaMorpheus [38] (https://github.com/smith-chem-wisc/MetaMorpheus/), MSFragger [39] (https://msfragger.nesvilab.org/), and Andromeda [40] (http://coxdocs.org/doku.php?id=maxquant:andromeda:start), among others, can be used to obtain peptide and protein identifications from the acquired mass spectra.
42. Depending on the search software used, there may be an option during search setup to prevent low-confidence identifications from being written to the output file. If selected, this would remove the need to perform this step manually after the search is complete.

43. Software programs capable of label-free peptide and protein quantification include FlashLFQ [41] (https://github.com/smith-chem-wisc/FlashLFQ) and MaxQuant [42] (http://coxdocs.org/doku.php?id=maxquant:start), among others.
44. If using a normalization algorithm that is built into the quantification software, ensure that the assumptions made by the algorithm are appropriate for the HyPR-MS experimental design.
45. Imputed values should maintain the normal distribution of the data and not create a bimodal distribution. If this occurs, adjust the parameters of how imputed values are determined.
46. We recommend applying a fold-change cutoff at this point, which enables one to specify a minimum difference in abundance that is required for a protein to be considered enriched in the target RNA capture sample. There is no single fold-change cutoff that is suitable for all experiments. In general, a smaller sample variance and more sample replicates enable one to apply a lower fold-change cutoff and still discover biologically significant changes.

Acknowledgments

This work was supported by NIH-NCI grant R01CA193481. K.B.H. was supported in part by the National Human Genome Research Institute grant to the Genomic Science Training Program, 5T32HG002760. R.M.M. was supported in part by the NIH Chemistry-Biology Interface Training Grant, T32GM008505. The authors would like to thank members of the Smith lab for helpful discussions and guidance in the development of HyPR-MS. The figures in this chapter were created with BioRender.com.

References

1. Moore MJ (2005) From birth to death: the complex lives of eukaryotic mRNAs. Science 309:1514–1518
2. Glisovic T, Bachorik JL, Yong J et al (2008) RNA-binding proteins and post-transcriptional gene regulation. FEBS Lett 582:1977–1986
3. Mitchell SF, Parker R (2014) Principles and properties of eukaryotic mRNPs. Mol Cell 54:547–558
4. Re A, Joshi T, Kulberkyte E et al (2014) RNA-protein interactions: an overview. In: Gorodkin J, Ruzzo WL (eds) RNA sequence, structure, and function: computational and bioinformatic methods, Methods in molecular biology, vol 1097. Humana Press, Totowa, NJ, pp 491–521
5. Matera AG, Terns RM, Terns MP (2007) Non-coding RNAs: lessons from the small nuclear and small nucleolar RNAs. Nat Rev Mol Cell Biol 8:209–220
6. Mayr C (2017) Regulation by 3′-untranslated regions. Annu Rev Genet 51:171–194
7. Marchese FP, Raimondi I, Huarte M (2017) The multidimensional mechanisms of long noncoding RNA function. Genome Biol 18:206
8. Allerson CR, Cazzola M, Rouault TA (1999) Clinical severity and thermodynamic effects of

iron-responsive element mutations in hereditary hyperferritinemia-cataract syndrome. J Biol Chem 274:26439–26447

9. Lukong KE, Chang KW, Khandjian EW et al (2008) RNA-binding proteins in human genetic disease. Trends Genet 24:416–425
10. Corbett AH (2018) Post-transcriptional regulation of gene expression and human disease. Curr Opin Cell Biol 52:96–104
11. Ramanathan M, Porter DF, Khavari PA (2019) Methods to study RNA–protein interactions. Nat Methods 16:225–234
12. Ule J, Jensen KB, Ruggiu M et al (2003) CLIP identifies Nova-regulated RNA networks in the brain. Science 302:1212–1215
13. Licatalosi DD, Mele A, Fak JJ et al (2008) HITS-CLIP yields genome-wide insights into brain alternative RNA processing. Nature 456:464–469
14. Hafner M, Landthaler M, Burger L et al (2010) Transcriptome-wide identification of RNA-binding protein and microRNA target sites by PAR-CLIP. Cell 141:129–141
15. König J, Zarnack K, Rot G et al (2010) iCLIP reveals the function of hnRNP particles in splicing at individual nucleotide resolution. Nat Struct Mol Biol 17:909–915
16. Van Nostrand EL, Pratt GA, Shishkin AA et al (2016) Robust transcriptome-wide discovery of RNA-binding protein binding sites with enhanced CLIP (eCLIP). Nat Methods 13:508–514
17. Kim B, Kim VN (2019) fCLIP-seq for transcriptomic footprinting of dsRNA-binding proteins: lessons from DROSHA. Methods 152:3–11
18. West JA, Davis CP, Sunwoo H et al (2014) The long noncoding RNAs NEAT1 and MALAT1 bind active chromatin sites. Mol Cell 55:791–802
19. Chu C, Zhang QC, da Rocha ST et al (2015) Systematic discovery of Xist RNA binding proteins. Cell 161:404–416
20. McHugh CA, Chen CK, Chow A et al (2015) The Xist lncRNA interacts directly with SHARP to silence transcription through HDAC3. Nature 521:232–236
21. Knoener RA, Becker JT, Scalf M et al (2017) Elucidating the in vivo interactome of HIV-1 RNA by hybridization capture and mass spectrometry. Sci Rep 7:16965
22. Spiniello M, Knoener RA, Steinbrink MI et al (2018) HyPR-MS for multiplexed discovery of MALAT1, NEAT1, and NORAD lncRNA protein interactomes. J Proteome Res 17:3022–3038
23. Spiniello M, Steinbrink MI, Cesnik AJ et al (2019) Comprehensive in vivo identification of the c-Myc mRNA interactome using HyPR-MS. RNA 25:1337–1352
24. Knoener R, Evans E III, Becker JT et al (2021) Identification of host proteins differentially associated with HIV-1 RNA splice variants. eLife 10:e62470
25. Erde J, Loo RRO, Loo JA (2014) Enhanced FASP (eFASP) to increase proteome coverage and sample recovery for quantitative proteomic experiments. J Proteome Res 13:1885–1895
26. Tyanova S, Temu T, Sinitcyn P et al (2016) The Perseus computational platform for comprehensive analysis of (prote)omics data. Nat Methods 13:731–740
27. Altschul SF, Gish W, Miller W et al (1990) Basic local alignment search tool. J Mol Biol 215:403–410
28. Schroeder A, Mueller O, Stocker S et al (2006) The RIN: an RNA integrity number for assigning integrity values to RNA measurements. BMC Mol Biol 7:3
29. Zuker M (2003) Mfold web server for nucleic acid folding and hybridization prediction. Nucleic Acids Res 31:3406–3415
30. Kibbe WA (2007) OligoCalc: an online oligonucleotide properties calculator. Nucleic Acids Res 35:W43–W46
31. Shipley GL (2013) Assay design for real-time qPCR. In: Nolan T, Bustin SA (eds) PCR technology: current innovations, 3rd edn. CRC Press, Taylor & Francis Group, Boca Raton, FL, pp 177–197
32. Jackson V (1978) Studies on histone organization in the nucleosome using formaldehyde as a reversible cross-linking agent. Cell 15:945–954
33. Kennedy-Darling J, Smith LM (2014) Measuring the formaldehyde protein-DNA cross-link reversal rate. Anal Chem 86:5678–5681
34. Livak KJ, Schmittgen TD (2001) Analysis of relative gene expression data using real-time quantitative PCR and the $2^{-\Delta\Delta Ct}$ method. Methods 25:402–408
35. The UniProt Consortium (2019) UniProt: a worldwide hub of protein knowledge. Nucleic Acids Res 47:D506–D515
36. Frankish A, Diekhans M, Ferreira AM et al (2019) GENCODE reference annotation for the human and mouse genomes. Nucleic Acids Res 47:D766–D773
37. O'Leary NA, Wright MW, Brister JR et al (2016) Reference sequence (RefSeq) database at NCBI: current status, taxonomic expansion, and functional annotation. Nucleic Acids Res 44:D733–D745

38. Solntsev SK, Shortreed MR, Frey BL et al (2018) Enhanced global post-translational modification discovery with MetaMorpheus. J Proteome Res 17:1844–1851
39. Kong AT, Leprevost FV, Avtonomov DM et al (2017) MSFragger: ultrafast and comprehensive peptide identification in mass spectrometry-based proteomics. Nat Methods 14:513–520
40. Cox J, Neuhauser N, Michalski A et al (2011) Andromeda: a peptide search engine integrated into the MaxQuant environment. J Proteome Res 10:1794–1805
41. Millikin RJ, Solntsev SK, Shortreed MR et al (2018) Ultrafast peptide label-free quantification with FlashLFQ. J Proteome Res 17:386–391
42. Tyanova S, Temu T, Cox J (2016) The MaxQuant computational platform for mass spectrometry-based shotgun proteomics. Nat Protoc 11:2301–2319

Part IV

The RNA Lifecycle

Chapter 13

Visualization and Quantification of Subcellular RNA Localization Using Single-Molecule RNA Fluorescence In Situ Hybridization

Ankita Arora, Raeann Goering, Pedro Tirado Velez, and J. Matthew Taliaferro

Abstract

Advancements in imaging technologies, especially approaches that allow the imaging of single RNA molecules, have opened new avenues to understand RNA regulation, from synthesis to decay with high spatial and temporal resolution. Here, we describe a protocol for single-molecule fluorescent in situ hybridization (smFISH) using three different approaches for synthesizing the fluorescent probes. The three approaches described are commercially available probes, single-molecule inexpensive FISH (smi-FISH), and in-house enzymatically labeled probes. These approaches offer technical and economic flexibility to meet the specific needs of an experiment. In addition, we provide a protocol to perform automated smFISH spot detection using the software FISH-quant.

Key words RNA localization, Fluorescence in situ hybridization, RNA transport, Fluorescence microscopy, RNA imaging, Single-molecule RNA quantification

1 Introduction

RNA plays a multifaceted role in the regulation of gene expression, and the subcellular localization of RNA contributes to the proper spatial patterning of proteins within cells. RNA localization has been shown to contribute to diverse processes across species including germline differentiation in Drosophila [1–3], embryonic development in Xenopus [4], mating-type switching in yeast [5], and neuronal development in mammals [6–9]. Advancements in imaging techniques have led to a better understanding of the spatial and temporal localization of RNA within cells. One method in particular, single-molecule fluorescence in situ hybridization (smFISH), has been widely used in the study of RNA localization.

smFISH allows for the direct visualization and quantification of individual RNA molecules. In short, cells are fixed, permeabilized,

Erik Dassi (ed.), *Post-Transcriptional Gene Regulation*, Methods in Molecular Biology, vol. 2404,
https://doi.org/10.1007/978-1-0716-1851-6_13,

and then treated with multiple small (~25–50 nt) fluorescently labeled DNA oligonucleotides which hybridize across the length of the RNA of interest (Fig. 1a). The hybridization of multiple labeled probes to a single RNA molecule increases the signal to noise ratio and results in a locally bright spot that can be detected by fluorescence microscopy. A 3D Gaussian fitting algorithm together with thresholding various parameters is used in image analysis tools to detect the spots in the resulting images (*see* data analysis for details).

Traditionally, smFISH probes are synthesized chemically and directly coupled to the fluorophore during synthesis. Although these reagents often give high quality images with a high signal-to-noise ratio, they suffer from two main disadvantages. First, because the fluor is directly coupled to the hybridizing probes, they have reduced flexibility in terms of multiplexing fluors to identify multiple RNA species in a single experiment. Second, these probes, while commercially available, are quite expensive. As an alternative, modifications of the smFISH technique have been developed that allow indirect labeling of relatively inexpensive unlabeled probes. One of the approaches, termed single-molecule inexpensive FISH (smiFISH) [10], uses primary probes which are nonfluorescent but carry a shared sequence at the 5′ end called a FLAP [10] (Fig. 1b). The FLAP sequence hybridizes to a secondary fluorescent oligonucleotide labeled with two fluorophores. This allows the more expensive fluorescent secondary probe to be easily used with multiple primary probe sets, greatly reducing the cost of the experiment. Further, since the primary probes are unlabeled, this gives greater flexibility in choosing the fluorescence channels that will be occupied by a given probe set when visualizing multiple RNA species in a single experiment.

Another modification to smFISH takes advantage of the enzymatic activity of terminal deoxynucleotidyl transferase (TdT) to add amino-allyl modified nucleotides at the 3′ end of the oligonucleotides [11]. The amine modification at 3′ end of the oligonucleotides enables coupling of fluorophore–succinimidyl (NHS)-ester conjugates in a single-step reaction rendering the probes fluorescent (Fig. 1c).

In this chapter, we present smFISH images from two cultured cell types, mouse neuronal CAD cells (Fig. 2) and human intestinal epithelial C2bbe1 cells (Fig. 3). Although in this protocol we use the example of a reporter messenger RNA (mRNA) encoding firefly luciferase, smFISH can also be used to detect and quantify endogenous mRNA [12], long noncoding RNAs (lncRNA) [13], and viral RNA genomes [14].

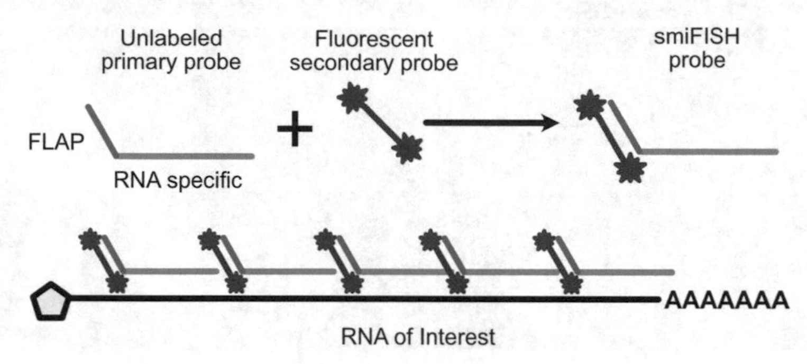

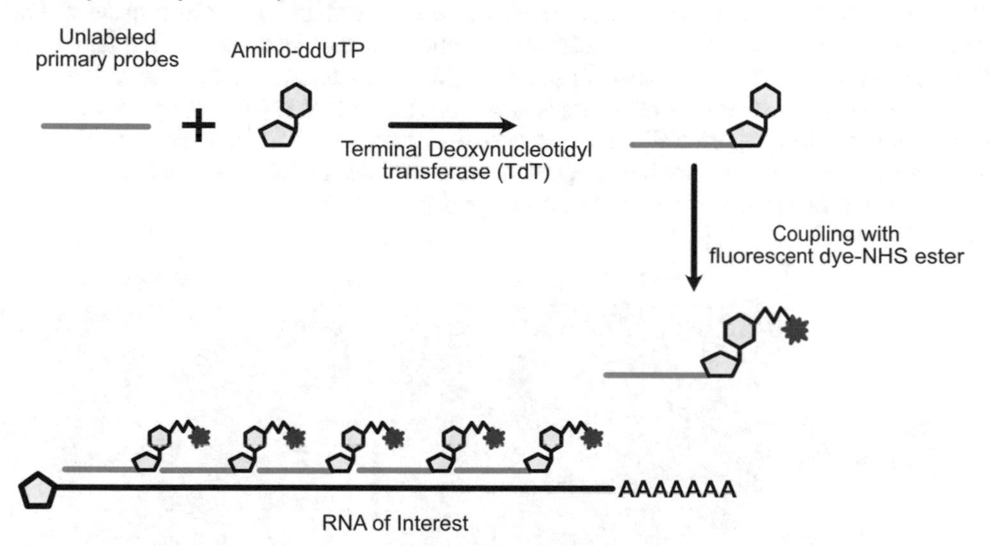

Fig. 1 Schematics showing the principles of various techniques to perform smFISH. (**a**) Commercially available probes, (**b**) smiFISH probes, and (**c**) enzymatically labeled probes

2 Materials

Prepare all solutions using nuclease and RNase-free water and molecular biology grade reagents, unless specified otherwise.

2.1 Stable or Transiently Transfected Cell Lines

This can be ignored if imaging endogenous RNA.

1. Plasmid expressing transcript of interest.
2. Cell culture media and supplements (e.g., DMEM supplemented with 10% FBS and penicillin–streptomycin).

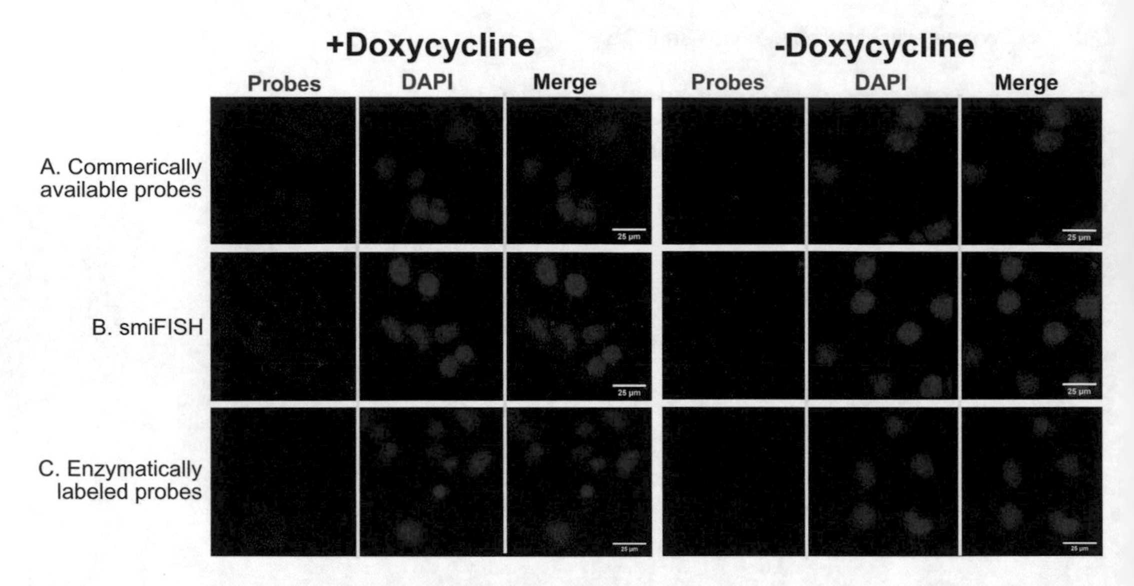

Fig. 2 Representative smFISH images from the three different strategies, commercially available probes from Stellaris, smiFISH probes and enzymatically labeled probes. Undifferentiated mouse neuronal CAD cells expressing a doxycycline-inducible firefly luciferase reporter construct were probed with (**a**) commercially available probes from Stellaris, (**b**) smiFISH probes or (**c**) enzymatically labeled probes. The addition of doxycycline ensures robust reporter mRNA expression while doxycycline negative images show background probe signal. Images are representative max projections through a z-stack. Large blotches of signal, most visible in the smiFISH signal, are likely aggregates of probe that, due to their large and irregular size, will be ignored during the computational calling of spots. Scale bar: 25 μm

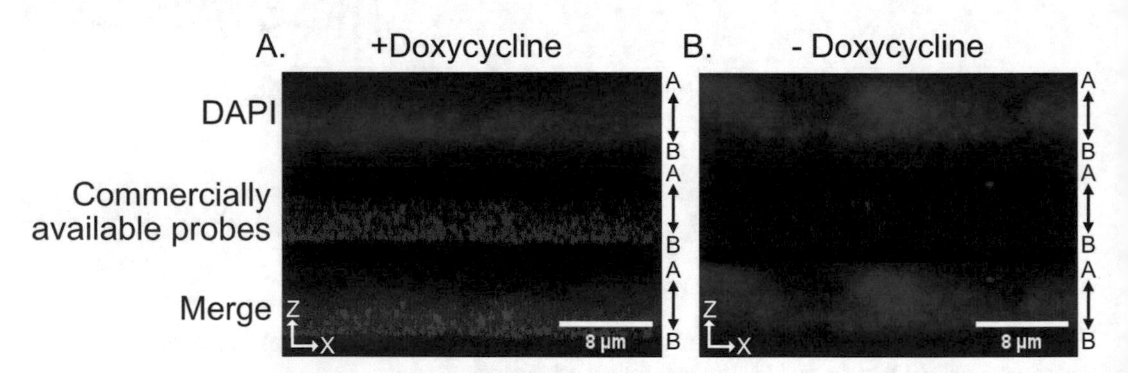

Fig. 3 Using smFISH to image apical and basal localization in intestinal epithelial C2bbe1 cells. Differentiated human C2bbe1 intestinal epithelial cell monolayers expressing (**a**) and not expressing (**b**) Firefly luciferase reporter constructs. Images are representative orthogonal max projections through a z-stack with the apical surface of the cells toward the top and the basal surface of the cells toward the bottom (denoted by vertical arrows). Scale bar: 8 μm

3. Doxycycline (2 mg/mL).
4. Puromycin (5 mg/mL).
5. Lipofectamine LTX (Invitrogen).
6. Opti-MEM media (Gibco).

2.2 Probe

2.2.1 Commercially Available Probes

1. Stellaris probes against RNA of interest (Biosearch Technologies) (*see* Subheading 3.3.1 for details).

2.2.2 smiFISH

1. Oligonucleotides complementary to the transcript of interest. Primary probes are produced in 96-well plates. For convenience, order the oligonucleotides resuspended in Tris–EDTA pH 8.0 (TE) buffer, at final concentration of 100 μM (*see* Subheading 3.3.1 for details).
2. Fluorescently labeled secondary probe: Resuspend in TE at 100 μM. Store at −20 °C in the dark (Can be ordered from multiple oligonucleotide synthesis companies).
3. 10× NEB Buffer 3.
4. 1× TE (Tris–EDTA) Buffer: 10 mM Tris–HCl and 1 mM EDTA at pH 8.0.

2.2.3 Enzymatically Labeled Probes

1. Oligonucleotides complementary to the transcript of interest (*see* Subheading 3.3.1 for details).
2. Amino-11-ddUTP: dissolve in nuclease-free water to 20 μM.
3. Terminal Deoxynucleotidyl transferase (TdT).
4. Oligo cleanup columns (Zymo).
5. 1 M $NaHCO_3$.
6. Quasar 570 Succinimidyl ester (Biosearch technologies): Dissolve in DMSO to 40 mM (Stock concentration).
7. 10× tris–borate–EDTA (TBE) buffer.
8. 40% (w/v) acrylamide–bis solution.
9. Urea.
10. TEMED (TMEDA, 1, 2-Bis(dimethylamino)cthanc).
11. 15% polyacrylamide (PA)–8 M Urea Stock: Mix 75 mL 40% (v/v) acrylamide–bis solution, 20 mL 10× TBE, and 96 g urea. Make up the volume to 200 mL with dH_2O. Stir and heat (max 100 °C) until urea completely dissolves. Store at RT away from light.
12. 10% (w/v) ammonium persulfate (APS).
13. Gel Green (Biotium).
14. Glass plates and spacers for casting Polyacrylamide (PAGE) gel.

2.3 smFISH

1. 20× saline–sodium citrate (SSC) buffer.
2. Hybridization buffer: 10% dextran sulfate, 10% formamide in 2× SSC. Store at −20 °C in 500 μL aliquots.
3. 1× phosphate buffer saline (PBS).
4. Fixation buffer: 3.7% formaldehyde in 1× PBS.

5. Wash buffer: 10% formamide in 2× SSC. Make fresh each time and keep at room temperature (RT) during the experiment.
6. Poly-D-Lysine–coated coverslips: 18 mm diameter, 1.5 thickness, although this may differ upon the microscope to be used for visualization.
7. Hybridization chamber (*see* protocol for details).
8. Fluoromount G.
9. Optically clear nail polish.

2.4 Microscopy and Image Analysis

1. DAPI (nuclear stain).
2. Fiji.
3. Fish-quant.
4. MATLAB.
5. R and R-studio for plotting.
6. Widefield DeltaVision microscope (GE).

3 Methods

3.1 Stellaris

Customized Stellaris smFISH probes can be designed and ordered using the freely available probe designer software by Biosearch technologies [15]. Representative smFISH images using the stellaris probes to image the firefly reporter construct in CAD cells is shown in Fig. 2a.

3.2 smiFISH

3.2.1 smiFISH Probe Design

The primary probes for RNA of interest are designed using the R script Oligostan that accompanied the paper describing smiFISH [10]. This script can be found at the following location: https://bitbucket.org/muellerflorian/fish_quant/src/master/Oligostan/ . The algorithm uses the following adjustable parameters for each probe:

1. Length: 26–32 nt.
2. GC content: 40–60%.
3. Similar delta G values across all the probes.
4. Avoid stretches of consecutive nucleotides.

   Three different FLAP sequences are available allowing to choose for multiple fluorophores based on the need of the experiment. The sequences of the FLAPs are as follows:

   FLAPX = "CCTCCTAAGTTTCGAGCTGGACTCAGTG".
   FLAPY = " TTACACTCGGACCTCGTCGACATGCATT".
   FLAPZ = " CCAGCTTCTAGCATCCATGCCCTATAAG".

Secondary probes that are fluorescently labeled and complementary to these FLAPs can be purchased from a variety of oligonucleotide synthesis companies. For increased brightness, fluors can be placed at both the 5′ and 3′ ends of the FLAP.

Representative smFISH images using the smiFISH probes to image the firefly reporter construct in CAD cells is shown in Fig. 2b.

3.2.2 smiFISH Probe Synthesis

1. Make an equimolar mixture of primary probes such that the final concentration of ***individual probes*** is 0.833 μM. For instance, if there were 24 probes in the probe set, this would be a final ***total probe*** concentration of 20 μM.
2. Prepare FLAP hybridization reaction as follows.

Component	Volume per reaction	Amount
Primary probe set (step 1)	2 μL	*x* pmol (*see* **Note 1**)
100 μM labeled secondary probe	*Y* μL	1.2× pmol
10× NEB buffer 3	1 μL	
RNAse-free water	7–*Y* μL	
Total volume	**10 μL**	(*see* **Note 2**)

Incubate the reaction mix in a thermocycler with lid temperatures of 99 °C and cycling conditions as below.

Temperature	Time
85 °C	3 min
65 °C	3 min
25 °C	5 min
4 °C	Hold

The hybridized flap duplexes can undergo multiple freeze-thaw cycles (2–5) without decrease of signal (*see* **Note 3**).

3.3 Enzymatically Labeled Probes

3.3.1 Probe Design

Follow the same design principles as the smiFISH primary probes, but do not add the FLAP hybridization sequences. It is often advised to have all enzymatically labeled probes be the same length. This will help with assessing the degree of labeling of the probes by gel electrophoresis. Representative smFISH images using the enzymatically labeled probes to image the Firefly luciferase reporter construct in mouse CAD cells are shown in Fig. 2c.

3.3.2 Probe Synthesis

1. Resuspend every probe to a concentration of 100 μM.
2. Prepare probe mix by combining equal volumes of every primary probe. The final *total* probe mix will therefore be 100 μM, while the concentration of each *individual* probe will be 100 μM divided by the number of probes.
3. Prepare the following reaction mix to conjugate amino-ddUTP (AA-ddUTP) to the primary probe mix.

Component	Stock concentration	Final concentration	Volume
Probe mix	100 μM	20 μM	20 μL
AA-ddUTP	1 mM	100 μM	10 μL
TdT reaction buffer	5×	1×	10 μL
TdT enzyme	20 U/μL	1 U/μL	2.5 μL
RNAse-free water			7.5 μL
Total volume			**50 μL**

4. Incubate the reaction at 37 °C in the thermocycler for 18 h with lid temperature set to 37 °C.
5. Cleanup the reaction with an oligo purification column (Zymo) as per manufacturer's instructions and elute in 20 μL elution buffer (*see* **Note 4**).
6. In order to quantify the oligo mix, the average extinction coefficient (at 260 nm) for all of the oligos is required. The extinction coefficient for each oligo can be found on the spec sheet that comes with the oligos (*see* **Note 5**).
7. Calculate the oligo concentration using the Nanodrop, giving it the extinction coefficient at 260 nm from **step 5**.
8. Using the average molecular weight of your oligos (assume a molecular weight of 330 g/mol per nucleotide) (*see* **Note 6**), convert the ng/μL concentration obtained by the Nanodrop into a molar concentration.
9. Prepare the following reaction mix to conjugate the amino-ddUTP labelled oligos with Q570:

Component	Stock	Final	volume
Probe mix	*X* μM	10 μM	*X* μL
Q570 NHS ester	2 mM	500 μM	25 μL
$NaHCO_3$	1 M	100 mM	4 μL
RNase-free water			*X* μL
Total volume			**100 μL**

10. Incubate in the dark (preferably in a thermocycler) at room temperature for 16 h or overnight.
11. Cleanup with oligo cleanup column (Zymo) as per manufacturer's protocol.
12. Repeat the clean-up with a second oligo cleanup column (Zymo).
13. Elute in 15 μL water.
14. Next for quantifying the labeled probe mix with spectrophotometry, take an absorbance measurement of the probe mix at 547 nm (the absorbance maximum of Q570) (A_{547}) and at 260 nm (A_{260}) (*see* the following link for Q570 info including extinction coefficients: http://www.sbsbio.com/download/Quasar%20570%20Amidite.pdf).
15. Using the formula below calculate the variable a and b as follows.

$$a = \left(\varepsilon_{\text{oligo\_260}} / \varepsilon_{\text{dye\_547}}\right) \times A_{547}$$

$$b = A_{260} - \left(\left(\varepsilon_{\text{dye\_260}} / \varepsilon_{\text{dye\_547}}\right) \times A_{547}\right)$$

where $\varepsilon_{\text{oligo\_260}}$ = average extinction coefficient of probe oligos

$\varepsilon_{\text{dye\_260}}$ = extinction coefficient for Q570 at 260 nm (9000 M^{-1} cm^{-1}).

$\varepsilon_{\text{dye\_547}}$ = extinction coefficient for Q570 at 547 nm (115,000 M^{-1} cm^{-1}).

A_{547} = Absorbance of the probe mix at 547 nm.

A_{260} = Absorbance of the probe mix at 260 nm.

16. Calculate the degree of labeling or labeling efficiency as follows:
   $[Q_{570}]/[\text{oligo}] = a/b$ (*see* **Notes 7** and **8**).
17. In addition, use the A_{260} and $\varepsilon_{\text{oligo\_260}}$ to calculate molar oligo concentration as before (**step 6**).

3.3.3 PAGE Analysis of the Labeled Oligonucleotides

An alternative to estimating the degree of labeling by spectrophotometry is using polyacrylamide gel electrophoresis (PAGE). In addition, PAGE analysis also provides information and quality control to check whether the ddUTP conjugation to the probe mix was efficient (*see* **Note 9**).

1. Clean the glass plates for casting the gel with distilled water and 70% ethanol. Let them dry and assemble the cassette with spacers on the sides and the bottom. Seal the assembly with large two fold-back clamps on the two sides of the cassette such that the clamps clip above the spacer between the two glass plates.

2. To cast a 10 × 8 cm acrylamide gel, add 10 mL of 15% polyacrylamide—8 M urea stock (*see* Subheading 2 for recipe), 50 μL 10% APS and 6.25 μL TEMED into an Erlenmeyer flask. Pour the contents into the previously assembled gel cast. Insert the comb and allow the gel to polymerize for 30 min.
3. Remove the clamps and the comb. Rinse the cassette with dH_2O to remove gel pieces polymerized on the outer surface. Rinse the wells of the polymerized gel with 1× TBE using a syringe.
4. Fill the PAGE chamber with 1× TBE buffer and prerun in 1× TBE for 30 min at 20 mA or ~200 V. Once done, rinse wells again to ensure excess urea is flushed.
5. In the meantime, prepare the loading samples by combining 100 ng of conjugated probe sample and 6× loading dye (final concentration of 1×). Make up the volume to 5 μL using dH_2O.
6. Load samples (~100 ng) and run at 160 V until xylene cyanol (blue) and bromophenol blue (purple) markers reach 2/3 of the gel length and the bottom respectively (~1.5 h).
7. Image the fluorescently labeled pool of molecules on a gel-imager with appropriate filter sets to excite and detect the incorporated fluorescent dyes.
8. Incubate the gel with Gel Green (diluted 1:10,000 in 1× TBE) for 15 min.
9. Reimage the gel to detect both the nonmodified and modified pools of ssDNA.
10. The addition of a bulky nucleotide (ddUTP) at the 3′-end of the probes makes the labeled oligonucleotides run slower than the nonmodified probes on the PAGE gel. Thus, the amount of nonmodified oligos can be estimated by comparing the loss of Gel Green fluorescence intensity to that of the control (Fig. 4).
11. Lastly, the addition of the Q570 to the conjugated ddUTP probe mix further reduces the migration speed of the oligo pool and the labeling efficiency can be quantified by the decrease in the fluorescent intensity of the ddUTP labeled probe (Fig. 4).

3.4 Hybridization of Probes

As stated above, this protocol is using an example where an mRNA encoding firefly luciferase is visualized using smFISH. Cells expressing this exogenous mRNA have been selected for stable expression, and expression of the reporter can be induced with doxycycline. If visualization of a transiently expressed RNA is desired instead, skip **steps 9–11**. If visualizing an endogenous

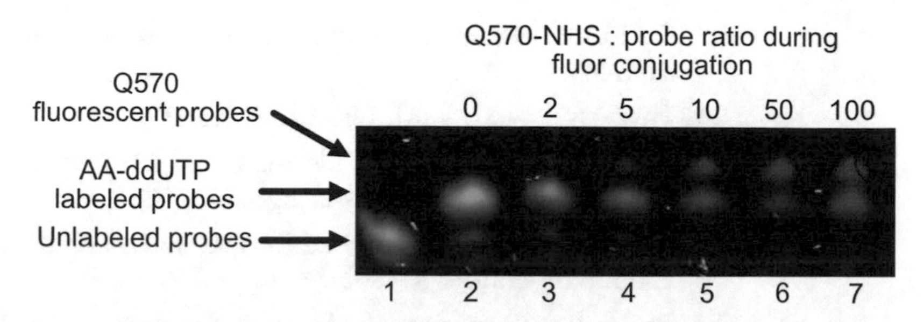

Fig. 4 Analyzing 3′ labeling of the oligonucleotide probes by PAGE. The unlabeled probes (Lane 1) are conjugated with amino-ddUTP at their 3′ end using the enzyme TdT (Lane 2). Q570–NHS ester is then reacted with NH_2-ddUTP labeled probes to yield fluorescent probes for smFISH imaging (Lane 3–7). Lanes 3–7 show increase in labeling efficiency with increasing molar excess of the Q570-NHS ester compared to the ddUTP-labeled probes. The labeling reaches the maximum at 50× molar excess of Q570-NHS ester (Lane 6). Gel Green staining is shown in green and Q570 fluorescence is shown in red

RNA, skip **steps 1–8**. The protocol is optimized for 12-well plate. Scale-up as needed.

1. For each well add 250 μL of confluent cells (~75,000 cells) in 1 mL of growth medium and allow to attach overnight. If not creating stably expressing lines, directly plate ~25,000 cells on a coverslip.
2. Dilute the plasmid DNA (1 μg) and 2 μL Plus reagent in 50 μL Opti-MEM per well.
3. Dilute the LTX reagent (4 μL) in 50 μL Opti-MEM per well.
4. Incubate at room temperature 5 min.
5. Mix the diluted DNA and LTX solutions together. Vortex briefly and incubate at room temperature for 15–20 min.
6. Add the transfection mix to the cells dropwise to prevent dislodging. Incubate the cells with the transfection mix for 24 h in a 37 °C incubator.
7. Change the media and allow expression for an additional 24 h.
8. If creating stable cell lines, split the cells, add puromycin at a final concentration of 5 μg/mL and select for integrants. If not creating stable lines, proceed to **Step 9**.
9. Place a coverslip into each well of a 12-well plate.
10. Plate 100 μL confluent puromycin selected cells (approximately 25,000 cells) in each well. Make sure to plate an additional well for a noninduced control (minus doxycycline). Add doxycycline (1 μg/mL) to express the firefly reporter mRNA. If visualizing an endogenous RNA instead of an exogenous reporter, simply plate 25,000 cells on a coverslip.
11. Incubate at 37 °C for 48 h.
12. Remove media and wash with 1 mL PBS.

13. Add 1 mL fixation buffer and incubate at room temperature for 10 min.
14. Wash twice with 1 mL PBS (*see* **Note 10**).
15. Permeabilize with 1 mL 70% ethanol at 4 °C. Incubate for 2 h at room temperature or overnight at 4 °C.
16. Remove 70% ethanol and add 1 mL wash buffer. Incubate at room temperature for 5–10 min.
17. In the meantime, prepare the hybridization chamber. Place wet napkins in the bottom of an empty pipette tip box. Place parafilm across the rack that once held tips. This is where the coverslips will be placed, so avoid any wrinkles in the parafilm (*see* **Notes 11** and **12**).
18. For each hybridization reaction, prepare 100 μL of hybridization buffer. Add 2 μL of each probe set for every 100 μL of hybridization buffer (*see* **Notes 13** and **14**).
19. Pipette the hybridization buffer on the parafilm avoiding bubbles. Place the coverslip directly on the drop of the hybridization buffer such that the cells side of the coverslip faces down into the hybridization buffer (*see* **Note 15**).
20. Seal the chamber with the lid of the pipette tip box and cover with aluminum foil to save from exposure to light. Incubate at 37 °C for 4–16 h in the dark.
21. Transfer the coverslip with cells side up to a fresh 12-well plate gently. Add 1 mL wash buffer.
22. Incubate in the dark at 37 °C for 30 min.
23. Remove the wash buffer and add 1 mL DAPI stain buffer (Wash buffer with 100 ng/mL DAPI).
24. Incubate in the dark at 37 °C for 30 min.
25. Remove the DAPI stain buffer and add 1 mL wash buffer. Incubate at room temperature for 5 min.
26. In the meantime, label one glass slide for each coverslip.
27. Add 6 μL fluoromount G to the glass slides for mounting and place the coverslip directly onto the fluoromount with cells facing down (*see* **Note 16**).
28. Wait for the mounting solution to dry in the dark. Once dried, seal the coverslips with optically clear nail polish.

3.5 Imaging

1. The slides are imaged using the Deltavision Elite microscope (GE) at 60× magnification with 1.4 numerical aperture and 1.15 refractive index oil. Similar microscopes that are available to the user may also be acceptable. Use the DAPI signal to focus on adherent cells.

2. For the fluorescence illumination, the following filter settings are used: for DAPI 350/50 nm excitation filter, 455/50 nm emission filter; for FITC (GFP) 490/20 nm excitation filter, 525/36 nm emission filter; and for TRITC (smFISH probes) 555/25 nm excitation filter, 605/52 nm emission filter.
3. For smFISH, it is preferred to use maximum (100%) transmission of the light source and an exposure time of 0.05–0.3 s for the probes channel. For DAPI (nucleus) and FITC (GFP), use a low transmission setting and very short exposure time (e.g., 5–10% transmission, 0.05–0.2 s).
4. For smFISH spot detection, more photons (exposure time and laser power) leads to better signal, but also increased background fluorescence and can lead to photobleaching of the probes. Therefore, the user should empirically decide the time and exposure parameters accordingly (*see* **Notes 17** and **18**).
5. In order to get a correct estimate of the smFISH spots, imaging across the volume of the cell is needed. Thus, multiple *z*-sections should be collected. It is recommended to use 0.2 μm steps and to collect sections until the probe signal becomes blurry on both ends. This might vary from cell-to-cell depending on its morphology.

3.6 Data Analysis: FISH-Quant

Quantification of smFISH spots is often required to answer the research questions of interest. An image alone is usually not enough to determine localization or quantity of smFISH spots, especially when comparing many conditions or different RNA transcripts. Software packages have been developed to aid in the quantification of RNA transcripts in smFISH images. Below a short user guide for the free software FISH-quant is provided.

FISH-quant is a free analysis tool for quantifying smFISH spots in three dimensional space [16]. While FISH-quant requires Matlab, it does not require any computational expertise as it operates through a graphical user interface (GUI). FISH-quant and its documentation are available for download here: https://bitbucket.org/muellerflorian/fish_quant.

1. FISH-quant operates through seven simple steps (Fig. 5). First, 3D smFISH image stacks and the properties of the microscope used to collect them must be provided. The type of microscope, pixel dimensions, refractive index, numerical aperture, and fluorescence wavelengths used to capture an image can be entered after pressing “Modify” within the “Experimental Parameters” box (Fig. 5).
2. Cells within an image can be outlined by pressing the “Define Outlines” button and opening the outline designer window (Fig. 6). Here, image brightness can be adjusted, additional images from other channels such as DAPI staining can be

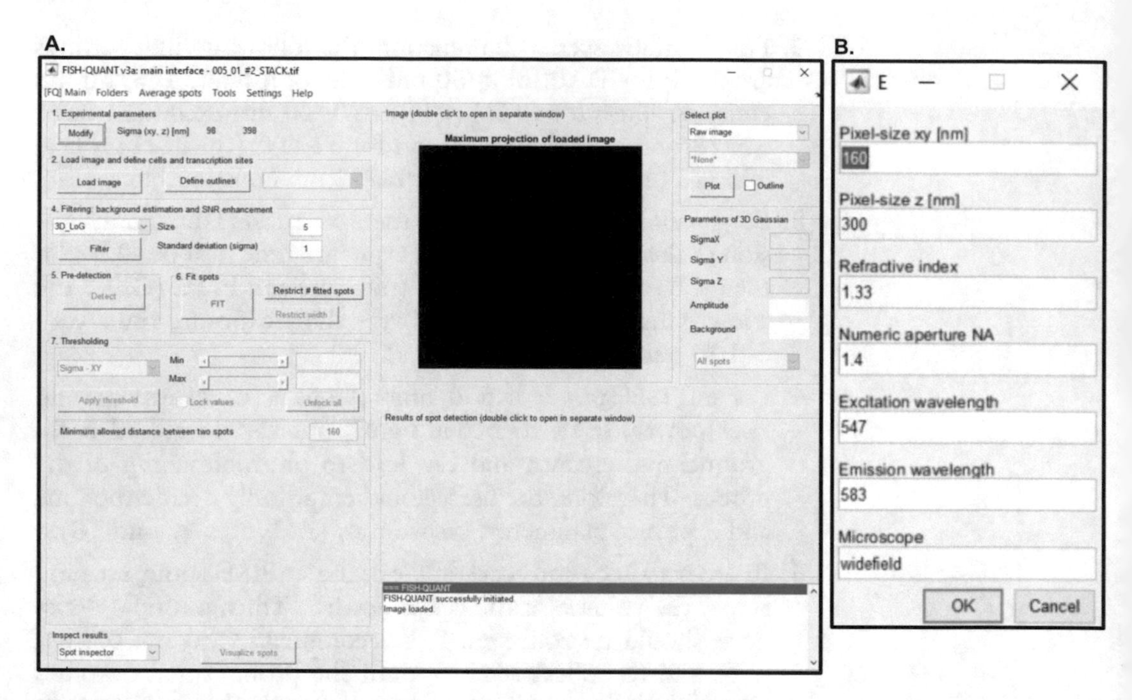

Fig. 5 Screenshots of FISH-quant GUI. (**a**) FISH-quant GUI after loading a single channel z-stack image. (**b**) GUI for managing experimental parameters

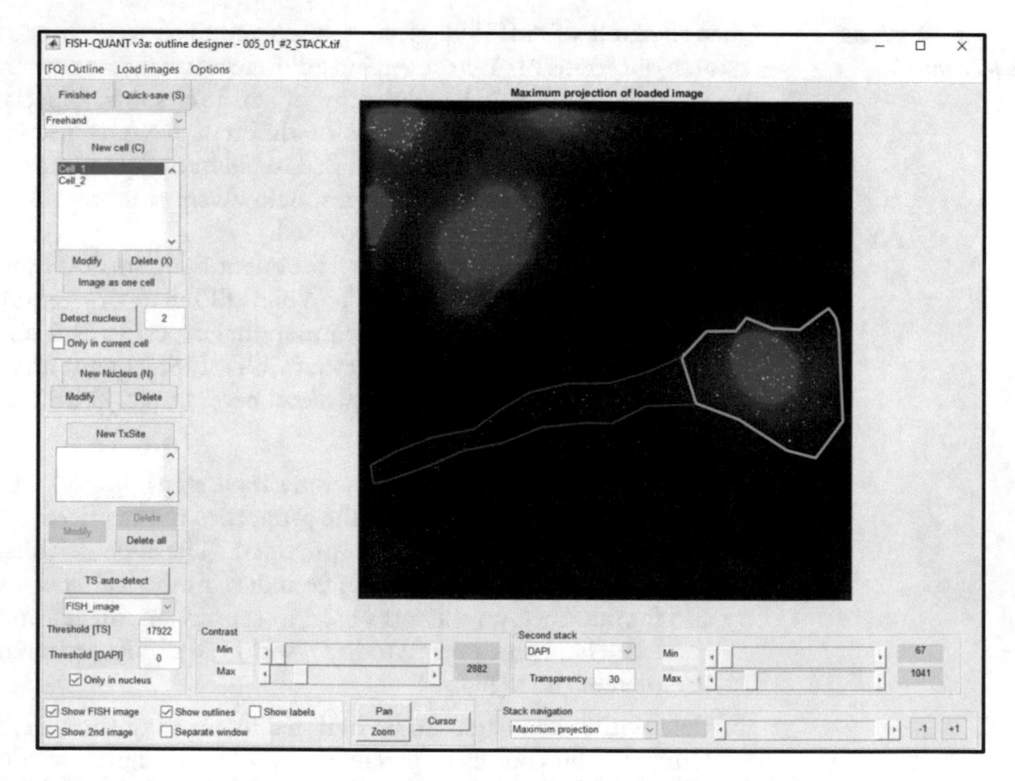

Fig. 6 Outline designer GUI. Screenshot of outline designer GUI with smFISH image with DAPI channel overlaid. The cell body and neurite have been outlined with the polygon feature

overlaid and outlines can be defined by freehand drawing or tracing a polygon shape. In Fig. 6, a neuronal cell's cell body and projection have been defined as two separate cells so that transcript abundance can be calculated in each part separately, allowing for a comparison of RNA abundance between the compartments.

3. Once cells within an image have been outlined as desired, the image is filtered using a 3D Gaussian method to eliminate image noise. Local maxima or spots can be detected by pressing "Detect". Testing a minimum intensity of 0 is often informative. Usually all other defaults are fine to leave unchanged (Fig. 7a).
4. Thresholding the intensity of local maxima to be considered as true spots is the first thresholding step. Overestimating the threshold intensity is very much preferred to underestimating. It is usually best to place the threshold just to the left of the pixel intensity histogram's "elbow" (Fig. 7b). Pressing "Perform Detections" will map thresholded spots onto your image and provide quality scores for the spots. With an intensity threshold of 28,1379 spots were detected within the neuronal cell's body (Fig. 7c).
5. Next, pressing "Apply detection to all cells in image" will close the predetection window. Spots can then be fitted in the entire image by pressing "FIT." Once spots with the set intensity threshold have been fit to the image, smFISH spots can be further thresholded based upon several parameters including Sigma or sphericity of the spot, Amplitude, Background, Intensity before and after filtering, and Position in Z. Thresholds are set by positioning the red min and max threshold lines around the peak of the histogram and checking the "Lock values" box. This will exclude outlier spots for any single parameter (Fig. 8).
6. These detection settings can be saved and applied to many cells and or images using Tools → "Batch Processing". Provide the detection settings and cell outlines for each image and click "PROCESS" (Fig. 9). When completed for every image, thresholding spots across all images can be done on the same parameters as before. Results can then be saved via the batch → save → summary: mature mRNA. This summary file includes total and thresholded spots detected for each cell. A more complete analysis for each spot can be saved (batch → save → results for each image [ALL spots]) which includes significantly more information for every single spot detected.
7. Usually, a summary of results will answer most smFISH questions about the number of transcripts in each cell. However,

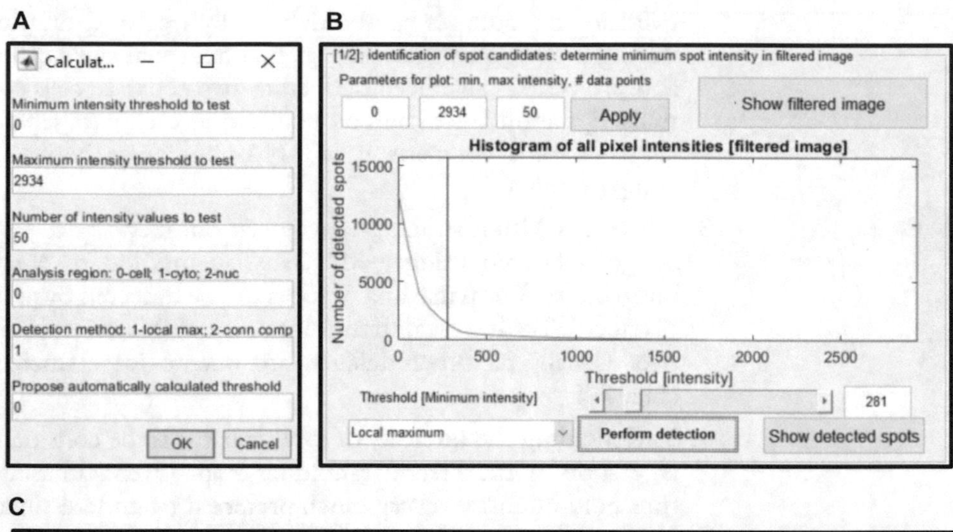

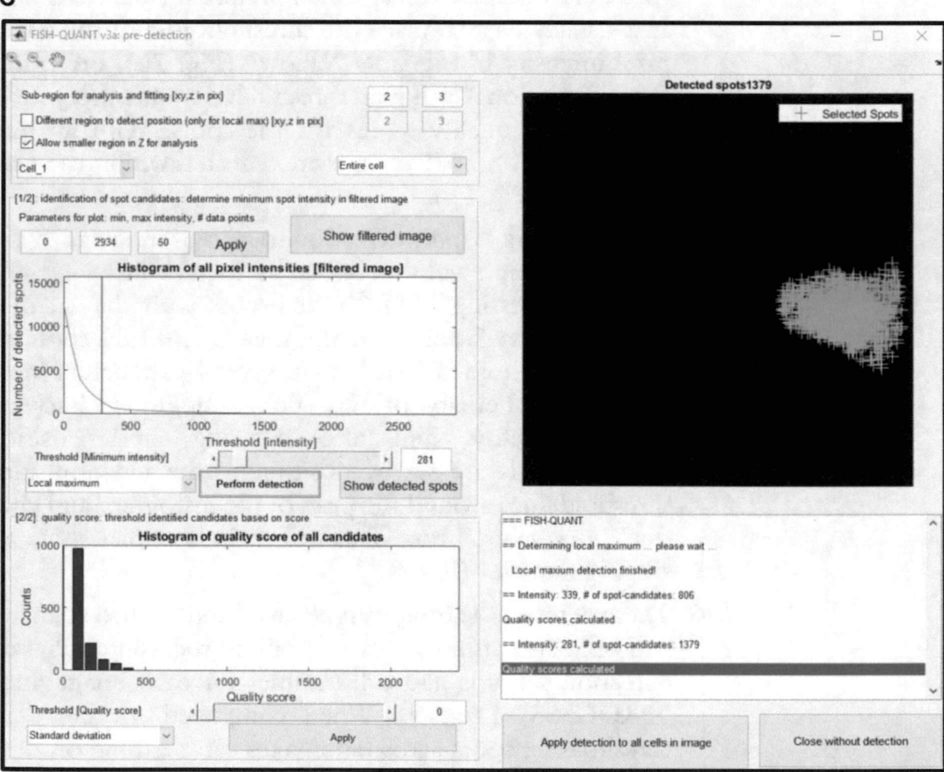

Fig. 7 Screenshot of predetection GUI. (**a**) GUI for managing predetection parameters. (**b**) GUI of pixel intensity thresholding. (**c**) Predetection of neuron cell body after thresholding

using the details of every spot detected allows for further custom thresholding of spots. For instance, a "Positive" image of a cell expressing a known transcript can be compared to a "Negative" image where a cell is known to not express the transcript. This can be used to separate true spots from noise in

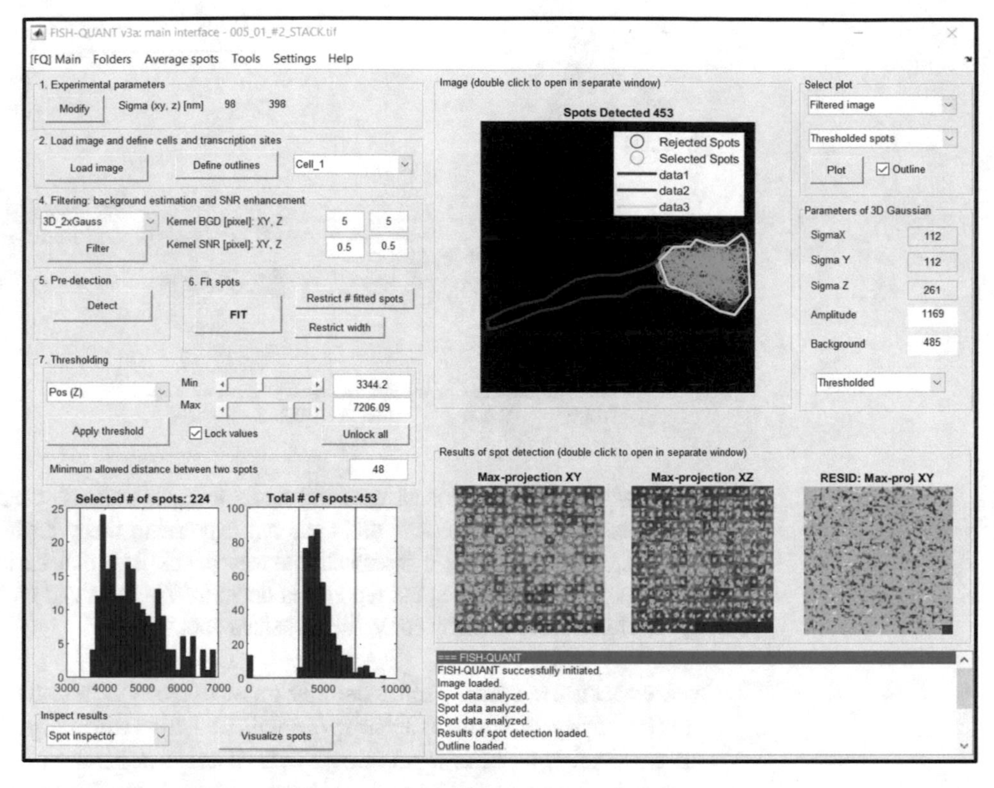

Fig. 8 Screenshot of FISH-quant main interface after fitting predetected spots. Results of detection after thresholding each spot characteristic are visualized

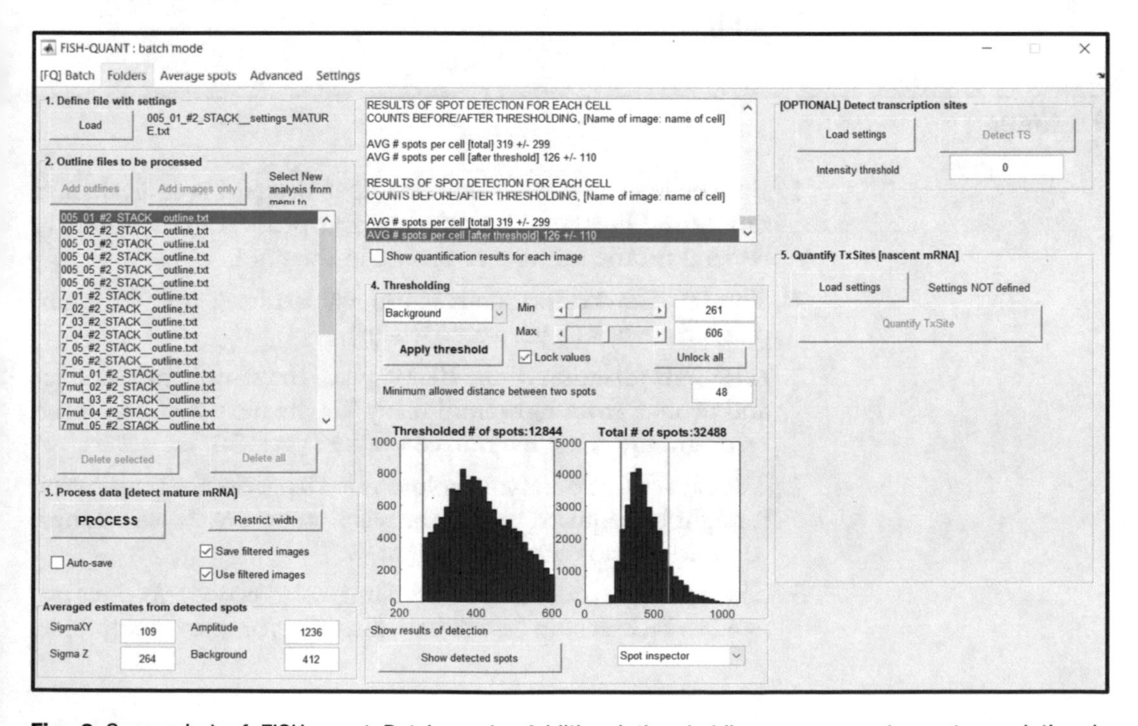

Fig. 9 Screenshot of FISH-quant Batch mode. Additional thresholding on aggregate spot population is available here

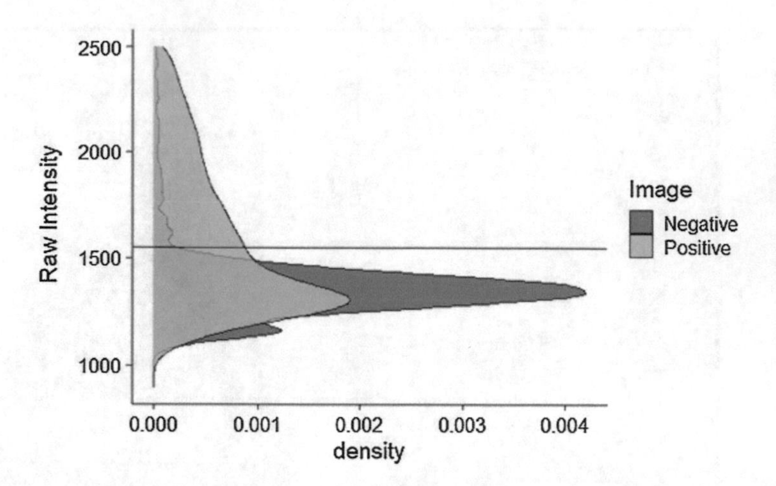

Fig. 10 Raw spot intensity for all detected spots in cells expressing Firefly luciferase reporter (green, positive) and cells not expressing firefly luciferase reporter (red, negative). Here, a threshold intensity (black line) that excludes almost all spots in cells lacking the reporter is defined. This threshold for spot calling therefore removes many noisy, false positive spots

your specific system. This is usually most useful when considering intensity of spots as false spots will be quite dim (Fig. 10). In this example, cells expressing the transcript of interest have a much higher density of true, bright spots where the negative sample, without expression of the transcript, mostly contains dim, false spots. These spots can be excluded from further study.

4 Notes

1. For the ease of conversion remember that μM is equivalent to pmol/μL. Further, the pmol of the probe mix is going to depend on the number of probes in the pool.
2. The 10 μL hybridization reaction will last for 5 smiFISH samples/coverslips (2 μL per coverslip).
3. After hybridization of the FLAP, place the sample tubes on ice and protect from light until ready for the next step. For long-term storage, keep hybridized duplexes at −20 °C.
4. The capacity of the Zymo columns is 10 μg. So, for purification it might be required to split the reaction over multiple columns to avoid overloading the column.
5. The extinction coefficient of the unlabeled probe mix does not need to be exact, an estimate within 5% error rate is acceptable.

6. While calculating the average molecular weight of the oligos remember to add the weight of the single extra nucleotide just added at the 3′ end.
7. For detailed information on how to calculate the degree of labeling (DOL) please refer to [11].
8. Labeled oligo mixtures with $0.8 < \text{DOL} \leq 1.0$ are considered good quality products that can be used in smFISH applications.
9. For accurate estimation of the labeling efficiency, only oligo mixtures containing ssDNA molecules of identical length (e.g., only 20 mers) should be analyzed by PAGE.
10. It is recommended to pipet liquids on the wall of the well and not directly onto the cells especially for cells with delicate structures (e.g., neurites, membranes). Also, be careful while aspirating the liquids between washes.
11. Ready to use commercial hybridization chambers are available as well (e.g., The ThermoBrite System from Abbott). Using a tip box as mentioned in the text is a cheaper but effective alternative.
12. The wet tissues are needed to maintain the humidity in the hybridization chamber and avoid evaporation of the hybridization buffer.
13. It is recommended to prepare the hybridization buffer with 10% excess than required as the buffer is very viscous.
14. In order to avoid formation of fluorescent aggregates sometimes observed in smiFISH imaging, we recommend to spin down the FLAP hybridization mix (Subheading 3.2) and the hybridization buffer with the probes added at 15,000 rpm for 5 min. Carefully pipette and use the supernatant, avoiding any pelleted aggregates on the bottom.
15. Sometimes due to surface tension, the coverslips are hard to pick up from the multiwell tissue culture plate. Using a bent syringe needle at the edges of the coverslip to break the surface tension and then holding the coverslip with a forceps can help with the process of transferring the coverslip. Hold the coverslip with the forceps gently in order to avoid cracking the glass.
16. Pipette the Fluoromount G slowly to avoid bubbles. Additionally, place the coverslip over the mounting solution gently to avoid bubbles.
17. Determine the suitable exposure time for different channels such that one obtains the highest fluorescent signal without saturating the pixels. This becomes extremely important when counting the FISH spots.
18. All samples that need to be compared should be captured using the same laser intensity and exposure times.

Acknowledgments

This work was funded by the National Institutes of Health (R35-GM133885) (JMT), a Predoctoral Training Grant in Molecular Biology (NIH-T32-GM008730) (RG) and the RNA Bioscience Initiative at the University of Colorado Anschutz Medical Campus (RG and JMT). It was further funded by the RNA Bioscience Initiative Summer Intern Program at the University of Colorado Anschutz Medical Campus (PTV).

References

1. Kim-Ha J, Smith JL, Macdonald PM (1991) Oskar mRNA is localized to the posterior pole of the drosophila oocyte. Cell 66:23–35
2. Zimyanin VL, Belaya K, Pecreaux J et al (2008) In vivo imaging of oskar mRNA transport reveals the mechanism of posterior localization. Cell 134:843–853
3. St Johnston D, Beuchle D, Nüsslein-Volhard C (1991) Staufen, a gene required to localize maternal RNAs in the drosophila egg. Cell 66:51–63
4. Mowry KL, Melton DA (1992) Vegetal messenger RNA localization directed by a 340-nt RNA sequence element in Xenopus oocytes. Science 255:991–994
5. Long RM, Singer RH, Meng X et al (1997) Mating type switching in yeast controlled by asymmetric localization of ASH1 mRNA. Science 277:383–387
6. Cajigas IJ, Tushev G, Will TJ et al (2012) The local transcriptome in the synaptic neuropil revealed by deep sequencing and high-resolution imaging. Neuron 74:453–466
7. Goering R, Hudish LI, Guzman BB et al (2020) FMRP promotes RNA localization to neuronal projections through interactions between its RGG domain and G-quadruplex RNA sequences. Elife 9:e52621
8. Zivraj KH, Tung YCL, Piper M et al (2010) Subcellular profiling reveals distinct and developmentally regulated repertoire of growth cone mRNAs. J Neurosci 30:15464–15478
9. Taliaferro JM, Vidaki M, Oliveira R et al (2016) Distal alternative last exons localize mRNAs to neural projections. Mol Cell 61:821–833
10. Tsanov N, Samacoits A, Chouaib R et al (2016) smiFISH and FISH-quant—a flexible single RNA detection approach with super-resolution capability. Nucleic Acids Res 44:e165
11. Gaspar I, Wippich F, Ephrussi A (2017) Enzymatic production of single-molecule FISH and RNA capture probes. RNA 23:1582–1591
12. Raj A, van den Bogaard P, Rifkin SA et al (2008) Imaging individual mRNA molecules using multiple singly labeled probes. Nat Methods 5:877–879
13. Cabili MN, Dunagin MC, McClanahan PD et al (2015) Localization and abundance analysis of human lncRNAs at single-cell and single-molecule resolution. Genome Biol 16:20
14. Chou Y-Y, Heaton NS, Gao Q et al (2013) Colocalization of different influenza viral RNA segments in the cytoplasm before viral budding as shown by single-molecule sensitivity FISH analysis. PLoS Pathog 9:e1003358
15. Stellaris Probe Designer. https://www.biosearchtech.com/support/tools/design-software/stellaris-probe-designer
16. Mueller F, Senecal A, Tantale K et al (2013) FISH-quant: automatic counting of transcripts in 3D FISH images. Nat Methods 10:277–278

Chapter 14

Single-Molecule RNA Imaging Using Mango II Arrays

Adam D. Cawte, Haruki Iino, Peter J. Unrau, and David S. Rueda

Abstract

In recent years, fluorogenic RNA aptamers, such as Spinach, Broccoli, Corn, Mango, Coral, and Pepper have gathered traction as an efficient alternative labeling strategy for background-free imaging of cellular RNAs. However, their application has been somewhat limited by relatively inefficient folding and fluorescent stability. With the recent advent of novel RNA-Mango variants which are improved in both fluorescence intensity and folding stability in tandem arrays, it is now possible to image RNAs with single-molecule sensitivity. Here we discuss the protocol for imaging Mango II tagged RNAs in both fixed and live cells.

Key words Fluorogenic RNA Aptamer, RNA-Mango, Single-Molecule Microscopy, Live cell imaging, RNA dynamics, Superresolution microscopy

1 Introduction

For over 20 years, the gold standard technique for imaging RNA molecules in live cells with single molecule resolution has been the MS2 bacteriophage imaging system [1, 2]. This method utilizes an exogenously introduced cassette of MS2 bacteriophage RNA stem-loops (MS2SL) into an RNA of interest and is subsequently labeled with fluorescent MS2 coat proteins (MCP). To date, there have been a number of improved iterations of the MS2 system [3, 4] and orthogonal imaging approaches. With the use of these tools studies have largely focused on RNA processing events, such as mRNA transcription dynamics [5, 6], mRNA splicing [7, 8], mRNA translation [9–12], and mRNA localization and transport [13–16]. However, one drawback to bacteriophage labeling strategies is the constitutive fluorescence of excess nuclear localized capsid protein [17]. This excess fluorescence can impede the imaging of nuclear RNAs and occasionally cytosolic mRNAs. Additionally, there have been reports of MS2SL-MCP labeling systems giving rise to capsid stabilized degradation products [18, 19]. To address excess background signal, split GFP methodologies taking advantage of orthogonal MS2/PP7 labeling strategies, can enable

Erik Dassi (ed.), *Post-Transcriptional Gene Regulation*, Methods in Molecular Biology, vol. 2404,
https://doi.org/10.1007/978-1-0716-1851-6_14,

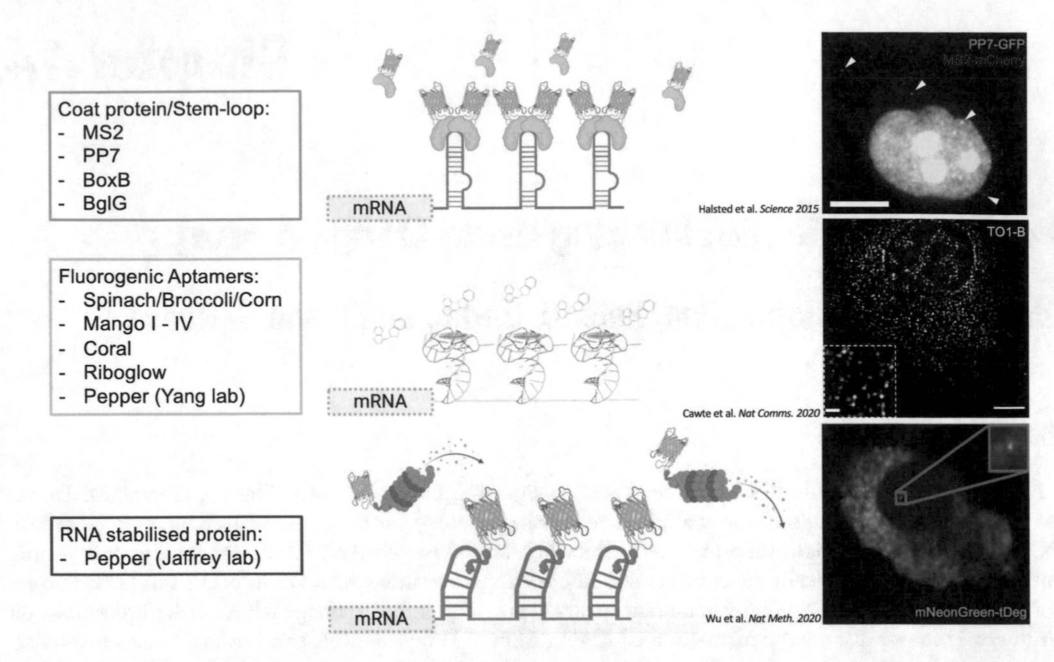

Fig. 1 Examples of fluorogenic imaging methodology: Top—Diagram of coat protein/stem-loop systems classically used for RNA imaging. Depicted are dimers of MCP-GFP binding MS2SL arrays. Representative image showing dually labeled RNAs with MS2 (red) and PP7 (green) systems adapted from Halstead et al. [9]. Excess nuclear localized CP can be seen as a saturating signal. Middle—Diagram of fluorogenic RNA aptamer imaging systems. Depicted is an array of RNA Mango aptamers binding their fluorogenic dye TO1-Biotin and inducing fluorescence (green). Representative image shows single-molecule imaging of RNA using fluorogenic Mango II arrays (yellow) adapted from Cawte et al. [38]. Bottom—Diagram of RNA stabilized imaging system where degron tagged fluorescent proteins are stabilized in the presence of Pepper RNA expression. Representative image shows imaging of RNA stabilized mNeonGreen-tDeg signal (green), adapted from Wu et al. [21]. Scale bars = 10 μm and inset 1 μm

background-free imaging [20]. However, these systems are more rarely used in the literature. Recent developments for background-free imaging of RNAs have opted either for RNA stabilized fluorescent proteins [21] or the use of fluorogenic RNA aptamers [22–31] (Fig. 1).

Here, we will focus on the recent advent of independently developed fluorogenic RNA aptamers. With such a large variety of fluorogenic RNA aptamers, it is difficult to decide which aptamer is best to use for your desired application. With a number of existing fluorogenic aptamers demonstrating some advantages and disadvantages in different cellular imaging contexts, it is best to understand the benefits and draw backs to your aptamer of choice.

Firstly, it is important to select the correct aptamer for the cell type being used. Most aptamers have been shown to work well in bacteria [24, 25, 28, 32–36], however obtaining single-molecule resolution seems to be limited when compared to mammalian cells. It has been reported that RNA G-quadruplex structures are globally unfolded in mouse embryonic stem cells (mESCs) [37]. This effect seems to be less significant in other immortalized mammalian cells that we and others have reported [23, 25–27, 38]. However, it may be safe idea to look into aptamers that are proposed not to contain G-quadruplex structures [22, 28, 29, 31, 34, 35] if using mESCs is essential.

Secondly, thought has to be given how to incorporate the aptamer into the chosen RNA. The Spinach family aptamers such as Spinach, Broccoli and Corn seem to show optimal performance when incorporated into a folding scaffold such as the F30 scaffold [39]. These sorts of folding scaffolds may not always be appropriate for the RNA of interest. Other aptamers such as Mango, Pepper, and Riboglow do not show this same requirement and similarly show the ability to be used either within internal loops or as multimeric arrays [23, 29, 38, 40]. Additionally, some aptamers seem to form dimeric interfaces or alternate structures which may affect their fluorescence in cells [27, 41].

Thirdly, understanding the physical properties of a fluorogenic aptamer of interest will help get the most out of that imaging technology. Remarkably, some aptamers such as Mango II and RhoBAST are resistant to formaldehyde fixation enabling both live and fixed cell imaging [23, 38, 42]. Fixation can also be a great tool to confirm an aptamers stability within the RNA of interest and also assess the localization of the tagged RNA with other complimentary applications such as immunofluorescence and RNA FISH.

One important factor that impacts the efficacy of imaging is a fluorogenic aptamers ability to exchange its ligand. Such exchange can drastically increase imaging times with minimal loss of signal in both standard and superresolution imaging techniques. The discovery of this photophysical property appears likely to revolutionize future superresolution imaging of RNA aptamers in both live and fixed cells [38, 42, 43]. Similarly, an RNA aptamers affinity to its ligand of interest is also an important property to be considered. With the recent developments of multiple high-affinity aptamers such as Mango, Coral, Pepper, and RhoBAST, it is evident that high affinity for their ligands enable more stable fluorescence whilst still retaining appropriate levels of dye exchange.

Given the varied brightness, photostability, and dye exchange of current aptamers, it is important to select an appropriate imaging technique to optimize the fluorescent signal. Most aptamers work well with widefield microscopes and pulsed illumination which allows for dye exchange. Only in a few instances has confocal or two-photon based imaging been used, for example with the Pepper-HBC RNA fluorophore pair [29]. Superresolution imaging approaches can be extremely useful but understanding which fluorogenic aptamer to use and their physical properties for a given superresolution technique are important considerations. Currently, Pepper, Mango, and RhoBAST aptamers which report high affinity for their ligands and a significant amount of dye exchange have benefited from the use of structured illumination microscopy (SIM) [29, 38, 42]. Furthermore, due to its flexibility using rhodamine-based dyes, RhoBAST has also been used with single-molecule localization microscopy (SMLM) and stimulated emission depletion (STED) microscopy. These recent developments show how new high-affinity fluorogenic aptamers can be used effectively with superresolution imaging approaches.

Finally, if single-molecule resolution is required to visualize RNAs dynamics and localization there are currently only a few fluorogenic aptamers that have demonstrated single-molecule detection [38, 40]. Our recent work with efficiently folding Mango II aptamer arrays and their high-affinity for TO1-Biotin has enabled single-molecule imaging of RNAs. Direct comparison with the MS2SL-MCP imaging systems (Fig. 1) show improvements in background fluorescence, most notably in the nucleus. Unlike other fluorogenic aptamer arrays, the stability of the Mango II aptamer array provides the ability to visualize single RNA molecules. Here we will describe a general method to image single molecules of RNA using fluorogenic Mango II arrays in both fixed and live cells.

2 Materials

2.1 Cell Lines and Media

1. HEK293T.
2. Cos-7.
3. Dulbecco Modified Eagle's Medium (DMEM), high glucose, L-glutamine, sodium pyruvate.
4. Fetal bovine serum (FBS).
5. Penicillin–streptomycin 10,000 U/ml.
6. Poly-D-lysine.
7. 1× PBS (-$MgCl_2$, -$CaCl_2$).
8. Trypsin–EDTA 0.05% phenol red.
9. EasYFlask™ T25 tissue culture treated flasks.
10. Doxycycline 1 mg/ml in sterile H_2O.
11. 8-well glass bottom slides.
12. 35 mm glass bottom dish.

2.2 Transfection and Fixation Reagents

1. FuGENE® HD Transfection reagent (Promega).
2. Opti-MEM reduced serum media (Gibco).
3. Plasmids (Addgene)—https://www.addgene.org/David_Rueda/

Controls:

(a) pSLQ-mgU2-47 (#127585) and pSLQ-mgU2-47 Mango II (#127586) [23].

Single-molecule mRNA imaging (*see* **Notes 1–2**)

(b) pLenti-mCherry-Mango II ×24 (#127587).

(c) pLenti-Halo-β-actin-Mango II ×24.

4. PKM buffer (10 mM sodium phosphate pH 7.2, 140 mM KCl, 1 mM $MgCl_2$).
5. Pierce™ 16% methanol-free paraformaldehyde (Thermo Fisher).
6. 10% Triton™ X-100 (Thermo Fisher).

2.3 Fluorescent Dyes

1. TO1-3PEG-Biotin.
2. Hoechst 33258.
3. Alexa Fluor™ 647 Phalloidin.
4. HaloTag® TMR ligand (Promega).
5. Vectashield Antifade mounting medium + DAPI (H1200).

2.4 Microscopy

1. Zeiss Elyra widefield/structural illumination S1 microscope and software.
2. Laser lines 405 nm (blue), 488 nm (green), 561 nm (red), and 642 nm (far-red).
3. 63× oil immersion objective Zeiss Plan-Apochromat NA = 1.4.
4. Band-pass filters used are 420–480 nm (blue), 495–550 nm (green), 570–640 nm (red), and a long-pass filter >650 nm (far-red).
5. Stage top incubator and CO_2 controller.

3 Methods

3.1 Cell Culture

1. Culture cells in tissue culture treated T-25 flasks, seeding at ~0.7×10^6 cells per flask.
2. Split cells every 2–3 days or when ~80% confluent.
3. To spilt cells, rinse once with 5 ml PBS.
4. Trypsinize cells with 1 ml Trypsin–EDTA.
5. Resuspend cells in 10 ml DMEM supplemented with 10% FBS and 100 U/ml Pen/Strep.
6. Count cells using a hemocytometer or cell counting machine and use appropriate dilutions for passaging cells and seeding into Ibidi imaging slides/dishes.

7. Poly-D-lysine–coat the 8-well glass bottom chamber slides.
8. Seed 15–20,000 cells per well (vol. 300 μl) and leave to adhere overnight prior to transfection.

3.2 Transfecting Cells

1. Equilibrate FuGENE® HD reagent and Opti-MEM to room temperature prior to transfection.
2. Briefly vortex the FuGENE® HD reagent before use.
3. Transfect 40–100 ng of plasmid into each well of an ibidi 8-well chamber slide.
4. Use a 3:1 FuGENE® HD:DNA ratio (μl:μg).
5. Add the sterile plasmid DNA to an appropriate volume of Opti-MEM (~40 μl per well—8-well chamber slide).
6. Add the FuGENE® reagent, mix thoroughly and incubate for 10 min.
7. Add the FuGENE®:DNA mix to the chamber slide containing fresh media. (Note: there is no need to remove serum or antibiotics when using FuGENE® reagents.)
8. Leave to express for 12–48 h depending on experiments.
9. Treat cells with 1 μg/ml doxycycline for 1–4 h if required for RNA expression prior to fixation or live-cell imaging.

3.3 Fixation and Immunostaining

1. Fix with 4% Paraformaldehyde diluted in PKM buffer at room temperature for 10 min. (*Note: Potassium is essential for the Mango G-Quadruplex formation; fluorescence is also quenched in high* Na^+ *conditions*).
2. Wash off fixative with 3× washes in PKM buffer which is essential to maintain Mango fluorescence. (*Note: Do not use amine-containing groups to quench formaldehyde,* e.g., *Tris or Glycine, as they can dramatically reduce Mango fluorescence.*)
3. Permeabilize the cells in 0.2% Triton X-100 at room temperature for 10 min.
4. Immunostain cells directly in the chamber slides. (*if required*).
   (a) Block cells in 2% BSA diluted in PKM buffer (wt/vol) for 30 min at room temperature.
   (b) Dilute primary antibody at the required concentration in 2% BSA and incubate for 2 h at room temperature or overnight at 4 °C.
   (c) Wash 3× 5 min with PKM buffer.
   (d) Incubate secondary antibody in 2% BSA for 1 h at room temperature.
5. Wash 3× 5 min with PKM buffer.
6. Incubate with 200 nM TO1-Biotin diluted into PKM buffer for 10 min at RT.

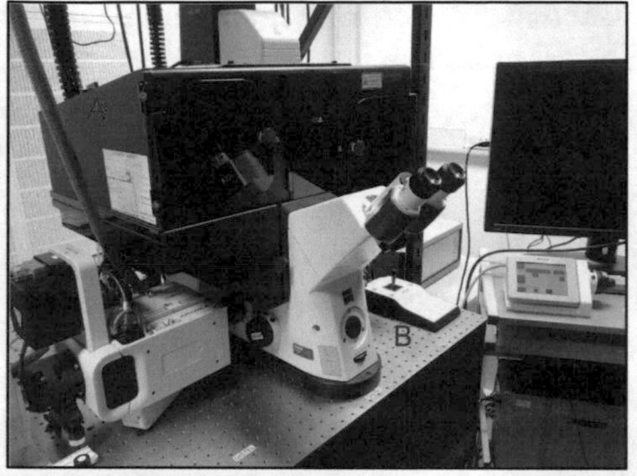

Zeiss Elyra S1	Specification	Legend
Main Objective	63x/1.4NA oil Plan-Apochromat	E
Laser lines	405nm (diode laser – 50 mW) 488nm (OPSL laser – 100 mW) 561nm (OPSL laser – 100 mW) 642 nm (diode laser – 150 mW)	N/A
Camera	PCO.edge 4.2 sCMOS	N/A
Incubation	Microscope incubation chamber Tokai Hit Temp/CO_2 controller Tokai Hit stage heater/water reservoir CO_2 input Heated stage lid	A C F G H
Stage	Motorized Piezo XY scanning stage and controller	D, B

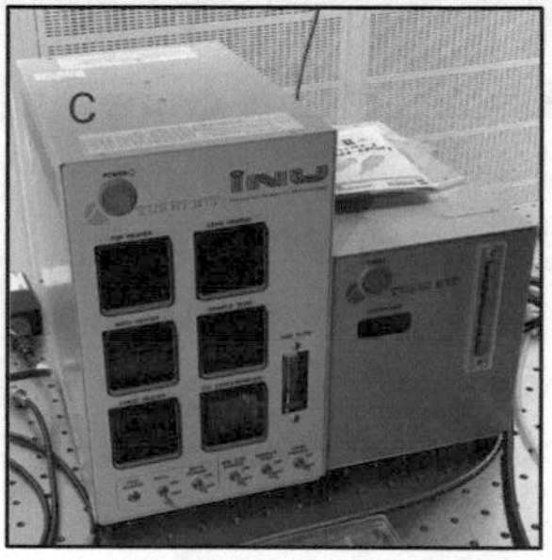

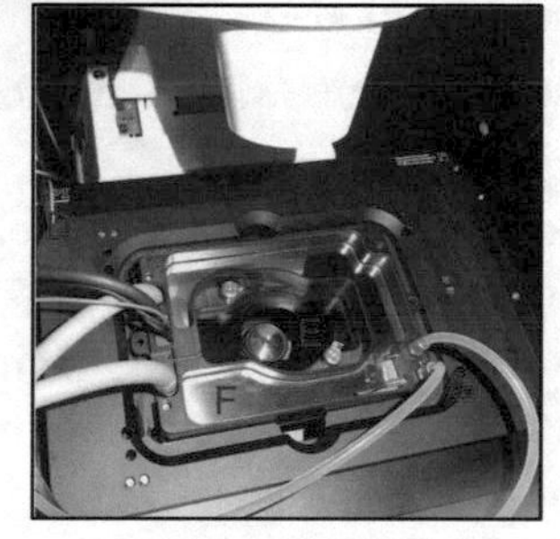

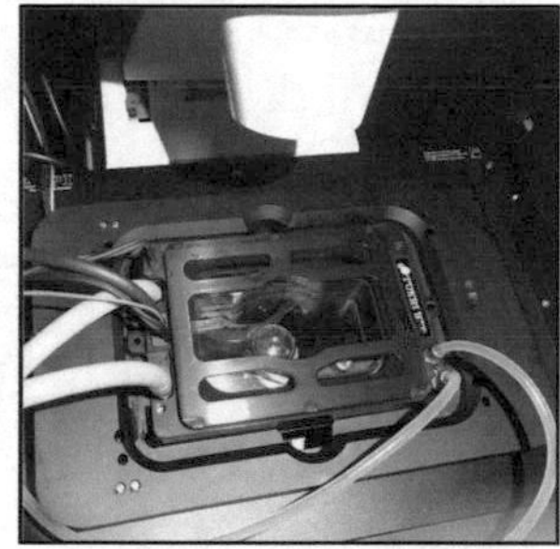

Fig. 2 Zeiss Elyra S1 widefield/structured illumination microscope: Microscope specifications and red highlighted annotations are described in the table above

7. Wash once with PKM buffer and image in sodium phosphate buffer +1 μg/ml Hoechst 33258.
8. If mounting medium is required, always use liquid mediums (e.g., Vectashield H1200). Hard-setting mediums reduce TO1-B fluorophore exchange.

3.4 Microscopy

3.4.1 Microscopy for Fixed Samples

1. Start up Zeiss Elyra microscope, lasers, and Zen software (Fig. 2).
2. Carefully add a small amount of 518F immersion oil (NA 1.518) to the 63× objective (Fig. 2e).
3. Place 8-well chamber slide onto slide holder.
4. Position XY stage on cells of interest by imaging in either the blue or red channels.
5. For widefield image acquisition of green (TO1-Biotin) and red (mCherry) channels, use 5% of total laser power (5 mW) and 200 ms exposure time.
6. Mango-TO1-Biotin signal is stabilized under pulsed illumination [38] (*see* **Notes 3–4**).
7. Therefore Z-stacks and time series experiments containing more than one colour should be acquired by alternating between each fluorescent channel for an individual frame or

Z-slice. This leads to sustained recovery and minimal loss of the Mango signal throughout acquisition.

8. For photobleaching-assisted quantification experiments use 10 mW laser power and 50 ms exposure time under constant illumination.
9. SIM images can also be easily acquired using a Zeiss Elyra microscope and the fluorescent recovery of Mango-TO1-Biotin lends itself well for superresolved imaging of RNA.
10. Sequentially alternate the excitation as before whilst retaining the same settings with respect to phase, rotation and use of a multicube filter-set.
11. Acquire images with 5 phases, 3 rotations, and a grating of 32 μm.
12. Process images using either Zen's automatic reconstruction settings or other settings that the user defines as appropriate.
13. Use a TetraSpek™ bead sample to create a bead map from which chromatic aberrations are corrected following manufacturer instructions.
14. Use the SIMcheck ImageJ plugin [44] to determine the quality of the reconstructed images, chromatic aberration, drift, and, most notably, the rates of bleaching throughout acquisition.

3.4.2 Microscopy for Live Samples

1. Start up Zeiss Elyra microscope, lasers, Tokai hit incubation chamber/CO_2, and Zen software (Fig. 2c).
2. Fill Tokai Hit heated incubation chamber water reservoir with sterile water to just below CO_2 input tubing (Fig. 2f, g).
3. Carefully add a small amount of 518 F immersion oil (NA 1.518) to the 63× objective (Fig. 2e).
4. Allow stage and objective to settle for at least 30–60 min (Fig. 2d–h).
5. Gently place the 35 mm dish into the incubation chamber.
6. Position XY stage on cells of interest by imaging in either the blue or red channels.
7. For widefield snapshots, acquire green (TO1-Biotin) and red (mCherry) channels as before, use 5% of total laser power (5 mW) and 200 ms exposure time.
8. For fast acquisition live-cell experiments use 10 mW laser power and 50 ms exposure time.
9. Alternate between 488 nm and 561 nm lasers with the use of the Zeiss multichannel filter cube (Filter Set 76 HE) to allow for fluorescence recovery of the Mango-TO1-Biotin signal.

3.5 Image Analysis

3.5.1 Image Processing

1. Process fixed cell images acquired as z-stacks using a 3D Laplacian of Gaussian (LoG) filter within the FISH-quant Matlab software [45].
2. 3D LoG settings can be optimized for the observed signal. For our data sets a foci size of 5 and a standard deviation of 2 were routinely used.
3. Foci detection was further limited to foci at least 1.5-fold brighter than background signal and ≤500 nm in diameter via thresholding the detected foci.
4. FISH-quant can be used to acquire foci intensity values for large numbers of single molecules over multiple fields of view using FISH-quant batch processing.

3.5.2 Polarization Index Calculation

The polarization index (PI) of a given RNA molecule can be determined to assess the directed accumulation of RNA in a specific cellular location; for example, RNA at the periphery of the cell gives a PI ~1, whereas RNA at the center has a PI ~0. In our recent publication we used this metric to quantify the difference between β-actin mRNA with and without its zip-code localization sequence. Here we describe a method to quickly calculate the PI of RNA foci in an unbiased manner. Further information on this method can be found in Park et al. and Cawte et al. [14, 38].

1. Download and install ImageJ plugin FociPicker3D for foci detection in both 2D and 3D (https://imagej.nih.gov/ij/plugins/foci-picker3d/index.html).
2. Use FociPicker3D to detect RNA foci and export their *x*, *y*, and *z* coordinates (Fig. 3).
3. For further tips on using FociPicker3D, follow the advice in the link above.
4. Create masks of each cell using a cell marker stain (e.g., Halo-β-actin or Alexa Fluor™ 647-Phalloidin) (Fig. 3).
5. Import cell masks and RNA localization coordinates into Matlab as individual matrices (Fig. 3).
6. Individual polarization indices (PI) can be simply calculated in Matlab using the script below to compute the following equation (Eq. 1).
7. Copy the text below into the Matlab "Editor" window, compile the function and simply enter *PIcalculate(Mask, RNA)* where Mask and RNA are the variables.
8. Our PI calculation is similar to that described in Park et al. [14]. However, for our Cos-7 cell lines, we found calculating the PI of each RNA localization individually then taking an average for each cell far more representative of the PI in this cell type (*see* **Notes 5–8** and Fig. 4).

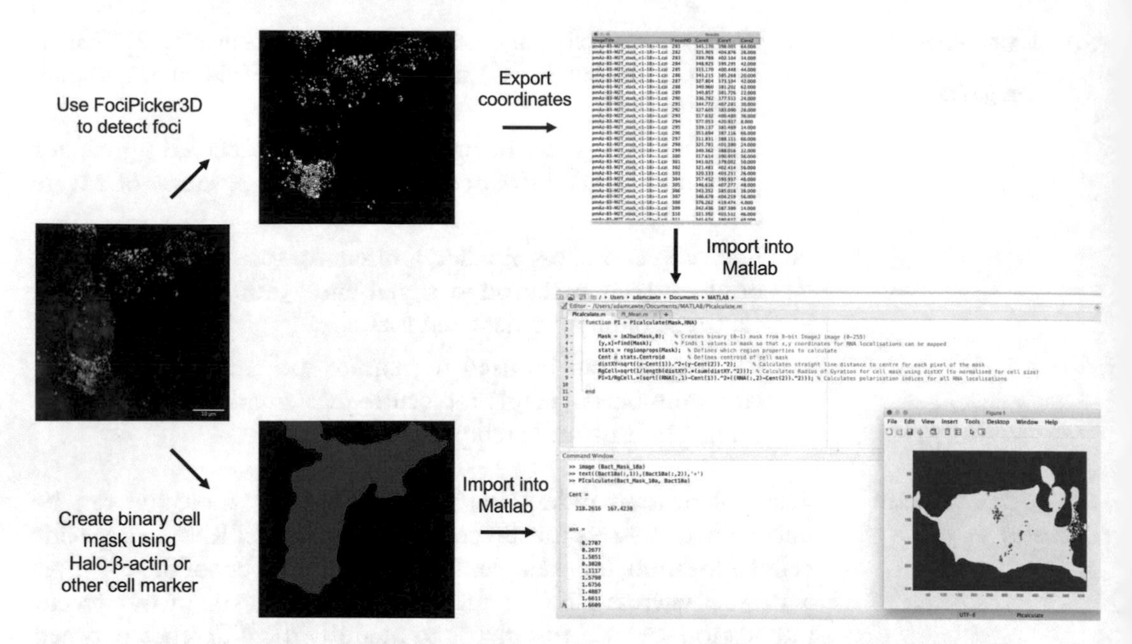

Fig. 3 Diagram of polarization index (PI) calculation using a combined ImageJ and Matlab workflow: Representative image shows Mango II labeled β-actin mRNA (green) in Cos-7 fibroblasts. Cell mask acquired using Alexa Fluor™ 647 Phalloidin staining (red). Scale bar = 10 μm. Following foci detection (white) using FociPicker3D localization coordinates and cell masks can be imported into Matlab for PI calculation

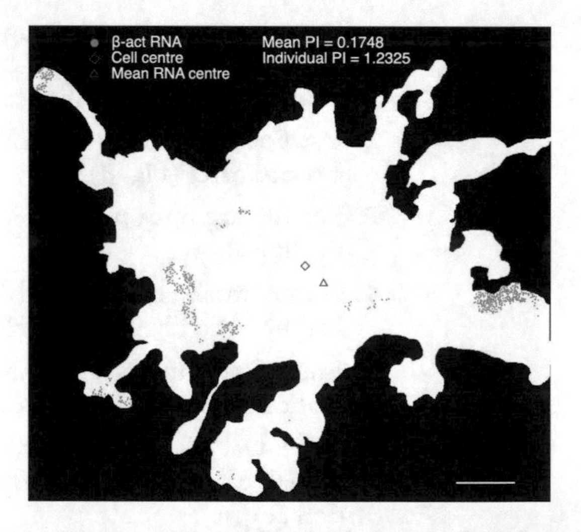

Fig. 4 Example of dampened PI quantification in Cos-7 cells: Cell mask shown in white with individual β-actin RNA localizations shown in green. Centroid of the cell mask and mean centroid of RNA shown as a red diamond and blue triangle respectively. Calculating the PI using the mean RNA centroid (Mean PI), the PI has a low value of 0.17. However, calculation using the average of individually calculated PIs (Individual PI) has a PI of 1.2. This is due to the more dispersed localizations of β-actin RNA in this cell type compared to primary fibroblasts

```
function PI = PIcalculate(Mask,RNA)

Mask = im2bw(Mask,0);    % Creates binary (0-1) mask from 8-bit
image (0-255)
[y,x]=find(Mask);         % Finds 1 values in mask so that x,y
coordinates for RNA localizations can be mapped within
stats = regionprops(Mask);  % Defines which region properties
to calculate
Cent = stats.Centroid        % Defines centroid of cell mask
distXY=sqrt((x-Cent(1)).^2+(y-Cent(2)).^2);  % Calculates
straight line distance to centre for each pixel of the mask
RgCell=sqrt(1/length(distXY).*(sum(distXY.^2))); % Calculates
Radius of Gyration for cell mask using distXY (normalization
for cell size)
PI=1/RgCell.*(sqrt((RNA(:,1)-Cent(1)).^2+((RNA(:,2)-Cent(2)).
^2))); % Calculates polarisation indices for all RNA localiza-
tions

end
```

$$PI = \frac{\sqrt{(x_{RNA} - x_{cell})^2 + (y_{RNA} - y_{cell})^2 + (z_{RNA} - z_{cell})^2}}{Rg_{cell}} \quad (1)$$

4 Notes

1. When transforming the pLenti-Mango II plasmids into bacteria, make sure a recombination stable strain is used. In the lab, we have seen best results using NEB Stable *E. coli* (Cat No. C3040H).
2. If amplifying the Mango II array by PCR, be sure to use KOD Hot start (71086) or Bioline Velocity (BIO-21098) polymerases as they are able to read through GC-rich region efficiently.
3. Where possible, always alternate or pulse Mango excitation to enable recovery of the Mango:TO1-Biotin signal.
4. This is particularly important when acquiring SIM images as fluorophore stability is imperative for efficient reconstruction.
5. For PI calculations, in Park et al. [14] the PI was calculated from the mean centroids for both RNA localizations and cell area. Due to our use of SV40 immortalized fibroblast cell lines such as Cos-7, the fibroblast motility is more sporadic and so too is the distribution of β-actin mRNA accumulation. This

means an average calculation of RNA distribution can lead to dampening of the PI (Fig. 4).

6. Therefore, modifications were made to this process using a custom Matlab image processing function. Individual polarization indices can be computed for each observed RNA localization and its relative distance from the centroid of the cell.
7. The mean of all RNA localizations in a cell can be calculated using the following command:

```
mean(PIcalculate(Mask, RNA))
```

8. To check the cell masks and RNA localizations in Matlab directly use the following commands.
   (a) `image(Mask)`
   (b) `text((RNA(:,1)),(RNA(:,2)),+)`

Acknowledgments

We would like to acknowledge members of the Rueda and Unrau labs for useful discussions and critically reading the manuscript. The Rueda lab is funded by a core grant of the MRC-London Institute of Medical Sciences (UKRI MC-A658-5TY10), a Wellcome Trust collaborative grant (P67153). The Unrau lab is funded by a National Sciences and Engineering Research Council of Canada Operating research grant.

Declaration of Interests: PJU is the founder of Aptamer Innovations Ltd., Chapel Hill, North Carolina, USA. ADC, HI, and DSR declare no competing interests.

References

1. Tutucci E, Livingston NM, Singer RH et al (2018) Imaging mRNA in vivo, from birth to death. Annu Rev Biophys 4:1–22
2. Bertrand E, Chartrand P, Schaefer M et al (1998) Localization of ASH1 mRNA particles in living yeast. Mol Cell 2:437–445
3. Tutucci E, Vera M, Biswas J et al (2018) An improved MS2 system for accurate reporting of the mRNA life cycle. Nat Methods 1:81–89
4. Wu B, Miskolci V, Sato H et al (2015) Synonymous modification results in high-fidelity gene expression of repetitive protein and nucleotide sequences. Genes Dev 8:876–886
5. Darzacq X, Shav-Tal Y, de Turris V et al (2007) In vivo dynamics of RNA polymerase II transcription. Nat Struct Mol Biol 9:796–806
6. Larson DR, Zenklusen D, Wu B et al (2011) Real-time observation of transcription initiation and elongation on an endogenous yeast gene. Science 6028:475–478
7. Martin RM, Rino J, Carvalho C et al (2013) Live-cell visualization of pre-mRNA splicing with single-molecule sensitivity. Cell Rep 6:1144–1155
8. Coulon A, Ferguson ML, de Turris V et al (2014) Kinetic competition during the transcription cycle results in stochastic RNA processing. Elife 3:e03939
9. Halstead JM, Lionnet T, Wilbertz JH et al (2015) An RNA biosensor for imaging the first round of translation. Science 347:1367–1371

10. Morisaki T, Lyon K, DeLuca KF et al (2016) Real-time quantification of single RNA translation dynamics in living cells. Science 6292:1425–1429
11. Wang C, Han B, Zhou R et al (2016) Real-time imaging of translation on single mRNA transcripts in live cells. Cell 4:990–1001
12. Wu B, Eliscovich C, Yoon YJ et al (2016) Translation dynamics of single mRNAs in live cells and neurons. Science 6292:1430–1435
13. Farina KL, Huttelmaier S, Musunuru K et al (2003) Two ZBP1 KH domains facilitate beta-actin mRNA localization, granule formation, and cytoskeletal attachment. J Cell Biol 1:77–87
14. Park HY, Trcek T, Wells AL et al (2012) An unbiased analysis method to quantify mRNA localization reveals its correlation with cell motility. Cell Rep 2:179–184
15. Yoon YJ, Wu B, Buxbaum AR et al (2016) Glutamate-induced RNA localization and translation in neurons. Proc Natl Acad Sci U S A 44:E6877–E6886
16. Katz ZB, Wells AL, Park HY et al (2012) Beta-actin mRNA compartmentalization enhances focal adhesion stability and directs cell migration. Genes Dev 17:1885–1890
17. Dolgosheina EV, Unrau PJ (2016) Fluorophore-binding RNA aptamers and their applications. Wiley Interdiscip Rev RNA 6:843–851
18. Garcia JF, Parker R (2015) MS2 coat proteins bound to yeast mRNAs block 5′ to 3′ degradation and trap mRNA decay products: implications for the localization of mRNAs by MS2-MCP system. RNA 8:1393–1395
19. Garcia JF, Parker R (2016) Ubiquitous accumulation of 3' mRNA decay fragments in *Saccharomyces cerevisiae* mRNAs with chromosomally integrated MS2 arrays. RNA 5:657–659
20. Wu B, Chen J, Singer RH (2014) Background free imaging of single mRNAs in live cells using split fluorescent proteins. Sci Rep 4:3615
21. Wu J, Zaccara S, Khuperkar D et al (2019) Live imaging of mRNA using RNA-stabilized fluorogenic proteins. Nat Methods 9:862–865
22. Braselmann E, Wierzba AJ, Polaski JT et al (2018) A multicolor riboswitch-based platform for imaging of RNA in live mammalian cells. Nat Chem Biol 10:964–971
23. Autour A, Jeng SC, Cawte AD et al (2018) Fluorogenic RNA mango aptamers for imaging small non-coding RNAs in mammalian cells. Nat Commun 1:656
24. Dolgosheina EV, Jeng SC, Panchapakesan SS et al (2014) RNA mango aptamer-fluorophore: a bright, high-affinity complex for RNA labeling and tracking. ACS Chem Biol 10:2412–2420
25. Filonov GS, Moon JD, Svensen N et al (2014) Broccoli: rapid selection of an RNA mimic of green fluorescent protein by fluorescence-based selection and directed evolution. J Am Chem Soc 46:16299–16308
26. Paige JS, Wu KY, Jaffrey SR (2011) RNA mimics of green fluorescent protein. Science 6042:642–646
27. Song W, Filonov GS, Kim H et al (2017) Imaging RNA polymerase III transcription using a photostable RNA-fluorophore complex. Nat Chem Biol 11:1187–1194
28. Sunbul M, Jaschke A (2018) SRB-2: a promiscuous rainbow aptamer for live-cell RNA imaging. Nucleic Acids Res 46(18):e110
29. Chen X, Zhang D, Su N et al (2019) Visualizing RNA dynamics in live cells with bright and stable fluorescent RNAs. Nat Biotechnol 37 (11):1287–1293
30. Autour A, Westhof E, Ryckelynck M (2016) iSpinach: a fluorogenic RNA aptamer optimized for in vitro applications. Nucleic Acids Res 6:2491–2500
31. Bouhedda F, Fam KT, Collot M et al (2020) A dimerization-based fluorogenic dye-aptamer module for RNA imaging in live cells. Nat Chem Biol 1:69–76
32. Jepsen MDE, Sparvath SM, Nielsen TB et al (2018) Development of a genetically encodable FRET system using fluorescent RNA aptamers. Nat Commun 1:18
33. Sunbul M, Arora A, Jaschke A (2018) Visualizing RNA in live bacterial cells using fluorophore- and quencher-binding aptamers. Methods Mol Biol 1649:289–304
34. Wirth R, Gao P, Nienhaus GU et al (2019) SiRA: a silicon rhodamine-binding aptamer for live-cell super-resolution RNA imaging. J Am Chem Soc 18:7562–7571
35. Yerramilli VS, Kim KH (2018) Labeling RNAs in live cells using malachite green aptamer scaffolds as fluorescent probes. ACS Synth Biol 3:758–766
36. Zhang J, Fei J, Leslie BJ et al (2015) Tandem spinach Array for mRNA imaging in living bacterial cells. Sci Rep 5:17295
37. Guo JU, Bartel DP (2016) RNA G-quadruplexes are globally unfolded in eukaryotic cells and depleted in bacteria. Science 353(6306):aaf5371
38. Cawte AD, Unrau PJ, Rueda DS (2020) Live cell imaging of single RNA molecules with fluorogenic mango II arrays. Nat Commun 1:1283

39. Filonov GS, Kam CW, Song W et al (2015) In-gel imaging of RNA processing using broccoli reveals optimal aptamer expression strategies. Chem Biol 5:649–660
40. Braselmann, E, Stasevich, TJ, Lyon, K, et al. (2019). Detection and quantification of single mRNA dynamics with the Riboglow fluorescent RNA tag bioRxiv: 701649
41. Trachman RJ 3rd, Cojocaru R, Wu D et al (2020) Structure-guided engineering of the Homodimeric mango-IV fluorescence turn-on aptamer yields an RNA FRET pair. Structure 7(776–785):e773
42. Sunbul M, Lackner J, Martin A et al (2021) Super-resolution RNA imaging using a rhodamine-binding aptamer with fast exchange kinetics. Nat Biotechnol 39 (6):686–690
43. Han KY, Leslie BJ, Fei J et al (2013) Understanding the photophysics of the spinach-DFHBI RNA aptamer-fluorogen complex to improve live-cell RNA imaging. J Am Chem Soc 50:19033–19038
44. Ball G, Demmerle J, Kaufmann R et al (2015) SIMcheck: a toolbox for successful super-resolution structured illumination microscopy. Sci Rep 5:15915
45. Mueller F, Senecal A, Tantale K et al (2013) FISH-quant: automatic counting of transcripts in 3D FISH images. Nat Methods 4:277–278

Chapter 15

Genome-Wide Identification of Polyadenylation Dynamics with TED-Seq

Yeonui Kwak and Hojoong Kwak

Abstract

Polyadenylation and deadenylation of mRNA are major RNA modifications associated with nucleus-to-cytoplasm translocation, mRNA stability, translation efficiency, and mRNA decay pathways. Our current knowledge of polyadenylation and deadenylation has been expanded due to recent advances in transcriptome-wide poly(A) tail length assays. Whereas these methods measure poly(A) length by quantifying the adenine (A) base stretch at the 3′ end of mRNA, we developed a more cost-efficient technique that does not rely on A-base counting, called tail-end-displacement sequencing (TED-seq). Through sequencing highly size-selected 3′ RNA fragments including the poly(A) tail pieces, TED-seq provides accurate measure of transcriptome-wide poly(A)-tail lengths in high resolution, economically suitable for larger scale analysis under various biologically transitional contexts.

Key words Poly(A) tail, TED-seq, RNA-seq, mRNA modification, Posttranscriptional regulation

1 Introduction

Poly(A) tail is one of the longest known classical mRNA modifications with multiple molecular functions. It is the binding site for poly(A) binding proteins (PABP) that protects the 3′ ends of mRNA from exonuclease mediated decay, and serves as a recruiter of translation initiation factors [1, 2]. Deadenylation—removal of poly(A) tail—is a critical process in mRNA decay mediated by CCR4-NOT deadenylases with 3′–5′ exonuclease activity, resulting in the shortening of poly(A) tails [3]. Polyadenylation, on the other hand, occurs co-transcriptionally during nascent RNA synthesis, coupled to 3′ cleavage polyadenylation site (CPS) formation [4]. Cytoplasmic polyadenylation can also occur later in the mRNA life span under specific biologically transitional contexts, most well known in early embryogenesis [5]. Therefore, monitoring the poly(A) tail lengths is increasingly becoming more important as one of the critical measures of posttranscriptional processes.

Erik Dassi (ed.), *Post-Transcriptional Gene Regulation*, Methods in Molecular Biology, vol. 2404,
https://doi.org/10.1007/978-1-0716-1851-6_15,

Traditionally, poly(A) tails have been measured through nuclease protection assays [6]. mRNA poly(A) tail hybridized to oligo-dT DNA probes is susceptible to RNase H degradation, and the comparison between the poly(A) intact and poly(A) degraded RNA will yield the poly(A) tail length. More recent procedures include 3′ ligation-mediated reverse transcription, and use of polymerase chain reaction (lmRT-PCR) to amplify the 3′ fragments of the mRNA including the full poly(A) tail of the specific transcript of interest [6]. However, these methods have limited throughput and require prior knowledge of the gene of interest.

Through the advances of the next-generation sequencing technologies, new transcriptome-wide poly(A) tail length methods have been developed. Measuring the poly(A) tail length through RNA sequencing is conceptually simple: count the number of A-bases. However, there are complications in accurately counting A-base homopolymer stretches using the currently prevailing next-generation sequencing technology that depends on fluorescent-base incorporation cycles. Incomplete quenching of the fluorescence signal from the previous base read cycle can result in a contaminated signal in the following read cycle, and this can become more problematic in reading through the repeat of the same bases. This sequencing ambiguity has been circumvented by methods such as PAL-seq or TAIL-seq [7, 8], where the poly (A) tails are either indirectly quantified using an additional fluorescent signal reporting A-base homopolymer abundance, or incorporating a custom fluorescence signal analysis of the ambiguous read cycles. However, these methods are device dependent, requiring modifications to the next-generation sequencing devices or analysis software. Direct long read sequencing methods using nanopores, such as FLAM-seq [9], can also be used to count poly (A) lengths, but currently has lower throughput than the previous methods, which makes larger scale applications cost-limited.

TED-seq is based on simple arithmetics that the length of an RNA fragment, including poly(A) tail and its flanking region, is equal to the addition of the poly(A) tail length and the distance from fragment-start to 3′ CPS (Fig. 1) [10]. Once the cDNA fragments are precisely size-selected, only sequencing and mapping the start of the cDNA fragment is necessary to calculate poly(A) tail length without the need to sequence through the poly(A) tail. The precise size-selection of the cDNA is performed by high-resolution polyacrylamide gel electrophoresis (PAGE). Knowing the selected size, and distance from the mapped 5′ end of the fragment to the 3′ CPS, poly(A) tail length is easily calculated for each mapped sequence read which represents each mRNA molecule. To map the 5′ end of the fragment to the genome, 30-40 base reads are sufficient rather than sequencing up to 250 bases of A-bases, which greatly enhance the cost efficiency of TED-seq.

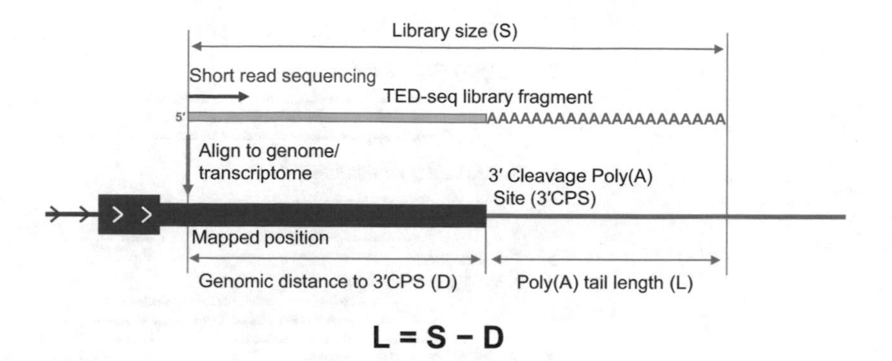

Fig. 1 Schematics of poly(A) tail length calculation in TED-seq. TED-seq library fragment encompassing poly (A) tail is aligned at the 3′ end of a gene. Typically, sequencing the first 40 bases is sufficient for the alignment (red arrow). Dark blue bar indicates the gene annotation: arrowheads pointing to the sense direction of transcription, thicker body reflect the coding sequence, thinner body reflect untranslated region (UTR), and line overlaid by arrowhead indicate spliced intron. From this diagram of poly(A) length (L), library size (S), and the distance from 3′ cleavage polyadenylation site (CPS) to TED-seq read (D), $L = S - D$

Outline of the major procedure: poly(A) RNAs are purified using Oligo d(T)-linked beads; purified poly(A) RNAs are ligated to 3′ adaptor; RNA fragmentation by base hydrolysis; the resultant is subjected to 5′ end repair and 5′ adaptor ligation; reverse transcription and PCR amplification selectively amplify cDNA fragments including poly(A) tail; high resolution native PAGE purification of the poly(A)-cDNA library at a specific length (300 base pair); next generation sequencing of the library from the 5′ end (Fig. 2). The sequence reads can be mapped to the reference genome or reference transcriptome sequences, and the distance from the mapped sites to the downstream 3′ CPS is subtracted from 300 base pairs to yield the poly(A) tail length.

TED-seq can perform as an RNA quantification method as well as the marker of posttranscriptional RNA regulation. Conceptually, it is similar to 3′ sequencing methods (3′-seq) [11] in that the reads are derived from regions near 3′ CPS, which is known to perform robustly for RNA quantification. Quantification pipelines established in 3′-seq can be adopted for TED-seq with minimal modification. Also, TED-seq can be versatile for conjugating with other modified RNA detection methods, such as after RNA immunoprecipitation [12] or metabolic RNA labeling [13, 14]. TED-seq is compatible with standard, unmodified next-generation sequencing platforms, which makes it easier to merge into existing RNA methods as a library preparation module. TED-seq can be even more powerful when used in combination with other RNA-seq methods, such as nascent RNA sequencing [15] and RNA stability measures [16], which will provide a complete set of mRNA regulation from its synthesis to decay. It will also be compatible with any upgrades in the next-generation sequencing devices, since TED-seq is device-independent.

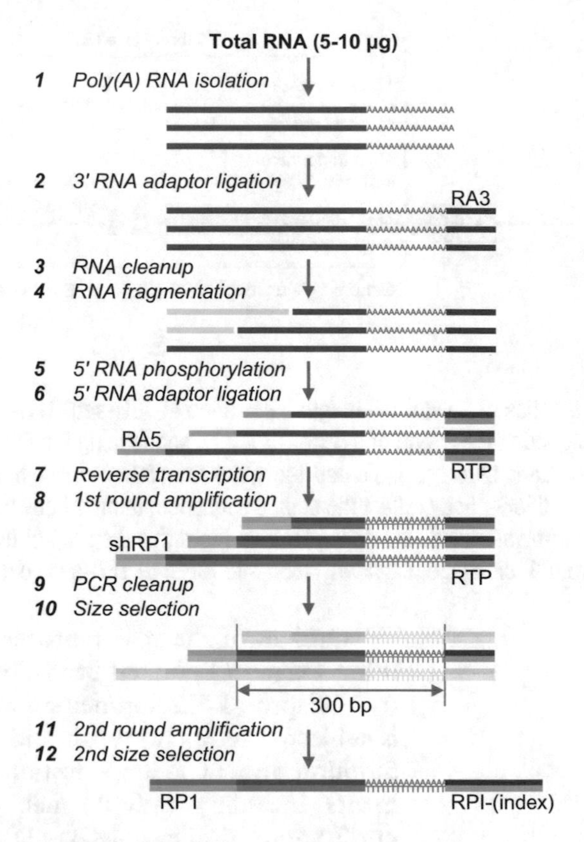

Fig. 2 Overview of TED-seq experimental procedures shows all the experimental steps in Subheading 3. Blue bars with poly(A) tail represent mRNA. 3′ and 5′ RNA adaptors are shown as red (RA3) and green (RA5) bars, respectively. Reverse transcription primer (RTP) is in orange after step 6, annealed to the ligated RA3. RTP is also used as a PCR primer with the short RP1 primer (shRP1; step 8). Full-length RP1 and RPI-index primers are used in the second amplification step (step 11)

TED-seq may have drawbacks in terms of resolution, isoform distinction, and the requirement of input material amount. The resolution of poly(A) tail length measurement is dependent on the precision of the library size selection by PAGE, typically about 20 bases. While it does not reach the single base resolution of poly(A) length, the 20 base resolution may be sufficient since the binding footprint of PABP encompasses about 20 bases, and mechanistically, the deterministic factor of the role of poly(A) length is dependent on the number of bound PABPs [17]. Also, TED-seq depends on preidentified 3′ CPS, and if the 3′ CPS is either ambiguous or multiple 3′ end isoforms are clustered, it will be difficult to assign the distance between the mapped reads and 3′ CPSs. However, the majority of the annotated transcripts have sharply defined 3′ CPSs within 10–20 bases, and the 3′ alternative polyadenylation or alternatively spliced isoforms are usually hundreds of bases apart

[11]. The amount of required input RNA may be in a rather higher range (5–10μg of total RNA or >200 ng of poly(A) RNA). This is due to the size selection step, where only a fraction of the cDNA from the fragmented RNA is recovered. However, adding an additional amplification and cDNA fragmentation step prior to the size selection can resolve the input requirement, which will make it compatible with less amount of input material. Overall, while limitations of TED-seq do exist, they are outweighed by its cost-effectiveness, versatility, and the potential for further improvements. Herein, we describe the experimental details of TED-seq and present a preliminary data processing pipeline.

2 Materials

2.1 Poly(A) RNA Isolation

1. Poly(A) RNA isolation kit (Ambion, Dynabeads mRNA purification kit): oligo-dT Dynabead, Binding buffer, Wash buffer (*see* **Note 1**).
2. Magnetic tube rack.
3. Tube rotator.
4. Heat block set to 65 °C.
5. Fluorometric nucleic acid quantification device and high sensitivity RNA detection reagent.
6. (Optional) poly(A) spike-in RNA (*see* **Note 2**).

2.2 3′ RNA Adaptor Ligation

1. 3′ RNA adaptor (RA3): 10μM of 5′-/phosphate/-rUrGrGrArArUrUrCrUrCrGrGrGrUrGrCrCrArArGrG-/inverted-dT/-3′ (*see* **Note 3**).
2. 10× T4 RNA ligase buffer.
3. 50% PEG-8000 (New England Biolabs).
4. 10 mM ATP.
5. RNase inhibitor (10 units/μl).
6. T4 RNA ligase I.
7. Heat block set to 65 °C.
8. Thermocycler.

2.3 RNA Cleanup

1. TRIzol reagent (Invitrogen).
2. Chloroform.
3. GlycoBlue (Ambion).
4. 100% isopropanol.
5. 75% ethanol.

2.4 RNA Fragmentation

1. 1 N NaOH.
2. 1 M Tris–HCl, pH 6.8.
3. Micro Bio-Spin P-30 Gel Column (Bio-Rad).
4. Poly(A) RNA isolation kit (Ambion, Dynabeads mRNA purification kit).
5. Heat block set to 65 °C.

2.5 5′ RNA Phosphorylation

1. Polynucleotide kinase (PNK).
2. 10× PNK buffer.
3. 10 mM ATP.
4. RNase inhibitor (10 units/μl).

2.6 5′ RNA Adaptor Ligation

1. 5′ RNA adaptor (RA5): 10μM of 5′-rGrUrUrCrArGrArGrUrUrCrUrArCrArGrUrCrCrGrArCrGrAr-UrCrNrNrNrNrNrNrNrN-3′ (*see* **Note 4**).
2. RT primer (RTP): 50μM of 5′-GCCTTGGCACCCGAGAATTCCA-3′.
3. 10× T4 RNA ligase buffer.
4. 50% PEG-8000 (New England Biolabs).
5. 10 mM ATP.
6. RNase inhibitor (10 units/μl).
7. T4 RNA ligase I.
8. Heat block set to 65 °C.
9. Thermocycler.

2.7 Reverse Transcription

1. Superscript II Reverse Transcriptase (Invitrogen).
2. 0.1 M DTT.
3. 5× First Strand buffer (Invitrogen).
4. RNase inhibitor (10 units/μl).
5. 12.5 mM dNTP mix: 12.5 mM dATP, 12.5 mM dCTP, 12.5 mM dGTP, 12.5 mM dTTP.
6. Thermocycler.

2.8 First Round Amplification of the Library

1. Short RP1 primer (shRP1 primer): 10μM of 5′-GTTCAGAGTTCTACAGTCCGA-3′.
2. RTP primer for PCR: 10μM of 5′-GCCTTGGCACCCGAGAATTCCA-3′.
3. High-fidelity hot-start PCR premix (2×).
4. Thermocycler.

2.9 PCR Cleanup Using SPRI Beads

1. Ampure XP beads (Beckman Coulter).
2. 75% ethanol.
3. Magnetic tube rack.

2.10 Size Selection of the Library

1. 5× TBE: 0.45 M tris–borate pH 8.3, 10 mM EDTA.
2. 6% PAGE gel, 16–20 cm of vertical height: 6% acrylamide, 0.5× TBE, 1% APS, 0.1% TEMED.
3. Vertical gel electrophoresis module, 16–20 cm height.
4. Power supply.
5. 6× gel loading dye, orange G.
6. 100 bp DNA ladder.
7. 25 bp DNA ladder.
8. (Optional) 10 bp DNA ladder.
9. Fluorescent DNA gel staining reagent.
10. Blue light gel illuminator.
11. TE-TW buffer: 10 mM tris, pH 8.0, 1 mM EDTA, 0.01% Tween 20.
12. 37 °C incubation chamber with rotator.
13. Spin X column (Sigma-Aldrich).
14. Ampure XP beads (Beckman Coulter).
15. 75% ethanol.
16. Magnetic tube rack.

2.11 Second Round Full Amplification of the Library

1. RP1 primer: 10μM of 5′-AATGATACGGCGACCACC GAGATCTACACGTTCAGAGTTCTACAGTCCGA-3′.
2. RPI-index primer: 10μM of 5′-CAAGCAGAAGACGGCAT ACGAGAT JJJJJJ GTGACTGGAGTTCCTTGGCACCCGA-GAATTCCA-3′ (*see* **Note 5**).
3. High-fidelity hot-start PCR premix (2×).
4. Thermocycler.

2.12 Second Size Selection and PCR Cleanup

1. 5× TBE: 0.45 M tris–borate pH 8.3, 10 mM EDTA.
2. 6% PAGE gel, 16–20 cm of vertical height: 6% acrylamide, 0.5× TBE, 1% APS, 0.1% TEMED.
3. Vertical gel electrophoresis module, 16–20 cm height.
4. Power Supply.
5. 6× gel loading dye, orange G.
6. 100 bp DNA ladder.
7. 25 bp DNA ladder.
8. (Optional) 10 bp DNA ladder.

9. Fluorescent DNA gel staining reagent.
10. Blue light gel illuminator.
11. TE-TW buffer: 10 mM tris, pH 8.0, 1 mM EDTA, 0.01% Tween 20.
12. 37 °C incubation chamber with rotator.
13. Spin X column (Sigma-Aldrich).
14. Ampure XP beads (Beckman Coulter).
15. 75% ethanol.
16. Magnetic tube rack.
17. Fluorometric nucleic acid quantification device and high-sensitivity DNA detection reagent.

2.13 TED-seq Data Analysis

1. UNIX compatible computing system with the following GNU software installed and accessible from the $PATH variable: bash shell, awk, sort, samtools [18], bedtools [19], STAR aligner [20].
2. Reference genome sequence file in a fasta format (./genome/genome.fasta).
3. Gene annotation file of the 3′ cleavage polyadenylation site (CPS) in a bed format (./gene/gene.bed) (*see* **Note 6**).

3 Methods

Prepare all solutions in ultrapure DNase and RNase free water. Prepare and store all reagents on ice unless indicated otherwise. Use 1.5 ml microcentrifuge tubes unless indicated to use 0.2 ml PCR tubes compatible with thermocyclers. Use heat blocks to incubate at temperatures higher than room temperature, except when indicated to use a thermocycler. Use DNase and RNase free plasticwares.

3.1 Poly-A RNA Isolation

1. Adjust RNA volume to 50μl in water (5–10μg of total RNA). Heat RNA at 65 °C for 2 min and place the tube on ice.
2. Prepare Oligo d(T) Dynabeads. Transfer 100μl of the beads from the kit to a microcentrifuge tube and place on a magnetic rack for 30 s. Discard the supernatant and wash the beads once with 50μl Binding Buffer. Pace the tube on the magnetic rack, and remove the supernatant after the supernatant gets clear. Add 50μl Binding Buffer to the beads, and mix beads thoroughly. Scale up accordingly if multiple samples are processed at once.
3. Mix the 50μl beads with 50μl RNA, followed by rotation on a mini rotator for 3 min at room temperature.

4. Wash the beads. Place the tube on the magnet for 30 s, and remove supernatant. Resuspend the beads in 100μl Washing Buffer B, and remove the buffer. Repeat once more for the total of two washes.
5. Add 11.5μl water to the bead for the elution. Mix thoroughly by gently pipetting. Heat the beads at 65 °C for 2 min in a heat block, and immediately place the tube on the magnetic rack. When the beads are clearly separated from the water, collect 10μl of supernatant containing eluted RNA.
6. Quantify poly-A isolated RNA using flourometer.
7. (Optional) Add poly(A) spike-in RNAs to the eluted RNAs (~1 ng spike-in per ~100 ng of poly-A isolated RNA; *see* **Note 2**).

3.2 3′ RNA Adaptor Ligation

1. Add 4μl of 10μM RA3 to 9μl of the poly-A RNA from Subheading 3.1. Heat the RNA mix at 65 °C for 40 s on a heat bock, then cool down on ice immediately.
2. Add the following reagents to the 13μl RNA mix in a PCR tube for the total reaction volume of 30μl: 3μl of 10× T4 RNA ligase buffer, 6μl of 50% PEG-8000, 3μl of 10 mM ATP, 2.5μl of RNase inhibitor, 2.5μl of T4 RNA ligase I (*see* **Note 7**).
3. Incubate at 20 °C for 6 h, followed by an infinite hold at 4 °C in a thermocycler. The reaction can remain at 4 °C up to overnight.

3.3 RNA Cleanup

1. Add 500μl TRIzol to the RNA ligation reaction and mix well (*see* **Note 8**).
2. Add 100μl chloroform and vortex for 30 s, followed by centrifugation at ≥15,000 × *g* for 5 min at 4 °C. Collect aqueous layer in a new microtube (~300μl).
3. Mix with 2μl GlycoBlue, then add 1× volume of 100% isopropanol (~300μl) and vortex. Centrifuge at ≥15,000 × *g* for 20 min at 4 °C, discard the supernatant, and wash pellet by gently pipetting 100μl 75% ethanol. The bluish gray RNA pellet in 75% ethanol can be stored in −80 °C for up to at least 2–3 weeks.
4. Completely remove supernatant and air dry for 5 min. Dissolve the RNA pellet in 20μl water.

3.4 RNA Fragmentation

1. Heat the RNA at 65 °C for 40 s for denaturation, then cool down on ice immediately.
2. Add ice cold 5μl 1 N NaOH to the RNA, and incubate on ice for 10 min. Add 25μl 1 M Tris–HCl, pH 6.8 to stop the base hydrolysis reaction.

3. Prepare a P-30 minicolumn by allowing the column storage buffer to flow by gravity. Centrifuge the column at 1000 × *g* for 2 min at room temperature to remove all buffers. Place the column on a new tube.
4. Transfer 50μl of base hydrolyzed RNAs to the column. Pass the RNA through the P-30 column by centrifuging at 1000 × *g* for 2 min at room temperature.
5. Using the poly(A) RNA isolation kit, repeat Subheading 3.1, and elute in 20μl water (*see* **Note 9**).

3.5 5′ RNA Phosphorylation

1. Add the following reagents to the 20μl of poly(A) RNA from the previous step: 3μl of 10× PNK buffer, 3μl of 10 mM ATP, 2μl of RNase inhibitor, and 2μl PNK enzyme (*see* **Note 10**).
2. Incubate the reaction at 37 °C for 1 h.
3. RNA cleanup by repeating Subheading 3.3 (*see* **Note 11**). Do not dissolve the RNA pellet in water. Proceed with the precipitated RNA pellet.

3.6 5′ RNA Adaptor Ligation

1. Add 3μl of 10μM RA5 adaptor and 1μl of 50μM RT primer to the RNA pellet. Incubate at RT for 1 min, then vortex and spin down to dissolve the RNA with the adaptor and the primer. Heat the sample at 65 °C in a heat block for 20 s, and cool down on ice immediately. Transfer all (4μl) of the mix to a PCR tube.
2. Add the following reagents to the sample (total reaction volume: 10μl): 1μl of 10× RNA ligase buffer, 2μl of 50% PEG-8000, 1μl of 10 mM ATP, 1μl of RNase inhibitor, and 1μl of T4 RNA ligase (*see* **Note 12**).
3. Incubate at 20 °C for 6 h, and hold out at 4 °C on a thermocycler overnight.

3.7 Reverse Transcription (RT)

1. Directly add the following reagents to the 10μl RNA ligation: 2.8μl of 5x FS buffer, 1.4μl of 0.1 M DTT, 0.8μl of 12.5 mM dNTP, 1μl of RNase inhibitor, and 1μl of Superscript II RTase (*see* **Note 13**).
2. Proceed with the RT reaction by incubating at 50 °C for 1 h.

3.8 First Round Amplification of the Library

1. Make a PCR mix by mixing the following reagents: 25μl of 2× PCR premix, 2.5μl of 10μM short RP1 primer, 2.5μl of 10μM RTP primer, and 3μl of water. Add 33μl of the PCR mix to the 17μl RT reaction (*see* **Note 14**).
2. Incubate in the thermocycler at 98 °C 2 min, 8 cycles of 98 °C for 30 s, 64 °C for 30 s, and 72 °C for 30 s, followed by 72 °C extension for 5 min (*see* **Note 15**).

3.9 PCR Cleanup Using SPRI Beads

1. Add 1× volume (50μl) of Ampure XP beads to the PCR reaction thoroughly by pipette mixing. Place the reaction at room temperature for 5 min.
2. Incubate on a magnetic rack for 5 min and discard the cleared solution.
3. Add 200μl of 75% ethanol and incubate for 30 s while the tube is placed on the magnetic rack. Discard the ethanol and wash once more with 75% ethanol.
4. Briefly spin down, place on the magnetic rack, and completely remove any trace of ethanol. Allow the bead to dry for 5 min on the magnetic rack with the cap open.
5. Add 12μl of water to each tube with dried magnetic SPRI beads and incubate at room temperature for 2 min. Place the mix on the magnet for 1 min and collect 10μl of the eluted supernatant.

3.10 Size Selection of the Library

1. Prepare a 6% native TBE polyacrylamide gel for a 16–20 cm length vertical electrophoresis unit (45 ml). Prerun the gel (Protean II xi gel) for 20 min at 40 mA in 0.5× TBE prepared from 5× TBE.
2. Mix the 10μl sample from the previous step with 2μl of 6× gel loading buffer. Load the sample with 25 bp and 100 bp DNA ladders at its both sides, which will be used later for measuring cDNA library size (Fig. 3).
3. Run PAGE for 1 h 30 min (40 mA), or until 10–15 min after the orange dye completely passes through the gel.
4. Disassemble the unit and take out the gel. Stain the gel with SYBR gold reagent for 3–5 min (*see* **Note 16**).
5. Precisely excise a rectangular gel piece between 350 and 360 base pairs according to the DNA ladder (Fig. 3; *see* **Note 17**). Place the excised gel piece into a 0.2 ml DNase-free PCR tube and grind it with a sterile pipette tip (Fig. 4a). Add 100μl TE-TW buffer to the gel and incubate it overnight at 37 °C with rotation.
6. Place the PCR tube with the cap removed, upside down on a Spin X column (Fig. 4b–g). Pass the eluate through a Spin X column using microcentrifuge for 5 min at 15,000 × *g*. Approximately 100μl of eluant will be collected in the microcentrifuge tube.
7. Add 1x volume (100μl) of Ampure XP beads to the pass-through, and repeat steps in Subheading 3.9 for the DNA cleanup. Elute in 17μl of water.

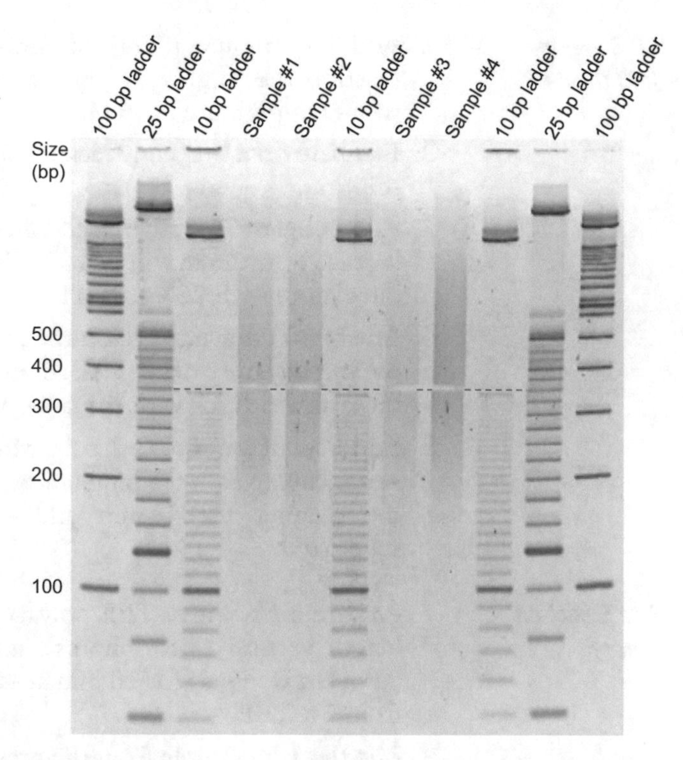

Fig. 3 Electrophoresis for the size selection. Shown is a post-excision polyacrylamide gel. DNA size markers and samples are labeled on the top, and DNA size labeled on the left size. Dashed lines indicate 350 bp. Note the excised region within the smear of the sample (modified from Woo et al. (2018) *Cell Rep*)

3.11 Second Round Full Amplification of the Library

1. Make PCR mix by mixing the following reagents: 25μl of 2× PCR premix, 2.5μl of 10μM RP1 primer, and 3μl of water (*see* **Note 14**).
2. Add 30.5μl PCR mix to the size selected cDNA library (356 bp) from Subheading 3.10.
3. Add 2.5μl of RPI-1 or other RPI-index primer (e.g., RPI-2, RPI-3, RPI-4) to each sample.
4. Perform PCR as follows: at 98 °C 2 min, 5 cycles of 98 °C for 30 s, 64 °C for 30 s, and 72 °C for 30 s, followed by 72 °C extension for 5 min (*see* **Note 15**).

3.12 Second Size Selection and Cleanup

1. Clean up the PCR products using 1.8× volume (90μl) of Ampure XP beads, otherwise proceed as Subheading 3.9; elute in 10μl of water.
2. Prerun the large 6% PAGE gel for 15–20 min.
3. Mix the 10μl sample with 2μl 6× loading dye. Load each sample into each well with the DNA size markers as described

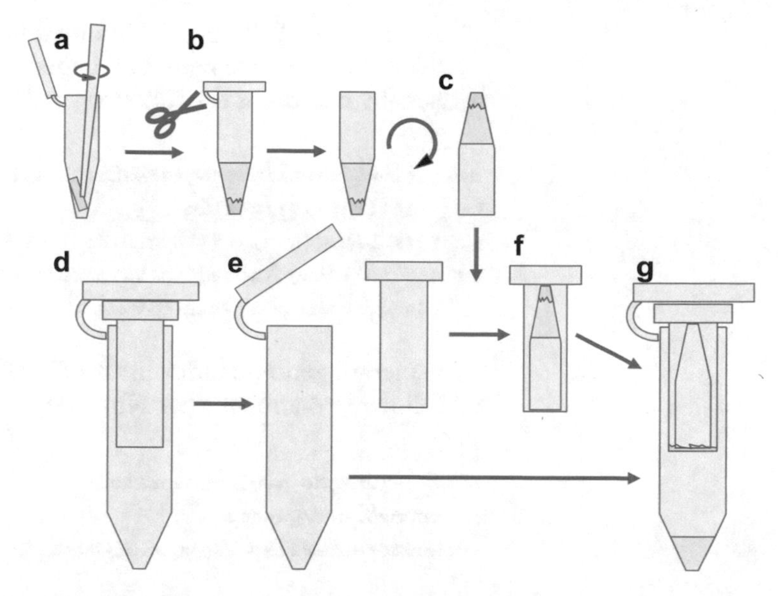

Fig. 4 DNA elution from a polyacrylamide gel. (**a**) Grinding a small gel piece in a 0.2 ml PCR tube (Subheading 3.10, **step 5**). (**b**) Cutting out the cap of the PCR tube in Subheading 3.10, **step 6**. (**c**) Inserting the decapped PCR tube in the microspin filter unit. (**d**) Microspin (Spin-X) filter unit. (**e**) Inner filter unit detached. (**f**) Decapped PCR tube in the filter unit. (**g**) Reassembly before centrifugation

in Subheading 3.10, **step 2**. Electrophorese for 90 min at 40 mA.

4. Stain the gel with 1× SYBR Gold reagent diluted in 1× TBE for 3–5 min.
5. Using sterile forceps and a cutting blade, excise out between 420 and 440 bp (*see* **Note 18**). Place the excised gel piece in a PCR tube and grind it with a pipette tip. Add 50μl of TE-TW buffer to the excised gel followed by incubation overnight at 37 °C on a rotator.
6. Pass the gel mixture through a Spin X column. DNA cleanup using 1× volume (50μl) of Ampure XP beads; repeat steps in Subheading 3.9 and elute in 10μl of water.
7. Quantify the library using a DNA fluorometer, and normalize the library to 2 ng/μl. Multiplex the libraries as needed. Proceed with the single-end Illumina sequencing compatible with TRU-seq small RNA adaptors.

3.13 TED-seq Data Visualization

1. Download the result Illumina sequencing fastq file (TEDseq. fastq) to the working directory (./) of the UNIX compatible system.

2. Extract first 8 bases of Unique Molecular Identifier (UMI) from the sequence reads and append it to the sequence identifiers by executing the following awk command:

```
awk '{id=$1;getline;tag=substr($1,1,8); \
seq=substr($1,9);getline; \
phred=substr($1,9);if(length(seq)>=16) \
printf id":"tag"\n"seq"\n+\n"phred"\n"}' \
./TEDseq.fastq > ./TEDseq.UMItag.fastq
```

3. Generate genome index file for STAR aligner by executing the following command (*see* **Note 19**):

```
STAR --runMode genomeGenerate \
--genomeDir ./genome \
--genomeFastaFiles ./genome/genome.fasta
```

4. Align TED-seq reads to the genome:

```
STAR -- genomeDir ./genome \
--readFilesIn ./TEDseq.UMItag.fastq
--outFilterMultimapMax 1 \
--outFileNamePrefix TEDseq
```

5. Collapse identical UMIs of the alignment file by executing the following lines (note that temporary files are created and deleted):

```
samtools view -S TEDseq.Aligned.out.sam | \
awk '{n=length($1);print substr($1,n-7,8)"\t"$0;}' \
> _sam.tmp
samtools view -SH TEDseq.Aligned.out.sam \
> _umi_unique_sam.tmp
sort -k4,4 -k5,5n -k1,1 -u _sam.tmp | \
cut -f2- >> _umi_unique_sam.tmp
samtools view -Sb _umi_unique_sam.tmp > TEDseq.uniqueUMI.bam
rm _sam.tmp _umi_unique_sam.tmp
```

6. Generate strand specific bedgraph files that can be loaded on genome browser softwares, such as Integrative Genomics Viewer (IGV) [21].

```
bedtools genomecov -ibam TEDseq.uniqueUMI.bam -bg -strand + -5 \
> TEDseq.pl.bedgraph
bedtools genomecov -ibam TEDseq.uniqueUMI.bam -bg -strand - -5 | \
awk '{print $1"\t"$2"\t"$3"\t"$4*-1}' > \
> TEDseq.mn.bedgraph
```

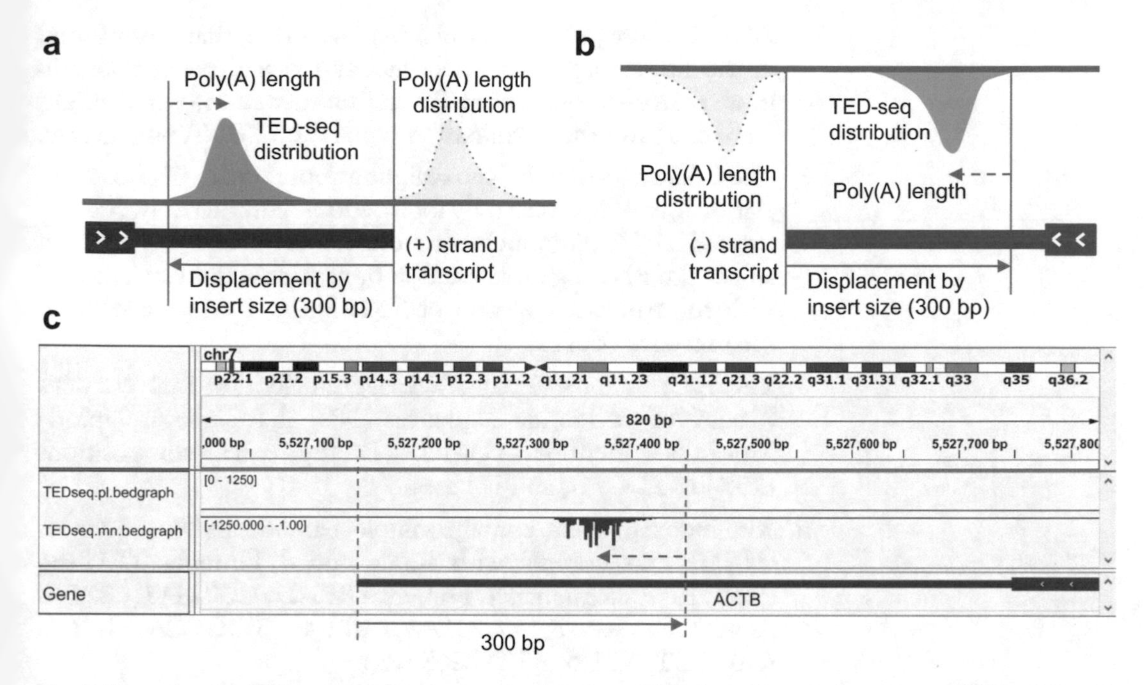

Fig. 5 Example of a TED-seq result in a genome browser. (**a**) Schematics of visualizing poly(A) tail length distribution by TED-seq on the genome browser. Red bar on the bottom indicates the gene annotation of a (+) strand transcript. Dotted distribution reflects the poly(A) tail length distribution of the transcripts. Orange filled distribution reflects the TED-seq distribution which is the poly(A) tail length distribution shifted upstream by the library insert size (300 bp). (**b**) Schematics of TED-seq browser view of a (−) strand transcript. Blue bar indicates the gene annotation, and the light blue filled distribution is the TED-seq distribution, otherwise as described in (**a**). Note that the (−) strand reads are inverted horizontally on the negative side of the y-axis for visualization. (**c**) Example of TED-seq at human *ACTB* gene (data from Woo et al. (2018) *Cell Rep*) on the Integrative Genomics Viewer [21]. Note that *ACTB* gene is on the (−) strand. TED-seq distribution relative to the position 300 bp upstream of the 3′CPS (dashed line) is the poly(A) tail length distribution (dashed arrow) of *ACTB* transcripts

7. Browse the plus and minus strand bedgraph files on a genome browser. (Fig. 5; *see* **Note 20**).

4 Notes

1. Oligo(dT) magnetic beads can be made custom by conjugating 3′-amino modified oligo-dT and carboxylic acid linked magnetic beads using *N*-(3-Dimethylaminopropyl)-*N*-′-ethylcarbodiimide (EDC).
2. Poly(A) spike-in RNA can be prepared by in vitro transcription of PCR amplified unique sequence template with T7 promoter sequence at the 5′ end of the forward primer and poly-dT sequence of desired length at the 5′ end of the reverse primer. The unique sequence template can be an arbitrary sequence of

700–800 base pairs from a plasmid backbone, that is not found in the target organism's genome, and that does not contain more than 4–5 consecutive T's on the sense strand which might serve as an internal termination signal for T7 RNA polymerase.

3. The RA3 adaptor is an RNA oligonucleotide with 5′ phosphorylation and 3′ inverted dT modifications. Alternatively, preadenylated DNA oligonucleotide can be used, but will need to adjust the RNA ligation reaction by replacing the RNA ligase I with the truncated version of RNA ligase II enzyme without the presence of ATP.
4. The RA5 adaptor is an RNA oligonucleotide that contains 8 random nucleotide sequences (N's) that serve as Unique Molecular Identifier (UMI). N's are equal compositions (25%) of A, C, G, and U bases.
5. RPI-index primers contain sample barcode index sequences (JJJJJJ) that comply with single ended Illumina TRU-seq small RNA sequencing primers (RPI-1: CGTGAT, RPI-2: ACATCG, RPI-3: GCCTAA, RPI-4: TGGTCA, RPI-5: CACTGT, RPI-6: ATTGGC; etc.)
6. Ideally, 3′ CPS annotation should be from a 3′ end sequencing data in the same biological sample. Alternatively, the last 300 bases of transcripts from reference gene annotation, such as RefSeq, can be used. The TED-seq reads will be positioned in the 300 bases region.
7. Mix thoroughly by gently pipetting as PEG-8000 is highly viscous. The reagents can be premixed when processing multiple samples at once. Scale up accordingly, and add 17μl of the premix to each reaction.
8. The remaining steps in Subheading 3.3 can be replaced by other column based or SPRI bead based RNA cleanup procedures. Adjust the final volume to 20μl in water. TRIzol procedure is preferred if the procedure cannot resume immediately, and extended storage of the material is needed at Subheading 3.3, **step 3**.
9. This step removes hydrolyzed RNA fragments that do not contain poly(A) tail regions, and remaining RA3 adaptors. It is possible to skip this step (but not currently recommended), which results in products with higher internal RNA reads and adaptor dimers.
10. The reagents can be premixed when processing multiple samples at once. Scale up accordingly, and add 10μl of the premix to each reaction.
11. The next step requires small volume reactions, and precipitated RNA suits better for this purpose. It is also possible to use other column-based or SPRI bead-based RNA cleanup

procedures. In these cases, use appropriate mixtures of RA5 adaptor and RTP primer in water for the final elution.

12. As in **Note 7**, make sure to mix thoroughly by gently pipetting. The reagents can be premixed when processing multiple samples at once. Scale up accordingly, and add 6μl to each reaction.
13. The reagents can be premixed when processing multiple samples at once. Scale up accordingly, and add 7μl to each reaction.
14. 2× PCR premix from commonly obtainable sources or custom mixes containing hot-start high-fidelity thermostable DNA polymerase can be used. For multiple samples, scale up accordingly.
15. Optimal number of PCR cycles may be determined empirically. Typically, 1μl of the input material is serially diluted by 4 folds, and subject to 20–25 cycles of PCR. Alternatively, 5μl of test PCR products in ~50μl PCR reaction can be taken out every 2 cycles. Over-amplified products will appear as an upshift of the smear due to self-priming. Determine the number of cycles by subtracting 4–6 cycles before over-amplification happens.
16. Make sure that the plate for gel staining is clean to minimize the possible contamination to the cDNA library.
17. Target product size is precisely at 356 bp (insert 300 bp + PCR primer and UMI 56 bp). 10 bp DNA ladder may not show up to this size (typically up to 330 bp). Using a 25 bp DNA ladder and 350 bp mark (Fig. 3, dashed line) as a guide, cut out a thin (~1 mm) slice of the gel. Loading the DNA ladder on both sides of the sample and cut along a straight line between the two 350 bp marks. The insert size can be other than 300 bp, such as 250 bp, as long as the products are precisely size-selected. The advantage of using a shorter size is that the on-gel manipulation is easier. The drawback of using shorter size (e.g., 250 bp) is that it will not be possible to detect poly (A) tails longer than the insert size, since L = S − D (Fig. 1). However, most eukaryotic poly(A) tails are known to be shorter than 250 bp, and may not affect many transcripts.
18. There will be a single band at 426 bp (cDNA insert: 300 bp, PCR primers: 126 bp), that may form a thicker smear due to self-priming. Use the 25 bp DNA ladder as a guide to cut out around 425 bp band. Try to include as much as possible, but cut out the tails of the smear to preserve the correct insert size.
19. This step needs to be done once, and the indexed genome files can be used for subsequent alignments. If the indexed genome file already exists, this step can be omitted.
20. Poly(A) tail length distribution appears as a displaced distribution of TED-seq reads by 300 bases upstream from the 3′CPS. For most transcripts, calculating the distance from the TED-

seq reads to its downstream 3′ CPS (D) is sufficient to yield poly(A) tail length (L) and using the formula L = S − D (Fig. 1). If there are multiple 3′ CPS within the 300 bp window, either use the most dominant 3′ CPS or exclude the transcript from the analysis. On rare occasions, there may be a spliced intron within the 300 bp window. On those genes, there may appear to be accumulations of TED-seq reads at the splice junction due to the partial alignment of TED-seq reads. In such cases, spliced reads need to be treated separately to be mapped to the correct 5′ end positions.

References

1. Eckmann CR, Rammelt C, Wahle E (2011) Control of poly(A) tail length. Wiley Interdiscip Rev RNA 2(3):348–361
2. Weill L, Belloc E, Bava FA et al (2012) Translational control by changes in poly(A) tail length: recycling mRNAs. Nat Struct Mol Biol 19(6):577–585
3. Mugridge JS, Coller J, Gross JD (2018) Structural and molecular mechanisms for the control of eukaryotic 5′-3' mRNA decay. Nat Struct Mol Biol 25(12):1077–1085
4. Colgan DF, Manley JL (1997) Mechanism and regulation of mRNA polyadenylation. Genes Dev 11(21):2755–2766
5. Richter JD (1999) Cytoplasmic polyadenylation in development and beyond. Microbiol Mol Biol Rev 63(2):446–456
6. Murray EL, Schoenberg DR (2008) Assays for determining poly(A) tail length and the polarity of mRNA decay in mammalian cells. Methods Enzymol 448:483–504
7. Subtelny AO, Eichhorn SW, Chen GR et al (2014) Poly(A)-tail profiling reveals an embryonic switch in translational control. Nature 508 (7494):66–71
8. Chang H, Lim J, Ha M et al (2014) TAIL-seq: genome-wide determination of poly(A) tail length and 3′ end modifications. Mol Cell 53 (6):1044–1052
9. Legnini I, Alles J, Karaiskos N et al (2019) FLAM-seq: full-length mRNA sequencing reveals principles of poly(A) tail length control. Nat Methods 16(9):879–886
10. Woo YM, Kwak Y, Namkoong S et al (2018) TED-Seq identifies the dynamics of poly (A) length during ER stress. Cell Rep 24 (13):3630–3641. e3637
11. Lianoglou S, Garg V, Yang JL et al (2013) Ubiquitously transcribed genes use alternative polyadenylation to achieve tissue-specific expression. Genes Dev 27(21):2380–2396
12. Köster T, Haas M, Staiger D (2014) The RIPper case: identification of RNA-binding protein targets by RNA immunoprecipitation. Methods Mol Biol 1158:107–121
13. Schwanhäusser B, Busse D, Li N et al (2011) Global quantification of mammalian gene expression control. Nature 473 (7347):337–342
14. Garibaldi A, Carranza F, Hertel KJ (2017) Isolation of newly transcribed RNA using the metabolic label 4-Thiouridine. Methods Mol Biol 1648:169–176
15. Mahat DB, Kwak H, Booth GT et al (2016) Base-pair-resolution genome-wide mapping of active RNA polymerases using precision nuclear run-on (PRO-seq). Nat Protoc 11 (8):1455–1476
16. Ross J (1995) mRNA stability in mammalian cells. Microbiol Rev 59(3):423–450
17. Mangus DA, Evans MC, Jacobson A (2003) Poly(A)-binding proteins: multifunctional scaffolds for the post-transcriptional control of gene expression. Genome Biol 4(7):223
18. Li H, Handsaker B, Wysoker A et al (2009) The sequence alignment/map format and SAMtools. Bioinformatics 25(16):2078–2079
19. Quinlan AR, Hall IM (2010) BEDTools: a flexible suite of utilities for comparing genomic features. Bioinformatics 26(6):841–842
20. Dobin A, Davis CA, Schlesinger F et al (2013) STAR: ultrafast universal RNA-seq aligner. Bioinformatics 29(1):15–21
21. Robinson J, Thorvaldsdóttir H, Winckler W et al (2011) Integrative genomics viewer. Nat Biotechnol 29:24–26

Chapter 16

In Vivo RNA Structure Probing with DMS-MaPseq

Paromita Gupta and Silvia Rouskin

Abstract

RNA has an extraordinary capacity to fold and form intrinsic secondary structures that play a central role in maintaining its functionality. It is crucial to have ways to study RNA structures and identify their functions in their biological environment. In the last few decades, a number of different chemical probing methods have been used to study RNA secondary structure. Here, we present a dimethyl sulfate–based (DMS) chemical probing method coupled with Next Generation sequencing (*DMS-MaPseq*) to study RNA secondary structure in vivo.

DMS modifies unpaired adenine and cytosine bases which are then converted to mutations/mismatches using a thermostable group II intron reverse transcriptase (TGIRT) and further analyzed using sequencing. We validated the technique in model systems ranging from Drosophila to human cell lines, thus increasing the technique's broad range of applications. DMS-MaPseq provides high quality data and can be used for both gene-targeted as well as genome-wide analysis.

Key words RNA, RNA structure, DMS, DMS-MaPseq, In vivo, Reverse transcription, Sequencing, TGIRT

1 Introduction

RNA is essential macromolecule required for all known forms of life. It is pivotal to fully understand how this molecule functions in nature, since it plays a major role in a wide variety of cellular processes. Many of these processes are influenced by RNA's unique ability to form complex secondary and tertiary structures. Understanding how RNA folds into these structures in vivo (i.e., in the context of a living cell) and how that folding controls RNA function is of vital importance.

Developing targeted and high-resolution techniques for RNA structure determination in vivo is crucial for understanding the role these structures play in different biological processes. While a number of already available structure determination methods provide valuable information for in vitro analysis, generating robust high-quality data for in vivo analysis is extremely valuable.

Erik Dassi (ed.), *Post-Transcriptional Gene Regulation*, Methods in Molecular Biology, vol. 2404,
https://doi.org/10.1007/978-1-0716-1851-6_16,

DMS-MaPseq (Fig. 1) is a structure specific chemical probing technique which utilizes a small molecule called dimethyl sulfate to "mark" unpaired adenine and cytosines with a methyl group at Watson–Crick base pairing positions N1 and N3 respectively [1]. Postmodification, a specialized reverse transcriptase, TGIRT is used to convert the modifications to a mismatch which is detected via sequencing. Unlike previously used DMS-based method [2] as well as Selective 2′-hydroxyl acylation analyzed by primer extension (SHAPE) [3, 4], *DMS-MaPseq* does not rely on truncation-based approach. The mutational profiling (MaP) approach provides a significant advantage by allowing detection of multiple modifications on the same molecule, thus allowing structure analysis on a global level for single molecules including genes with relatively low levels of expression. Thus, our method allows the in vivo detection of novel RNA structures by combing RNA sequence information with experimental data analyzed at single-nucleotide resolution.

Using *DMS-MaPseq*, we have probed Drosophila ovaries as well as human cell lines and carried out both an amplicon-targeted RT-PCR approach and a genome-wide analysis using library generation strategy [5].

2 Materials

2.1 Buffers

1. **STOP solution**—30% solution of beta-mercaptoethanol (BME) IN 1× PBS.
2. **Solution A**—200 μl DMS (undiluted) to 10 ml of prewarmed cell culture media.
3. **Solution B**—1 ml of TRIzol containing 10 μl BME.
4. **Solution C**—3 mM sodium acetate (1/10th total volume), 3 μl GlycoBlue and 600 μl 100% isopropanol.
5. **Solution D**—70% ethanol solution.
6. **rRNA subtraction mix**—designed from Adiconis et al.
7. **Hybridization mix**—1 M NaCl, 500 mM Tris–HCl pH 7.5.
8. **DNase treatment buffers and enzymes**—as mentioned in the Turbo DNase kit.
9. **RT reaction mix**—2 μl 5× FS Buffer (M-MLV RT buffer), 1 μl dNTP (10 mM mix), 1 μl Reverse Primer (10 μM stock), 0.5 μl each of DTT (0.1 M stock), RNaseOUT, and 0.5 μl RT enzyme (TIGRT).
10. **Fragmentation Buffers**—Zn^{2+} Fragmentation solution and EDTA stop solution.
11. **Zymo Column Buffers**—All dilutions are prepared according to the instructions provided in the Zymo kit.

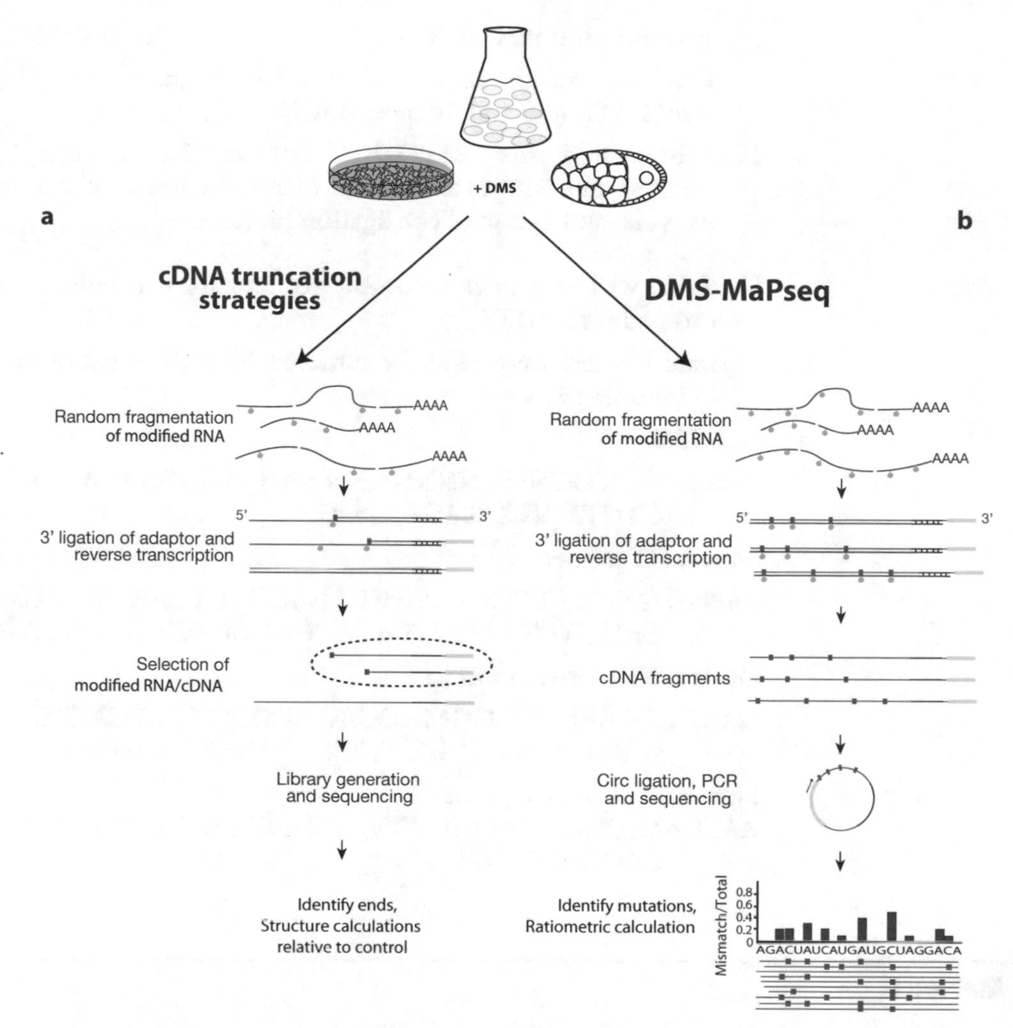

Fig. 1 Sequencing library generation for RNA structure probing techniques. Schematic of library preparation strategies for cDNA truncation approaches (**a**) and for DMS-MaPseq (**b**), which has a higher DMS modification level, no selection for modified molecules, and ligations that are no longer directly proximal to the structure information-containing positions. The structure signal for DMS-MaPseq is inherently ratiometric and calculated per nucleotide as the number of mismatches divided by base sequencing depth. 5′ to 3′ orientation noted relative to RNA fragment

12. **Dephosphorylation Buffer**—1 μl 10× CutSmart Buffer, 1 μl SUPERase Inhibitor, and 1.5 μl rSAP.
13. **Ligation Mix**—6 μl 50% PEG 8000, 1 μl Linker, 2.2 μl 10× T4 Ligase buffer, and 2 μl 1:20 truncated T4 RNA ligase.
14. **Linker Degradation Mix**—2 μl 10× Rec J Buffer and 1 μl RNase OUT.
15. **Library RT Mix**—4 μl 5× First Strand buffer, 1 μl dNTP, 1 μl DTT, 1 μl RNaseOUT, and 1 μl TGIRT enzyme.
16. **Gel extraction solution**—300 mM NaCl.

17. **Precipitation mix**—600 μl isopropanol and 3 μl GlycoBlue.
18. **Circular ligation mix**—2 μl 10× Circ ligation buffer, 1 μl 1 mM ATP, and 1 μl 50 mM $MnCl_2$.
19. **Library PCR mix**—26 μl DEPC water, 8 μl 5× HF buffer, 1 μl forward and 1 μl reverse primer (10 mM), 0.4 μl of Phusion enzyme, and 1–2 μl of circ ligation product.

2.2 Oligos

Linker as well as primer used only for Library generation were ordered from IDT.

Separate primers were used for targeted RT-PCR mentioned in Subheading 3.6.

N12 Linker.

/5rApp/ TCNNNNNNNNNNNNAGATCGGAAGAGCGTC GTGTAGGGAAAGA/3ddC/.

Library RT primer.

/5Phos/AGA TCGGAAGAGCACACGTCTGAACTCCAG / iSp18/TCTTTCCCT ACACGACGCTCTTCC GATCT.

Library PCR forward primer.

CAAGCAGAAGACGGCATACGAGATXXXXXXGTGACTG-GAGTTCAGACGTGTGCTC XXXXXX = Index.

Library PCR reverse primer.
AATGATACGGCGACCACCGAGATCTACACTCTTTCCCTA-CACGACGCTC.

3 Methods

All reactions using DMS must be carried out in chemical hoods. Appropriate protective gear-lab coats, gloves must be used while handling DMS. All tips, tubes, and so on exposed to DMS must be discarded as hazardous material.

3.1 In Vivo Dimethyl Sulfate (DMS) Modification in Drosophila Ovary

1. Collect flies fed on yeast paste for a day.
2. Dissect ovaries (slightly tease ovaries while dissection) from ~100 flies in 250 μl 1× PBS. Keep samples on ice till you add DMS (*see* **Note 1**).
3. Add 250 μl DMS to 250 μl 1× PBS containing dissected ovaries. Incubate for 5 min at 26 °C with shaking at 500 rpm.
4. To stop the reaction, add 1 ml of **STOP solution** and transfer the oocytes using a pipette onto a sieve.
5. Wash the DMS treated ovaries three times in **STOP solution** and two times with sterile water.

6. Finally, collect the ovaries and resuspend in **Solution B** (*see* **Note 2**).
7. Grind ovaries manually using a sterile pestle (*see* **Note 3**).
8. Proceed with Subheading 3.3.

3.2 In Vivo Dimethyl Sulfate (DMS) Modification in HEK293T Cells (See Note 4)

1. Use only confluent cells for this treatment (*see* **Note 5**).
2. Mix **Solution A** well before use.
3. Add **Solution A** to the plate (*see* **Note 6**). Make sure that the solution covers entire plate.
4. Incubate at 37 °C for 5 min (*see* **Note 7**).
5. Post incubation, discard all media (*see* **Note 8**).
6. Stop the reaction by adding 10 ml of **STOP solution**. Resuspend cells in the solution and transfer to a 50 ml falcon tube on ice.
7. Spin cells for 3 min at 3000 rpm. Discard supernatant (*see* **Note 9**).
8. Wash pellet in 10 ml dWater by spinning again at 3000 rpm for 3 min.
9. Discard most of the supernatant. Resuspend pellet in remaining (~2 ml) water and transfer to 2 ml Eppendorf.
10. One final spin at 3000 rpm for 3 min.
11. Discard supernatant and resuspend pellet in 1 ml of **Solution B**.
12. Proceed with Subheading 3.3.

3.3 TRIzol RNA Extraction

1. Grind up ovaries/homogenized cells in **Solution B**.
2. Add 200 μl chloroform and mix tube by quickly inverting 5–10 times.
3. Incubate at room temperature for 2–3 min.
4. Spin at 12,000 × g for 15 min at 4 °C.
5. Carefully collect clear supernatant avoiding interphase layer (*see* **Note 10**).
6. Add **Solution C** to the supernatant and incubate on dry ice for 15 min (*see* **Note 11**).
7. Spin at maximum speed for 45 min at 4 °C.
8. Wash the pellet once in **Solution D** and air-dry pellet (*see* **Note 12**).
9. Resuspend pellet in 30–50 μl 10 mM Tris pH 7.0 and store at −20 °C or −80 °C.

3.4 RNase H–Based Ribosomal RNA Extraction

1. Add 1.5 μl of **rRNA subtraction mix** (3 μg/μl) and 2.5 μl of 5× **Hybridization buffer** to 9.5 μl of RNA (*see* **Note 13**).
2. Incubate at 68 °C (−1 °C/min) for 23 cycles till temperature is lowered to 45 °C.
3. Add 30.5 μl of DNase/RNase-free water, 5 μl of RNase H buffer and 2 μl of **thermostable** RNase H enzyme to the mix and incubate at 45 °C for 30 min (*see* **Note 14**).
4. Proceed with Zymo column cleanup for >200 nt as mentioned after DNase treatment step.
5. Elute twice in 22.5 μl RNase-free water (*see* **Note 15**).

3.5 DNase Treatment (See Note 16)

1. Add 0.1 volume (5 μl) of DNase buffer to 45 μl of subtracted RNA (total vol-50 μl).
2. Add 1 μl of Turbo DNase enzyme and incubate at 37 °C for 20 min.
3. Add 0.1 volume (5.1 μl) of inactivation buffer and incubate at RT for 5 min (Mix manually intermittently).
4. Spin at 12,000 × *g* for 1 min 30 s.
5. Transfer supernatant to a fresh tube (*see* **Note 17**).
6. For Zymo Column cleanup refer Subheading 3.7.2 (for >200 nt). For elution volume *see* **Note 18**.

The subtracted RNA from Subheading 3.5 can now be used for both Subheading 3.6 as well as Subheading 3.7 depending upon gene targeted vs library generation strategy.

3.6 Reverse Transcription Polymerase Chain Reaction (RT-PCR)

3.6.1 Reverse Transcription

1. Prepare **RT reaction mix** on ice.
2. After adding required volume of subtracted RNA (50–100 ng), add required volume of DNase/RNase free water to reach a final volume of 10 μl.
3. Incubate at 57 °C for 1.5 h followed by deactivation at 85° for 5 min.
4. Add 1 μl RNase H to the reaction and incubate at 37 °C for 20 min.
5. Proceed to PCR reaction.

3.6.2 Polymerase Chain Reaction (See Note 19)—For Advantage HF Kit

1. PCR Program:
   (a) 94 °C for 1 min.
   (b) 94 °C for 30 s.
   (c) 60 °C for 30 s.
   (d) 68 °C for 2 min (1 min works for shorter templates).
   Go to **step 2** for 27×.
   (e) hold at 4 °C.

2. Check PCR products on 1% agarose gel.
3. Clean up PCR product using Zymo PCR cleanup kit.

3.6.3 Polymerase Chain Reaction (See **Note 19**)—For Phusion Kit

1. PCR Program
   (a) 98 °C for 30 s.
   (b) 98 °C for 10 s.
   (c) 62 °C for 30 s.
   (d) 72 °C for 2 min.
   Go to **step b** for 10×.
   (e) 98 °C for 10 s.
   (f) 57 °C for 30 s.
   (g) 72 °C for 30 s.
   Go to **step d** for 15×.
   (h) 72 °C for 1 min.
   (i) Hold at 4 °C.
2. Check PCR products on 1% agarose gel.
3. Clean up PCR product using Zymo PCR cleanup kit.

3.7 Library Preparation Strategy

3.7.1 Fragmentation (See **Note 20**)

1. Start with 1–2 μg of preselected mRNA in 9 μl 10 mM Tris pH 7.0.
2. Melt RNA structures at 95 °C for 1 min.
3. Move PCR tubes to ice block. Change program on PCR machine to 70 °C.
4. Add 1 μl buffered **Fragmentation solution** (*see* **Note 21**).
5. Move tubes back to 70 °C for 45 s.
6. Move PCR tubes to ice block.
7. Quickly add 1.1 μl **EDTA Stop Solution**. Again, pipet and swirl to mix.
8. Proceed to Zymo column cleanup (choose according to size of fragmented template). (See **Note 22**).

3.7.2 ZYMO Column Cleanup (see **Note 23**)

1. Add RNA binding buffer (using the Zymo kit instructions) based on expected RNA fragment size.
2. Spin at 18,000 × *g* for 30 s. Discard flow through.
3. For cleaning up **17–200 bp fragments**, include **additional step** of adding 1 volume of 100% ethanol to supernatant, mix and transfer to new Zymo column. Repeat **step 2**.
4. Add 400 μl RNA prep buffer and spin for 30 s at 18,000 × *g*. Discard flow through.
5. Add 700 μl wash buffer and spin for 30 s at 18,000 × *g*. Discard flow through.

6. Add 400 μl wash buffer and spin for 2 min at 18,000 × *g*. Discard flow through.
7. Spin column for 30 s to remove additional wash buffer.
8. Place column in an RNase free tube and elute in 6.5 μl RNase-free water.

3.7.3 Dephosphorylation

1. To 6 μl RNA, add **Dephosphorylation buffer** and mix well.
2. Incubate 1 h, 37 °C.
3. Proceed directly to ligation.

3.7.4 Linker Ligation (See Note 24)

1. Prepare **Ligation mix** for each reaction on ice, add and mix well.
2. Incubate 18 h at 22 °C.
3. Proceed with Zymo column cleanup for >200 bp cleanup as in Subheading 3.7.2.
4. Elute in 15.5 μl RNase-free water. (Save 0.5 μl of eluted RNA in 3 μl of Tris pH 7.0 for Bioanalyzer analysis.)

3.7.5 Linker Degradation (See Note 25)

1. Add 3 μl **Linker degradation mix** to 15 μl RNA for each sample. Mix well.
2. Add 1 μl 5′ deadenylase (Epicentre DA11101K 10 U/μl).
3. Add 1 μl RecJ exonuclease (Epicentre RJ411250 10 U/μl).
4. Incubate 30 °C for 1 h.
5. Proceed with Zymo column cleanup for >200 bp cleanup as mentioned in Subheading 3.7.2.
6. Elute in 11.5 μl RNase-free water. (Save 0.5 μl of eluted RNA in 3 μl of RNase-free water for Bioanalyzer analysis.)

3.7.6 Reverse Transcription (See Note 26)

1. To the 11 μl RNA, add 1 μl 10 μM RT primer (PAGE purified).
2. Add **Library RT mix** to each tube containing RNA and primer. Mix well.
3. Move sample to PCR tubes.
4. Incubate 65 °C for 1 h 30 min.
5. Add 1 μl 4 M NaOH, pipetting to mix, then incubate 95 °C 3 min.
6. Add 10 μl of commercial 2× Sample Buffer to each sample before loading on gel.

3.7.7 Size Selection on Gel (See Note 27)

1. Use 10% TBU gel for size selection post RT. Prerun the gel and flush out the wells properly before loading the samples.

2. After loading samples, run the gel for 1 h 15 min (depending on size of fragments-for 150 bp fragments run for 1 h) at 200 V.
3. Cut full-length fragments from gel. If using blue light and NOT UV, take picture before cutting the gel for future reference.
4. Cut ~170–240 nt fragments or around desired fragment range.

3.7.8 Gel Extraction (Using 300 mM NaCl to Extract)

1. Resuspend crushed gel in 400–500 μl of **Gel extraction solution**.
2. Melt gel at 70 °C with shaking at 1500 rpm for 10 min.
3. Transfer entire volume including smaller gel pieces to a Costar Spin column.
4. Transfer supernatant to a fresh tube and add **Precipitation mix**.
5. Store at −20 °C for a few days or keep on dry ice for 10–15 min to continue processing.
6. Spin frozen samples at 4 °C for 45 min.
7. Wash pellet in 500 μl of **Solution D** and air dry or discard supernatant completely with pipette after wash.
8. Resuspend pellet in 15 μl 10 mM Tris pH 8.0.
9. Leave on ice.

3.7.9 Circular Ligation

1. Prepare **Circular ligation reaction mix**.
2. Add 1 μl Circ Ligase ssDNA Ligase.
3. Add entire volume of gel extracted RT product to the mix.
4. Incubate 60 °C for 2 h.
5. Incubate 80 °C for 10 min to heat inactivate enzyme.
6. Keep on ice and proceed to amplification or store at −20 °C.

3.7.10 PCR to Add Sequencing Handles

1. Make **Library PCR mix** on ice.
2. PCR Conditions:
   (a) 98 °C, 30 s.
   (b) 94 °C, 15 s.
   (c) 55 °C, 5 s.
   (d) 65 °C, 10 s.
   Repeat **steps b–d** for *X* number of cycles, removing fewer cycle sample at right time.
   (e) 4 °C HOLD.
   * Collect half volume (~18 μl) after first cycle and the other half at the end of the PCR program (*see* **Note 28**).

3.7.11 PCR Gel

1. To 18 μl PCR reactions (2 tubes—e.g., one collected at the -end of 8 and the other at the end of 10 cycles), add 3 μl of commercial 6x DNA Loading Dye.
2. Load all 21 μl of sample on 12-lane, 8% TBE gel (no need to prerun gel).
3. Run 55 min at 180 V.
4. Cut band from gel.
5. There should be 3 populations on the gel.
6. Empty vector at bottom of gel.
7. Vector with product at size 200–300 nt.
8. May see larger smear if overamplified. In this case, do fewer PCR cycles.
9. Gel extraction (Subheading 3.7.10).
10. Resuspend in 10 μl 10 mM Tris pH 8.0.

3.8 Sequencing

1. Bioanalyze to quantify and check for library quality.
2. Proceed for sequencing using the chosen Illumina machine.

4 Notes

1. Start DMS treatment as soon as dissection is over. Do not keep ovaries on ice for a long period of time.
2. At this step make sure you that there is no leftover DMS. If required add an additional wash step.
3. Make sure the ovaries are ground up well and the solution looks homogenous.
4. Prewarm media at 37 °C and Prepare STOP solution containing 30% BME (beta-mercaptoethanol) (v/v) in 1× PBS.
5. It is crucial to start with at least 90% confluent cells since loss occurs during DMS treatment and wash steps. It is always helpful to treat replicates together. Do not treat more than 2 plates at the same time.
6. While adding Solution A make sure to add slowly from the edge to minimize dislodging cells.
7. This includes the time required to walk the plates to and from the incubator. Ideally 3 min 30 s to 4 min incubation at 37 °C gives well modified samples.
8. Collect the DMS containing media from the edge of the plate and avoid collecting modified cells. It is really important to get rid of almost all of the DMS to prevent cells from getting overmodified.

9. At this step make sure to remove DMS entirely. There should be some phase separation between DMS and BME.
10. Do not collect any interphase since it interferes with proceeding steps. If required respin and then collect supernatant.
11. If required the samples can be stored at −80 °C if a break is needed.
12. Instead of air drying a pipette can be used to get rid of the ethanol and the pellet can be subsequently resuspended.
13. Oligo mix or rRNA subtraction mix is a mix of rRNA-aligning oligos designed by Adiconis et al.
14. The thermostable RNase H enzyme and buffer are obtained from Illumina (Inc.#H39500). The oligo mix can be ordered from IDT according to the model organism.
15. Before proceeding to the next step, save 0.5 μl of eluted RNA in 3 μl of Tris pH 7.0 for bioanalyzer analysis.
16. The DNase used for this step is obtained from ThermoFisher. The buffer needs to be thawed completely before use.
17. Be careful not to transfer any inactivation buffer with the supernatant. If required, repeat the spin step.
18. Elute in ~3–4 μl more than desired volume. Before proceeding to the next step, save 0.5 μl of eluted RNA in 3 μl of DNase/RNase-free water for Bioanalyzer analysis.
19. Advantage HF kit was purchased from Takara Clontech, cat# 639123 and Phusion HF from NEB cat # M0530L. Buffers are used as recommended in the protocol.
20. Set two heat blocks/PCR machines to 95 °C and 70 °C before starting the reaction.
21. Fragmentation reagents are obtained from Ambion. It is important to carry out this step as quickly as possible making sure that the reagents are mixed well by using a pipette.
22. Based on the size of desired fragment, choose cleanup method using Zymo column. If standardization for fragmentation is required, it might be helpful to send some of the cleaned up fragmented product to bioanalyzer to determine the size range.
23. Zymo columns used for cleanup are obtained from Zymo Research, cat # R1015. For either cleanup always elute in a slightly higher volume than mentioned in the protocol (e.g., for 6.5 μl elute sample in about 8 μl) to counteract loss during spinning step.
24. Carry out incubation step in a thermomixer. The buffer mix used for ligation should be made fresh and try to maintain a 1:1 ratio for linker to fragments if possible.

25. No need to run gel to remove Linker!
26. Keep all reagents for RT reaction as fresh as possible, that is, aliquot 5× FS buffer, dNTPs, and TGIRT enzyme into single-use aliquots. Make fresh 0.1 M DTT from scratch before every RT reaction.
27. Using a control here helps a lot since RT product might not be clearly visible on the gel. Thus, running a control of known size would help in identifying which region to extract.
28. For PCR cycle number choose 8 and 10 cycles, 10 and 12 cycles, or 13 and 15 cycles depending on amount of material in RT gel. Fewer number of cycles reduce the risk of introducing PCR bias.

References

1. Wells SE, Hughes JM, Igel AH, Ares M Jr (2000) Use of dimethyl sulfate to probe RNA structure in vivo. Methods Enzymol 318:479–493
2. Rouskin S, Zubradt M, Washietl S, Kellis M, Weissman JS (2014) Genome-wide probing of RNA structure reveals active unfolding of mRNA structures in vivo. Nature 505:701–705
3. Mortimer SA, Weeks KM (2007) A fast-acting reagent for accurate analysis of RNA secondary and tertiary structure by SHAPE chemistry. J Am Chem Soc 129:4144–4145
4. Smola MJ, Rice GM, Busan S, Siegfried NA, Weeks KM (2015) Selective 2′-hydroxyl acylation analyzed by primer extension and mutational profiling (SHAPE-MaP) for direct, versatile and accurate RNA structure analysis. Nat Protoc 10:1643–1669
5. Zubradt M, Gupta P, Persad S, Lambowitz AM, Weissman JS, Rouskin S (2017) DMS MaPseq for genome-wide or targeted RNA structure probing in vivo. Nat Methods 14:75–82

Chapter 17

Transcriptome-Wide Profiling of RNA Stability

Nina Fasching, Jan Petržílek, Niko Popitsch, Veronika A. Herzog, and Stefan L. Ameres

Abstract

Gene expression is controlled at multiple levels, including RNA transcription and turnover. But determining the relative contributions of RNA biogenesis and decay to the steady-state abundance of cellular transcripts remains challenging because conventional transcriptomics approaches do not provide the temporal resolution to derive the kinetic parameters underlying steady-state gene expression.

Here, we describe a protocol that combines metabolic RNA labeling by 4-thiouridine with chemical nucleoside conversion and whole-transcriptome sequencing followed by bioinformatics analysis to determine RNA stability in cultured cells at a genomic scale. Time-resolved transcriptomics by thiol (SH)-linked alkylation for the metabolic sequencing of RNA (SLAMseq) provides accurate information on transcript half-lives across annotated features in the genome, including by-products of transcription, such as introns. We provide a step-by-step instruction for time-resolved transcriptomics, which enhances traditional RNA sequencing protocols to acquire the temporal resolution required to directly measure the cellular kinetics of RNA turnover under physiological conditions.

Key words RNA stability, Metabolic RNA sequencing, 4-Thiouridine, Gene regulation, SLAMseq

1 Introduction

Gene expression is the essential process that controls genome function in all living organisms, impinging on most biological processes from organismal development to cellular differentiation and physiological responses to external stimuli and pathogens [1]. Among the molecular processes that control gene expression, the regulated decay of RNA acts as a key step to control the quality and quantity of cellular transcripts [2]. To this end, a variety of cis- and trans-acting factors adjust the stability of RNA according to its encoded function [3]. Hence, robust measurements of gene expression kinetics represent important means by which to systematically determine the biological processes that control RNA turnover.

Erik Dassi (ed.), *Post-Transcriptional Gene Regulation*, Methods in Molecular Biology, vol. 2404,
https://doi.org/10.1007/978-1-0716-1851-6_17,

With the advent of next-generation sequencing technologies and the development of diverse cDNA library preparation protocols, measurements of global and transcript-specific changes in the steady-state abundance of cellular RNA in response to environmental or genetic perturbations became readily accessible [4]. But given the lack in temporal resolution, most of these protocols refrain from addressing the cause of gene expression changes that may arise from alterations in the rates of transcription or RNA stability, or both [2, 5]. To this end, various protocols have been developed to specifically address the kinetics of transcript turnover. Among those, the global inhibition of transcription by conditional mutants or small molecule inhibitors, such as actinomycin D or α-amanitin, frequently induce cellular stress responses and deliver imprecise results, most certainly due to the intrinsic coupling of transcription and decay [6–9]. Less invasive methods employ nucleoside analogs (e.g., 4-thiouridine, 5-ethyniluridine or 5-bromouridine) that possess unique physicochemical properties that can be coupled to biochemical enrichment procedures (e.g., by reversible biotinylation or classical immunoprecipitation) to capture label-containing transcripts following metabolic labeling in cultured cells [8]. When coupled to gene expression profiling, such protocols have been widely used to assess transcript stabilities, despite labor-intensive procedures, a need for excessive starting material, and the implementation of sophisticated kinetic modeling that account for enrichment procedures. Most of these limitations were recently overcome with the implementation of simple chemical protocols that convert s^4U into cytosine analogs [10–12]. Here, total RNA prepared from cultured cells subjected to s^4U metabolic labeling is exposed to chemical treatment that changes the base-pairing properties of uridine, resulting in specific T to C conversions at the sites of s^4U incorporation upon cDNA library preparation. Such protocols detect metabolically labeled transcripts in the context of total RNA and therefore enable the rapid assessment of gene expression dynamics. Among the available s^4U conversion chemistries, SLAMseq (thiol[SH]-linked alkylation for the metabolic sequencing of RNA) employs a rapid (15 min reaction time) and efficient (>94% detection efficiency) nucleophilic substitution reaction that transfers a carboxyamidomethyl-group from iodoacetamide (IAA) to the thiol group of s^4U, which causes the incorporation of guanine across alkylated s^4U during reverse transcription (Fig. 1). When coupled to well-established metabolic RNA labeling protocols in cultured cells, SLAMseq has been applied to systematically dissect gene expression kinetics, revealing immediate changes in transcriptional programs [12, 13], regulatory principles in global and gene-specific RNA decay rates [12], and basic concepts underlying cellular RNA homeostasis [12, 14].

We recently established a SLAMseq protocol that probes the half-life of polyadenylated RNA species by mRNA 3′ end

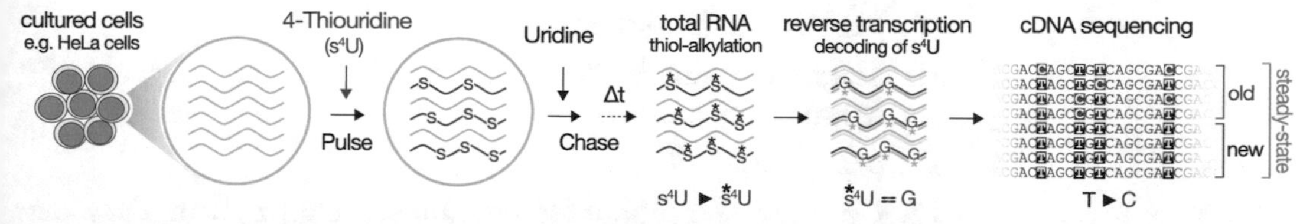

Fig. 1 Schematic overview of a protocol for the transcriptome-wide measurement of RNA stability by thiol(SH)-linked alkylation for the metabolic sequencing of RNA (SLAMseq). Cultured cells (e.g., HeLa cells) are subjected to metabolic RNA labeling by the addition of 4-thiouridine (s^4U) to the culture medium. Following total RNA extraction at several timepoints after uridine chase, treatment with iodoacetamide results in thiol-specific alkylation of s^4U, prompting the site-specific incorporation of guanine across chemically modified s^4U during cDNA synthesis by reverse transcription. As a consequence, SLAMseq specifically identifies s^4U incorporations at single-nucleotide resolution, enabling to distinguish preexisting from newly transcribed RNA in the context of steady-state transcript abundance measurements performed by conventional whole-transcriptome RNA sequencing. Subsequent bioinformatic analyses enable to quantify transcript decay kinetics in a global and transcript-specific manner

sequencing [12, 15]. Here, we describe a follow-up procedure that combines SLAMseq with whole transcriptome sequencing based on cDNA generation from rRNA-depleted total RNA of a human cervical carcinoma cell line (HeLa cells). We provide a stepwise instruction for experimental conditions and basic bioinformatics tools to acquire a broader view on RNA decay kinetics (e.g., of pre-mRNA processing products such as exons and introns) that allows to address most transcriptional products provided they are robustly detected at steady-state.

2 Materials

General considerations: Ensure nuclease-free working conditions. Use reagents and labware, which are free of nucleases and nucleic acid contaminations. Regularly clean working surfaces and follow manufacturer's safety recommendations. Solutions and reagents should be stored at room temperature unless otherwise specified by the manufacturer.

2.1 General Equipment

1. CO_2-controlled and humidified cell culture incubator.
2. Laminar flow hood.
3. Cell counter instrument (e.g., Nexcelom Cellometer Auto 1000).
4. Fume hood.
5. Benchtop refrigerated centrifuge (up to 20,000 × g).
6. Microvolume Spectrophotometer (e.g., Nanodrop).
7. Shaking heat block.
8. Thermal cycler.

9. Real-time PCR system.
10. Magnetic Stand for PCR and 1.5 mL tubes (e.g., DynaMag™ Magnet, Invitrogen).
11. pH meter.
12. Capillary electrophoresis instrument (e.g., Agilent Fragment Analyzer).
13. Qubit™ Fluorometer.
14. Next generation sequencing machine (e.g., Illumina HiSeq 2500).
15. Sterile tissue culture plasticware.
16. Volumetric pipettes (5, 10, 25 mL).
17. Pipette controller (e.g., Integra PIPETBOY pro).
18. 500 mL Nalgene® Rapid-Flow™ Filter Units and Bottle Top Filters (0.2 μm).
19. Set of laboratory pipettes (1–1000 μL).
20. Safe-lock 1.5 mL tubes.
21. Falcon tubes (15 mL and 50 mL).
22. 0.2 mL 8-tube PCR strips and ultraclear flat cap strips.
23. Filter tips (1–1000 μL).
24. Water bath.

2.2 Reagents and Buffers

1. Dulbecco's Modified Eagle (DMEM) high glucose medium.
2. Fetal bovine serum.
3. Sterile phosphate buffered saline (PBS) 1×.
4. 0.5% trypsin–EDTA.
5. Trypan blue staining solution for cell culture.
6. 4-Thiouridine.
7. Iodoacetamide; stock in ethanol (final conc. 100 mM); prepare fresh before use.
8. TRIzol™ Reagent (Ambion, Life Technologies).
9. Chloroform–isoamyl alcohol 24:1.
10. Liquid nitrogen.
11. Ethanol absolute.
12. Ethanol 80% (v/v).
13. Ethanol 75% (v/v).
14. 2-Propanol.
15. 3 M NaOAc (pH 5.2).
16. Glycogen (20 mg/mL).
17. OmniPur® DTT (Merck).

18. DMSO.
19. NaH_2PO_4 (monobasic).
20. Na_2HPO_4 (dibasic).
21. Nuclease-free water.
22. QuDye dsDNA HS Assay Kit (Qubit™).
23. TURBO DNA-free™ Kit (Thermo Fisher Scientific).
24. RiboCop rRNA Depletion kit for Human/Mouse/Rat (Lexogen).
25. NEBNext® UltraTM II Directional RNA Library Prep Kit for Illumina® (#E7765, New England Biolabs).
26. EvaGreen® Dye, 20 x in Water (Biotium).
27. NEBNext® Multiplex Oligos for Illumina 96 Index Primers (New England Biolabs).
28. DNF-474 High Sensitivity NGS Fragment Analysis Kit (1–6000 bp) (AATI).
29. Alkylation Master Mix; prepare directly before use.

Reagent	Volume per sample	Final concentration
IAA (100 mM) in EtOH 100%	5 μL	10 mM
$NaPO_4$, pH 8 (500 mM)	5 μL	50 mM
DMSO	25 μL	50% (v/v)
Final volume	**35 μL**	–

2.3 Cell Lines

HeLa cells (e.g., obtained by ATCC®).

3 Methods

Thiol-linked alkylation for the metabolic sequencing of RNA (SLAMseq) is based on metabolic RNA labeling by 4-thiouridine (s^4U) that can in principle be applied to any cultured metazoan cell type (Fig. 1). 4-thiouridine is biologically inert when applied at appropriate doses. When added as a supplement to cell culture medium, metazoan cells readily take up s^4U via nucleoside equilibrate transporters and incorporate it into newly synthesized RNA. Notably, the efficiency of s^4U incorporation into nascent RNA depends on cell type-specific variables, such as cellular uptake kinetics and overall transcriptional activity. Thus, sensitivity to s^4U and efficiency of s^4U incorporation into RNA should be assessed prior to a SLAMseq experiment in order to obtain meaningful results [15]. Experimental parameters, such as s^4U concentration, s^4U labeling time or cDNA library sequencing depth, need to be adjusted accordingly. Here, we describe an experimental setup to perform s^4U labeling in HeLa cells.

3.1 HeLa Cell Culture Procedures

Operate under a laminar flow hood and maintain sterile environment. Culture HeLa cells in DMEM + 10% FCS in a humidified incubator at 37 °C and 5% CO_2. Passage cells at a confluency of <85% every 2–3 days by diluting 1:6 in fresh media. Before passaging or medium exchange, preheat medium in a water bath to 37 °C. Sterile filter medium using a 500 mL 0.2 μm bottle top filter into a glass or plastic bottle and store at 4 °C.

3.2 Assessment of s^4U-Induced Cytotoxicity

Sensitivity of cultured cells to s^4U differs between cell lines. In order to ensure unperturbed physiological settings, it is essential that s^4U is administered at non-toxic concentrations. Detailed experimental designs to address optimal s^4U labeling conditions have been described previously [15].

3.3 General Considerations When Measuring Transcriptome Stability in HeLa Cells

The following protocol describes a pulse–chase experiment to directly measure RNA stability across the transcriptome in HeLa cells. To this end, RNA transcripts are first labeled to saturation with s^4U for 12 h. Note, that pulse labeling periods may be adjusted by taking specific expression kinetics of transcripts of interest into account (i.e., short-lived transcripts will be labeled to saturation within 1 h or less, while long-lived transcripts may require >12 h of labeling to reach saturation). Following s^4U labeling, samples are harvested at multiple time-points after chasing with uridine (0 h, 0.5 h, 1 h, 3 h, 6 h, and 12 h after chase onset) and total RNA is isolated, followed by alkylation, rRNA depletion and RNAseq library preparation. To ensure robust measurements of transcript stabilities, we recommend performing three technical replicates of each timepoint. Note, that depending on the expected half-lives of the transcripts of interest, a reduced number of timepoints may be sufficient to estimate RNA stability. To minimize effects of experimental handling, we recommend using a different plate for each timepoint. The experiment is described for three technical replicates across one unlabeled, one fully labelled and five chase timepoints in a 6-well–plate format. If downstream applications require higher RNA yield, the experiment can be scaled up accordingly.

3.4 Pulse–Chase Metabolic RNA Labeling

Limit light-exposure of s^4U-containing medium and samples to a minimum throughout all of the following steps.

1. Seed 3×10^5 cells (1.5×10^5 per mL) on 21 wells of 6-well plates (using 3 wells per plate) in standard medium the day before the labeling experiment (*see* **Notes 1** and **2**). Prior to seeding, assess the cell viability by Trypan blue staining in the process of cell counting. To guarantee reproducible results, ensure that the cell viability exceeds 90%.

2. Incubate cells at 37 °C over-night in standard culturing conditions.
3. On the next day, prepare 40 mL culturing medium supplemented with 100 μM s^4U (*see* **Note 3**).
4. Aspirate medium from cultured cells and add 2 mL of pre-warmed culturing medium without s^4U to three wells ("unlabeled"). To the rest of the plated cells add 2 mL of pre-warmed s^4U labeling medium per well.
5. Incubate for 12 h under standard culturing conditions.
6. Prior to the end of a 12 h labeling time, prepare 32 mL of chase medium consisting of culturing medium supplemented with 10 mM uridine.
7. When 12 h labeling concluded, harvest pulse samples (3 × unlabeled and 3 × 0 h chase sample) by aspirating medium and adding 500 μL TRIzol™ per well (*see* **Note 2**). Incubate 2–3 min until cell lysis is complete, pipet up and down a few times, transfer lysate to 1.5 mL tubes and freeze samples at −80 °C. For all other wells, remove s^4U labeling medium, wash twice with 1× PBS and add 2 mL uridine-containing chase medium. Incubate at standard culturing conditions.
8. Collect chase samples after the desired timepoints (i.e., 0.5 h, 1 h, 3 h, 6 h, and 12 h after chase onset) by aspiring medium and lysing the cells in 500 μL TRIzol™ per well as described in **step** 7 (*see* **Note 2**). Transfer lysate to 1.5 mL tubes. Freeze samples at −80 °C or directly proceed to RNA extraction.

3.5 RNA Extraction

Limit light exposure of s^4U-labeled samples to a minimum. Work under a fume hood at least until the EtOH addition step.

1. Thaw TRIzol™ lysate and incubate 5 min at room temperature.
2. In the meantime, prepare 6 mL 2-propanol containing 0.2 mM DTT and 25 mL 75% EtOH containing 0.1 mM DTT (*see* **Note 4**).
3. Add 100 μL chloroform:isoamyl-alcohol 24:1 per 0.5 mL of TRIzol™.
4. Vortex tube for 15 s and incubate at room temperature for 3 min.
5. Centrifuge at 16,000 × *g* for at least 15 min at 4 °C.
6. Transfer aqueous phase (~200 μL) to new tube. Do not disturb interphase.
7. Add 1 μL glycogen (20 mg/mL) and 200 μL 2-propanol (containing 0.1 mM DTT). Mix well by vortexing.
8. Incubate 10 min at room temperature.

9. Spin down at 16,000 × *g* for at least 20 min at 4 °C.
10. Take off supernatant with a pipette and discard. Handle carefully not to disturb pellet.
11. Add 500 μL DTT-containing 75% EtOH (DTT at 0.1 mM final concentration) and vortex well to wash pellet.
12. Spin down at 7500 × *g* for 5 min at 4 °C.
13. Remove the supernatant completely with a pipette and let the pellet dry for 5 min. Be careful not to over-dry as it might become difficult to resuspend the RNA pellet.
14. Resuspend in 25 μL of 1 mM DTT.
15. Incubate at 55 °C for 10 min to fully dissolve the RNA pellet.
16. Measure the concentration and purity of extracted RNA by Nanodrop. We typically obtain >7 μg of total RNA from a single well of a 6-well plate.
17. Control the integrity of the RNA, for example, on a capillary electrophoresis system like Fragment Analyzer (AATI, kit DNF-471), as judged by discrete 28S and 18S rRNA signal.
18. Store RNA at −80 °C (*see* **Note 5**) or proceed to s^4U alkylation.

3.6 s^4U-Alkylation (Iodoacetamide Treatment)

1. Prepare Sodium Phosphate buffer (500 mM $NaPO_4$, pH 8) as follows: Prepare 1 M stocks solutions of monobasic NaH_2PO_4 (138 g in 1 L H_2O) and dibasic Na_2HPO_4 (142 g in 1 L H_2O). To prepare 200 mL of 0.5 M sodium phosphate buffer, mix 93.2 mL of 1 M Na_2HPO_4 and 6.8 mL of 1 M NaH_2PO_4 and add 100 mL of H_2O. Adjust buffer to pH 8 using a pH meter.
2. Freshly prepare 100 mM iodoacetamide in EtOH. Protect from light.
3. Prewarm heating block at 50 °C.
4. Prepare 3.5 μg total RNA in 15 μL nuclease-free water (*see* **Note 6**). Keep RNA on ice.
5. Prepare alkylation mastermix (*see* Subheading 2 and **Note 7**).
6. Add 35 μL of the mastermix to 15 μL RNA, mix well.
7. Incubate the reaction at 50 °C for 15 min. Do not exceed incubation time.
8. Transfer samples on ice and immediately add 1 μL 1 M DTT to quench the reaction (*see* **Note 8**). Vortex briefly.
9. Precipitate RNA by adding 1 μL glycogen (20 mg/mL), 5 μL NaOAc (3 M, pH 5.2), and 125 μL EtOH 100%, and vortexing and incubating for 30 min at −80 °C or for >2 h at −20 °C.
10. Centrifuge samples at 16,000 × *g* for 30 min at 4 °C.

11. Remove supernatant carefully with a pipette and discard. Then add 1 mL 75% EtOH and vortex briefly.
12. Spin down at 16,000 × *g* for 10 min at 4 °C.
13. Completely remove supernatant carefully with a pipette and discard. Let the pellet airdry for 5–10 min. Do not overly dry as it might become difficult to resuspend the RNA pellet.
14. Resuspend RNA pellet in 10 μL H_2O.
15. Measure RNA concentration on Nanodrop. At this point, samples can be stored at −80 °C.
16. Control the integrity of the RNA, for example, on a capillary electrophoresis system like Fragment Analyzer (AATI, kit DNF-471), as judged by discrete 28S and 18S rRNA signal.

3.7 DNase Treatment

To remove genomic DNA contaminations, we recommend DNase treatment of RNA samples prior to library preparation (e.g., using TURBO DNA-free™ Kit, Thermo Fisher Scientific) according to manufacturer's instructions:

1. Prepare reaction as follows.

Reagent	Volume per sample
10× TURBO™ DNase buffer	4 μL
H_2O	30 μL
RNA sample	10 μL
TURBO™ DNasc	1 μL
Final volume	**45 μL**

2. Incubate at 37 °C for 20 min.
3. Add 4 μL of DNase Inactivation reagent.
4. Incubatc 5 min at room temperature; mix occasionally.
5. Centrifuge at 10,000 × *g* for 1.5 min at room temperature.
6. Transfer supernatant to a new tube.
7. Measure the concentration of RNA on microvolume spectrophotometer (e.g., Nanodrop). At this point, samples can be stored at −80 °C.

3.8 Depletion of Ribosomal RNA

To avoid adverse overrepresentation of abundant ribosomal RNA (rRNA) in RNA sequencing libraries, removal of rRNA by standard depletion protocols is essential. For cultured cells derived from human, mouse, and rat tissues, we recommend using the oligonucleotide-based RiboCop™ rRNA Depletion kit v1.3 (Lexogen GmbH) according to manufacturer's instructions, which removes cytoplasmic 28S, 18S, 5.8S, 45S, 5S, as well as mitochondrial mt16S, mt12S ribosomal RNA sequences. A brief outline of this protocol is as follows.

3.8.1 Hybridization

1. Prepare Hybridization Mix by combining 1 μg of total RNA (in a volume of 26 μL; dilute using RNase-free water if required), 4 μL hybridization solution (HS, Lexogen) and 5 μL Probe Mix (HMR V2, Lexogen) and mix thoroughly until homogenous.
2. Denature Hybridization Mix for 5 min at 75 °C in a thermomixer with agitation at 1250 rpm.
3. Incubate Hybridization Mix for 30 min at 60 °C in a thermomixer with agitation at 1250 rpm.

3.8.2 Bead Washing

1. Transfer 75 μL Depletion Beads (DB, Lexogen) per reaction to a fresh tube, place it onto a magnetic stand and let the beads collect for 5 min or until the supernatant is completely clear. Remove and discard the supernatant.
2. Add 75 μL Depletion Solution (DS, Lexogen) to the beads. Remove the tube from the magnet and resuspend the beads. Then place the tube back onto the magnetic stand and let the beads collect for 5 min or until the supernatant is completely clear. Remove and discard the supernatant. Repeat this washing step once, for a total of two washes.
3. Resuspend the beads in 30 μL Depletion Solution (DS, Lexogen).

3.8.3 Depletion

1. Spin down the Hybridization Mix (from Subheading 3.8.1) and add 30 μL of freshly prepared beads (from Subheading 3.8.2). Mix by pipetting up and down at least eight times, or until homogeneous.
2. Put the sample back to the thermomixer and incubate at 60 °C for 15 min with agitation at 1250 rpm.
3. Spin down briefly and place the sample on the magnet and let the beads collect for 5 min.
4. Recover and transfer 60 μL of the supernatant containing the rRNA-depleted RNA a fresh tube. Avoid disturbing the collected beads to prevent rRNA contamination (*see* **Note 9**).

3.8.4 Purification

1. Add 24 μL Purification Beads (PB, Lexogen) and 108 μL Purification Solution (PS, Lexogen) to the supernatant. Mix well by pipetting. Incubate for 5 min at room temperature.
2. Place the sample onto a magnetic stand and let the beads collect for 5–10 min or until the supernatant is completely clear.
3. Remove and discard the clear supernatant without removing the sample from the magnet. Make sure that accumulated beads are not disturbed.
4. Add 120 μL of 80% Ethanol (ensure that beads are fully covered) and incubate the beads for 30 s. Leave the sample on the magnet as beads should not be resuspended during this washing step. Afterward, remove and discard the supernatant.

5. Repeat this washing step once for a total of two washes. Make sure the supernatant is removed completely.
6. Leave the sample in contact with the magnetic stand at room temperature and let beads dry for 5–10 min or until all ethanol has evaporated. Do not dry extensively, since this may have negative impact on elution and RNA recovery.
7. Add 12 μL of Elution Buffer (EB, Lexogen), remove the sample for the magnet and resuspend the beads properly in EB. Incubate for 2 min at room temperature.
8. Place the sample onto the magnetic stand and let the beads collect for 2–5 min or until the supernatant is completely clear.
9. Transfer 10 μL of the supernatant into a fresh tube. Depleted RNA is now ready for optional further quality control and downstream processing. At this point, RNA can be stored at −80 °C.

3.9 RNA Library Preparation

SLAMseq is in principle compatible with any RNA sequencing library preparation protocol that converts RNA into cDNA. To determine global and transcript-specific stability of cellular RNA, we employ NEBNext® Ultra™ II Directional RNA Library Prep Kit for Illumina® (New England Biolabs). A brief outline of this protocol is as follows:

3.9.1 RNA Fragmentation and Priming

1. Concentrate purified rRNA-depleted total RNA in a SpeedVac vacuum concentrator to a total volume of 5 μL.
2. Add 4 μL NEBNext First Strand Synthesis Reaction Buffer (NEB) and 1 μL Random Primers (NEB) to 5 μL purified RNA (Subheading 3.8) and mix thoroughly by pipetting up and down ten times.
3. Place the sample on a thermocycler and incubate at 94 °C for 15 min (optimized for a final library insert size of ~200 nt) (*see* **Note 10**).
4. Immediately transfer the tube on ice upon completion of fragmentation and proceed to First Strand cDNA Synthesis.

3.9.2 First Strand cDNA Synthesis

1. Assemble the first strand synthesis reaction on ice by adding the following components to the fragmented and primed RNA (from Subheading 3.9.1):

Reagent	Volume per sample
Fragmented and primed RNA (from **step 1**)	10 μL
NEBNext Strand specificity reagent	8 μL
NEB next first Strand synthesis enzyme mix	2 μL
Final volume	**20 μL**

2. Mix thoroughly by pipetting up and down ten times.
3. Incubate the sample in preheated thermal cycler with the heated lid set at ≥80 °C.

Temperature	Time
25 °C	10 min
42 °C	15 min
70 °C	15 min
4 °C	Hold

4. Proceed directly to Second Strand cDNA Synthesis.

3.9.3 Second Strand cDNA Synthesis

1. Assemble the second strand cDNA synthesis reaction on ice by adding the following components into the first strand synthesis reaction from previous step.

Reagent	Volume per sample
First Strand synthesis reaction (Subheading 3.9.2)	20 μL
NEBNext second Strand synthesis reaction buffer with dUTP mix (10×)	8 μL
NEBNext second Strand synthesis enzyme mix	4 μL
Nuclease-free water	48 μL
Final volume	**80 μL**

2. Mix thoroughly by pipetting up and down several times while keeping the tube on ice.
3. Incubate in a thermocycler for 1 h at 16 °C with the heated lit set at ≤40 °C.

3.9.4 Purification of Double-Stranded cDNA

1. Let NEBNext Sample Purification Beads equilibrate to room temperature and vortex to resuspend immediately prior to use.
2. Add 144 μL (1.8×) of resuspended beads to the second strand synthesis reaction (80 μL). Mix well by vortexing or by pipetting up and down at least ten times.
3. Incubate for 5 min at room temperature.
4. Briefly spin the tube in a microcentrifuge to collect sample. Place the tube on a magnetic stand and let the beads collect for 2–5 min or until the supernatant is completely clear. Then carefully remove and discard the supernatant. Be careful not to disturb the beads.

5. Add 200 μL of 80% Ethanol to the tube while in the magnetic rack (ensure that beads are fully covered) and incubate the beads for 30 s. Leave the sample on the magnet as beads should not be resuspended during this washing step. Afterward, remove and discard the supernatant.
6. Repeat this washing step once for a total of two washes. Make sure the supernatant is removed completely.
7. Leave the sample in contact with the magnetic stand at room temperature and let beads dry for 5–10 min or until all ethanol has evaporated. Do not dry extensively, since this may have negative impact on elution and cDNA recovery.
8. Remove the tube from the magnetic stand and elute the DNA from the beads by adding 53 μL 0.1× TE Buffer (NEB). Mix well on a vortex mixer or by pipetting up and down at least ten times. Quickly spin the tube and incubate for 2 min at room temperature. Place the tube on the magnetic stand until the solution is clear.
9. Remove 50 μL of the supernatant and transfer to a clean nuclease-free PCR tube. At this point, cDNA can be stored at −20 °C.

3.9.5 End Prep of cDNA Library

1. Assemble the end prep reaction on ice by adding the following components to the purified double-stranded cDNA (Subheading 3.9.4):

Reagent	Volume per sample
Purified double-stranded cDNA (Subheading 3.9.4)	50 μL
NEBNext ultra II end prep reaction buffer	7 μL
NEBNext ultra II end prep enzyme mix	3 μL
Final volume	**60 μL**

2. Mix the reaction thoroughly by pipetting the entire volume up and down at least ten times. Spin down reaction briefly in a microcentrifuge.
3. Incubate the sample in a thermal cycler with the heated lid set at ≥75 °C.

Temperature	Time
20 °C	30 min
65 °C	30 min
4 °C	Hold

4. Proceed immediately to adaptor ligation.

3.9.6 Adaptor Ligation

1. Dilute the NEBNext Adaptor (NEB) 1/25 in Adaptor Dilution Buffer (NEB) prior to setting up the ligation reaction in ice-cold Adaptor Dilution Buffer and keep the adaptor on ice.
2. Assemble the ligation reaction on ice by adding the following components, in the order given, to the end prep reaction product from Subheading 3.9.5.

Reagent	Volume per sample
End prepped DNA (Subheading 3.9.5)	60 μL
Diluted adaptor	2.5 μL
NEBNext ligation enhancer	1 μL
NEBNext ultra II ligation master mix	30 μL
Final volume	**93.5 μL**

3. Mix the reaction thoroughly by pipetting the entire volume up and down at least ten times. Spin down reaction briefly in a microcentrifuge.
4. Incubate 15 min at 20 °C in a thermal cycler.
5. Add 3 μL USER Enzyme (NEB) to the ligation reaction (final total volume of 96.5 μL).
6. Mix well and incubate in a thermal cycler at 37 °C for 15 min with the heated lid set to ≥45 °C.
7. Proceed immediately to Purification of the Ligation Reaction.

3.9.7 Purification of the Ligation Reaction

1. Add 87 μL (0.9×) resuspended NEBNext Sample Purification Beads and mix well on a vortex mixer or by pipetting up and down at least ten times.
2. Incubate for 10 min at room temperature.
3. Briefly spin the tube in a microcentrifuge to collect sample. Place the tube on a magnetic stand and let the beads collect for 2–5 min or until the supernatant is completely clear. Then carefully remove and discard the supernatant. Be careful not to disturb the beads.
4. Add 200 μL of 80% Ethanol to the tube while in the magnetic rack (ensure that beads are fully covered) and incubate the beads for 30 s. Leave the sample on the magnet as beads should not be resuspended during this washing step. Afterward, remove and discard the supernatant.
5. Repeat this washing step once for a total of two washes. Make sure the supernatant is removed completely.

6. Leave the sample in contact with the magnetic stand at room temperature and let beads dry for 5–10 min or until all ethanol has evaporated. Do not dry extensively, since this may have negative impact on elution and adaptor-ligated cDNA recovery.
7. Remove the tube from the magnetic rack. Elute DNA target from the beads by adding 17 μL 0.1× TE (NEB) to the beads. Mix well on a vortex mixer or by pipetting up and down. Briefly spin the tube in a microcentrifuge incubate for 2 min at room temperature. Put the tube in the magnetic stand and let the beads collect for 2–5 min or until the supernatant is completely clear.
8. Without disturbing the bead pellet, transfer 15 μL of the supernatant to a clean PCR tube and proceed to PCR enrichment. At this point, adapter-ligated cDNA can be stored at −20 °C.

3.9.8 PCR Enrichment of Adaptor-Ligated DNA

1. Set up the following dual-barcode amplification reaction in ultra-clear flat cap PCR tubes. Use each one separate barcoded i7 and i5 primer per sample for unique dual barcoding.

Reagent	Volume per sample
Adapter-ligated DNA (from Subheading 3.9.7)	7.5 μL
NEB next ultra II Q5 master mix	25 μL
20× EvaGreen	2.5 μL
i7 primer (index X)	5 μL
i5 primer (index Y)	5 μL
H_2O	5 μL
Final volume	**50 μL**

2. Mix the reaction gently by pipetting up and down ten times. Quickly spin the tube in a microcentrifuge. Avoid introducing bubbles, as it will disturb the signal detection during the qPCR run.
3. Thaw KAPA fluorescent standards 3 and 4. Mix well and pipet 50 μL of each standard in each one ultra-clear flat cap PCR tube.
4. Place the dual-barcode amplification reactions and the standards into a real-time thermal cycler with the heated lid set to 105 °C and perform PCR amplification using the following PCR cycling conditions:

Temperature	Time	Cycles
98 °C	30 s	1
98 °C	30 min	5–20[a]
65 °C	30 min	
Signal detection		
65 °C	Hold	

[a]Amplification should be monitored by real-time amplification signal. As soon as signal reaches a range between Standard 3 and 4, briefly interrupt the cycling, remove sample from the machine and place on ice. Then continue cycling with the rest of the samples. At a starting total RNA input of 1 μg amplification should complete in less than 15 cycles

3.9.9 Purification of the PCR Reaction

1. Vortex NEBNext Sample Purification Beads to resuspend.
2. Add 45 μL (0.9×) of resuspended beads to the PCR reaction (~50 μL). Mix well on a vortex mixer or by pipetting up and down at least ten times.
3. Incubate for 5 min at room temperature.
4. Briefly spin the tube in a microcentrifuge to collect sample. Place the tube on a magnetic stand and let the beads collect for 2–5 min or until the supernatant is completely clear. Then carefully remove and discard the supernatant. Be careful not to disturb the beads.
5. Add 200 μL of 80% Ethanol to the tube while in the magnetic rack (ensure that beads are fully covered) and incubate the beads for 30 s. Leave the sample on the magnet as beads should not be resuspended during this washing step. Afterward, remove and discard the supernatant.
6. Repeat this washing step once for a total of two washes. Make sure the supernatant is removed completely.
7. Leave the sample in contact with the magnetic stand at room temperature and let beads dry for 5–10 min or until all ethanol has evaporated. Do not dry extensively, since this may have negative impact on elution and adaptor-ligated cDNA recovery.
8. Remove the tube from the magnetic rack. Elute DNA target from the beads by adding 23 μL 0.1× TE (NEB) to the beads. Mix well on a vortex mixer or by pipetting up and down. Briefly spin the tube in a microcentrifuge incubate for 2 min at room temperature. Put the tube in the magnetic stand and let the beads collect for 2–5 min or until the supernatant is completely clear.

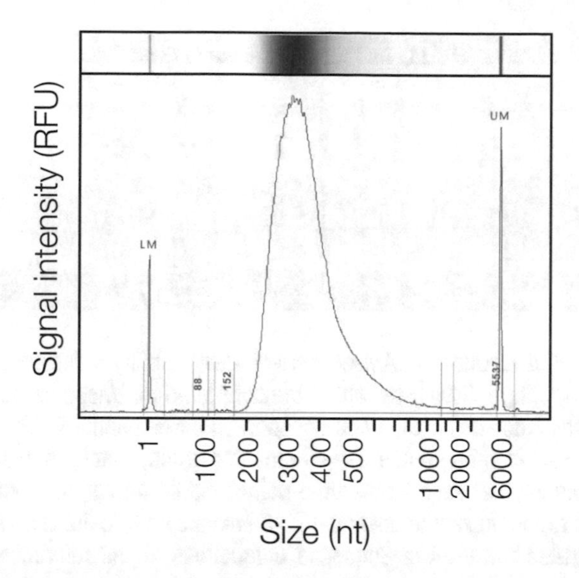

Fig. 2 Expected electropherogram of a representative cDNA library prepared from rRNA-depleted total RNA extracted from HeLa cells following the described protocol for metabolic RNA sequencing. RNA library was prepared using NEBNext® Ultra™ II Directional RNA Library Prep Kit for Illumina (New England Biolabs). Fragment size is indicated in nucleotides (nt). RFU, relative fluorescence units. Raw data image is shown on top

9. Without disturbing the bead pellet, transfer 20 μL of the supernatant to a clean PCR tube and proceed to PCR enrichment. At this point, adapter-ligated cDNA can be stored at −20 °C.

3.9.10 Assess Library Quantity and Quality

1. Assess the quality of the library on a capillary electrophoresis system such as Fragment Analyzer (AATI, kit DNF-474) following manufacturer's instructions (Fig. 2).
2. Determine the concentration, for example by using QuDye dsDNA HS Assay Kit (Qubit) following the instructions of the manufacturer.
3. For multiplexing samples before sequencing mix equimolar quantities of libraries at the molarity and volume required by the sequencing provider.

3.10 High-Throughput Sequencing

As a standard approach for SLAMseq libraries generated to monitor transcriptome-wide RNA stability, we recommend >20 million reads per library for the assessment of >5.000 transcripts. We recommend sequencing with single read 100 (SR 100) mode, which typically enables to recover the vast majority of labeled transcripts (>70% at a labeling efficiency of ~2.3%) [12].

3.11 Data Analysis

For data analysis, standard workflows for adapter trimming, genome alignment, and nucleotide conversion counting can be

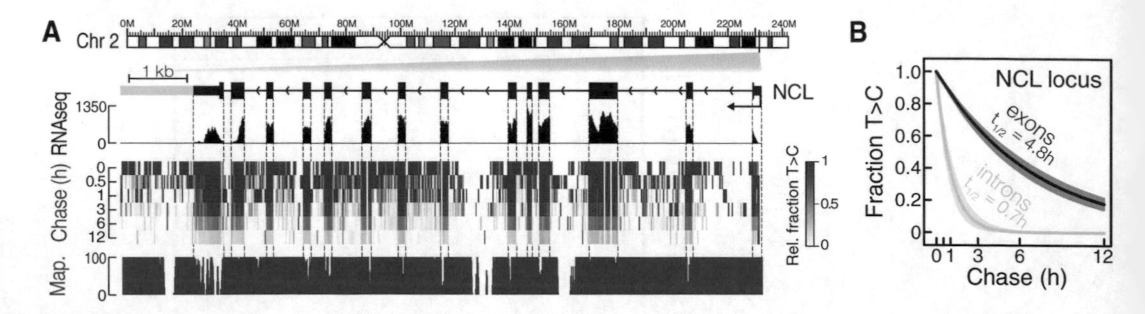

Fig. 3 Expected results of SLAMseq-derived RNA stability measurements in HeLa cells subjected to s^{4}U-pulse–chase labeling followed by whole-transcriptome SLAMseq as described in this protocol. (**a**) Sequencing tracks for the Nucleolin (NCL) gene encoded on chromosome 2 are displayed. Average steady-state RNAseq signal across the time-course experiment (top track, black), and relative T>C conversion rate for 20 bp counting windows at each chase-time point (middle tracks, red), where signal of each counting window is normalized to the maximum observed T>C conversion rate are shown. Mappability track (Map., bottom track, blue) illustrates the level of sequence uniqueness of the reference genome assembly, where high signal represents areas where the sequence is unique. (**b**) T>C conversion rates ± confidence interval of counting windows overlapping with exons (black) or introns (grey) of the NCL gene (shown in **a**) were determined for each timepoint of a s^{4}U pulse/chase experiment and fit to a single-exponential decay model to derive half-life ($t_{½}$) in hours (h)

employed [16]. A standard workflow employs the following processing steps: Raw reads are demultiplexed, reverse complemented (fastx-toolkit v0.0.13; http://hannonlab.cshl.edu/fastx_toolkit/), and adapter- and quality trimmed by trim_galore v0.3.7 (typically 10 nt were clipped from the 5′-end of each read) [17]. Preprocessed reads are mapped to GRCh38 with STAR_2.6.1d [18]. Read alignments are then processed column-wise (pileup) at selected genomic regions using python scripts based on pysam v0.15.4 [19]. Briefly, we count strand-specific T>T and T>C conversions in the respective pileup and calculated T/C fractions as frac_tc=#tc / (#tt +#tc). T/C SNPs are filtered from the data by either providing a list of SNP positions called from DNA-seq data of the respective cell line or, if this data is not available, by excluding genomic positions with an overall frac_tc across all timepoints above a given threshold (e.g., 80% for haploid and 30% for diploid cell lines). Fractions of T>C conversions are then calculated per tiling genomic window (window size = 20 bp) and resulting signals are normalized to the maximum fraction across all timepoints. Windows are classified as exonic/intronic/other according to gencode.v32 annotations. A single exponential decay model can be used to fit the normalized T>C conversions over time in order to obtain half-life information for each transcript or genomic feature. For representative examples of a SLAMseq RNA half-life measurement in HeLa cells, as described in this protocol, *see* Fig. 3.

4 Notes

1. The cell seeding density for s^4U-labeling should be determined by considering the growth rate of the cell type and the duration of the pulse–chase experiment. We recommend optimizing seeding conditions that allow exponential growth for the duration of the experiment.
2. To minimize disturbances of the cells during the s^4U labeling time and to avoid toxic effects of TRIzol® vapor arising from neighboring wells while harvesting, seed only three wells per 6-well plate, corresponding to the triplicates harvested at the same time.
3. Different cell types may vary in their sensitivity to s^4U. Thus, determine nontoxic s^4U concentrations by the cell viability assay (*see* Subheading 3.2).
4. Low DTT concentrations are added during the RNA extraction protocol to ensure the reduced state of the thiol group.
5. We recommend to first perform the alkylation of s^4U by treating the RNA with iodoacetamide before long-term storage of the RNA sample.
6. In case less RNA is available and/or downstream library preparations require less input material, input RNA amounts can be decreased for the IAA treatment. The reaction was tested to perform with equal efficiency with an input of up to 5 μg total RNA.
7. $NaPO_4$ can form a salt aggregate in DMSO. Always prepare >10% additional master-mix.
8. Alkylation of s^4U is completed at this point. It is not necessary to protect samples from conventional light sources from this point onward.
9. Do not throw away the supernatant, which contains the rRNA depleted total RNA sample required for downstream transcriptome library preparation.
10. Fragmentation conditions were optimized for rRNA-depleted human total RNA. RNA from other sources may require different fragmentation times and/or input amounts. Note, that adequate fragmentation relies on intact and undegraded input RNA.

Acknowledgments

HTP sequencing was performed at the VBCF NGS Unit (www.vbcf.ac.at). This work was supported in part by the European Research Council grants ERC-CoG-866166 (RiboTrace) and ERC-PoC-825710 (SLAMseq) to S.L.A.

References

1. Mata J, Marguerat S, Bähler J (2005) Post-transcriptional control of gene expression: a genome-wide perspective. Trends Biochem Sci 30:506–514. https://doi.org/10.1016/j.tibs.2005.07.005
2. Ghosh S, Jacobson A (2010) RNA decay modulates gene expression and controls its fidelity. Wiley Interdisciplinary Reviews RNA 1:351–361. https://doi.org/10.1002/wrna.25
3. Schwanhäusser B, Busse D, Li N, Dittmar G, Schuchhardt J, Wolf J, Chen W, Selbach M (2011) Global quantification of mammalian gene expression control. Nature 473:337–342. https://doi.org/10.1038/nature10098
4. Stark R, Grzelak M, Hadfield J (2019) RNA sequencing: the teenage years. Nat Rev Genet 20:631–656. https://doi.org/10.1038/s41576-019-0150-2
5. Cramer P (2019) Organization and regulation of gene transcription. Nature 573:45–54. https://doi.org/10.1038/s41586-019-1517-4
6. Singer RH, Penman S (1972) Stability of HeLa cell mRNA in actinomycin. Nature 240:100–102
7. Wang Y, Liu CL, Storey JD, Tibshirani RJ, Herschlag D, Brown PO (2002) Precision and functional specificity in mRNA decay. Proc Natl Acad Sci U S A 99:5860–5865. https://doi.org/10.1073/pnas.092538799
8. Tani H, Akimitsu N (2014) Genome-wide technology for determining RNA stability in mammalian cells. RNA Biol 9:1233–1238. https://doi.org/10.4161/rna.22036
9. Haimovich G, Medina DA, Causse SZ, Garber M, Millán-Zambrano G, Barkai O, Chávez S, Pérez-Ortín JE, Darzacq X, Choder M (2013) Gene expression is circular: factors for mRNA degradation also Foster mRNA synthesis. Cell 153:1000–1011. https://doi.org/10.1016/j.cell.2013.05.012
10. Riml C, Amort T, Rieder D, Gasser C, Lusser A, Micura R (2017) Osmium-mediated transformation of 4-Thiouridine to cytidine as key to study RNA dynamics by sequencing. Angew Chem Int Ed 56:13479–13483. https://doi.org/10.1002/anie.201707465
11. Schofield JA, Duffy EE, Kiefer L, Sullivan MC, Simon MD (2018) TimeLapse-seq: adding a temporal dimension to RNA sequencing through nucleoside recoding. Nat Methods 15:221–225. https://doi.org/10.1038/nmeth.4582
12. Herzog VA, Reichholf B, Neumann T, Rescheneder P, Bhat P, Burkard TR, Wlotzka W, von Haeseler A, Zuber J, Ameres SL (2017) Thiol-linked alkylation of RNA to assess expression dynamics. Nat Methods 14:1198–1204. https://doi.org/10.1038/nmeth.4435
13. Muhar M, Ebert A, Neumann T, Umkehrer C, Jude J, Wieshofer C, Rescheneder P, Lipp JJ, Herzog VA, Reichholf B, Cisneros DA, Hoffmann T, Schlapansky MF, Bhat P, Haeseler A, von Köcher T, Obenauf AC, Popow J, Ameres SL, Zuber J (2018) SLAM-seq defines direct gene-regulatory functions of the BRD4-MYC axis. Science 360:800–805. https://doi.org/10.1126/science.aao2793
14. Reichholf B, Herzog VA, Fasching N, Manzenreither RA, Sowemimo I, Ameres SL (2019) Time-resolved small RNA sequencing unravels the molecular principles of MicroRNA homeostasis. Mol Cell 75:756–768.e7. https://doi.org/10.1016/j.molcel.2019.06.018
15. Herzog VA, Fasching N, Ameres SL (2020) Determining mRNA stability by metabolic RNA labeling and chemical nucleoside conversion. Methods Mol Biol 2062:169–189. https://doi.org/10.1007/978-1-4939-9822-7_9
16. Neumann T, Herzog VA, Muhar M, von Haeseler A, Zuber J, Ameres SL, Rescheneder P (2019) Quantification of experimentally induced nucleotide conversions in high-throughput sequencing datasets. BMC Bioinformatics 20:258. https://doi.org/10.1186/s12859-019-2849-7
17. Martin M (2011) Cutadapt removes adapter sequences from high-throughput sequencing reads. EMBnet J 17:10. https://doi.org/10.14806/ej.17.1.200
18. Dobin A, Davis CA, Schlesinger F, Drenkow J, Zaleski C, Jha S, Batut P, Chaisson M, Gingeras TR (2012) STAR: ultrafast universal RNA-seq aligner. Bioinformatics 29:15–21. https://doi.org/10.1093/bioinformatics/bts635
19. Li H, Handsaker B, Wysoker A, Fennell T, Ruan J, Homer N, Marth G, Abecasis G, Durbin R, Subgroup 1000 Genome Project Data Processing (2009) The sequence alignment/map format and SAMtools. Bioinformatics 25:2078–2079. https://doi.org/10.1093/bioinformatics/btp352

Chapter 18

High-Throughput Quantitation of Yeast uORF Regulatory Impacts Using FACS-uORF

Gemma E. May and C. Joel McManus

Abstract

Eukaryotic *u*pstream *O*pen *R*eading *F*rame*s* (uORFs) are short translated regions found in many transcript leaders (Barbosa et al. PLoS Genet 9:e1003529, 2013; Zhang et al. Trends Biochem Sci 44:782–794, 2019). Modern transcript annotations and ribosome profiling studies have found thousands of AUG-initiated uORFs, and many more uORFs initiated by near-cognate codons (CUG, GUG, UUG, etc.). Their translation generally decreases the expression of the main encoded protein by preventing ribosomes from reaching the main ORF of each gene, and by inducing *n*onsense *m*ediated *d*ecay (NMD) through premature termination. Under many cellular stresses, uORF containing transcripts are de-repressed due to decreased translation initiation (Young et al. J Biol Chem 291:16927–16935, 2016). Traditional experimental evaluation of uORFs involves comparing expression from matched uORF-containing and start-codon mutated transcript leader reporter plasmids. This tedious process has precluded analysis of large numbers of uORFs. We recently used FACS-uORF to simultaneously assay thousands of yeast uORFs in order to evaluate the impact of codon usage on their functions (Lin et al. Nucleic Acids Res 2:1–10, 2019). Here, we provide a step-by-step protocol for this assay.

Key words mRNA translation, uORFs, Massively parallel reporter assay

1 Introduction

Eukaryotic mRNA translation is regulated primarily at the initiation stage [5]. Most eukaryotic translation is cap-dependent under unstressed conditions. Translation initiation requires recruitment of the ribosomal subunit and numerous eukaryotic initiation factors (eIFs) to mRNA 5′ caps to form a preinitiation complex (PIC). The assembled PIC then scans directionally, from 5′ to 3′, until a start codon is recognized, the complete ribosome is assembled, and protein synthesis is initiated. When translation initiates at uORFs, this typically precludes translation of the main protein encoded in the mRNA transcript. uORFs can also induce nonsense mediated decay, due to translation termination in the 5′ UTR [1]. Under stress conditions, phosphorylation of the alpha subunit of eIF2

Erik Dassi (ed.), *Post-Transcriptional Gene Regulation*, Methods in Molecular Biology, vol. 2404,
https://doi.org/10.1007/978-1-0716-1851-6_18,

decreases the rate of translation initiation. This allows stress-responsive mRNAs to be rapidly derepressed in response to numerous cellular stresses [3].

Genomic analyses have revealed that uORFs are very common, occurring in ~15% of yeast and ~50% of mammalian transcripts [2, 6]. Despite their pervasive nature, the functions of most uORFs have not been tested. Traditional assays to test uORF functions typically involve comparing expression from reporter genes with wild-type 5′ transcript leaders to expression from reporters with transcript leaders in which the uORF start codon has been mutated. Typically, these reporters are individually cloned and assayed one at a time. This low throughput approach has precluded high-throughput analyses of uORF activity.

Over the past decade, several groups have combined technological improvements in DNA synthesis and sequencing to create high-throughput reporter assays for transcription [7, 8], mRNA splicing [9–11], and translation [12–15]. These experimental systems are collectively known as massively parallel reporter assays, or MPRAs. Here we provide a detailed protocol for FACS-uORF, an MPRA system we generated to assay thousands of uORFs in parallel in yeast [4]. Our protocol covers all steps of the process, including cloning a designer library of reporter plasmids, FACS-sorting cells, and generating high-throughput sequencing libraries to evaluate uORF function (Fig. 1).

2 Materials

1. Pipettes with filter tips (P1000, P200, P20, P2).
2. 1xTE, 10 mM Tris, 0.1 mM EDTA, pH 8.0.
3. 10 pmole custom designed oligo pool (Agilent). Design an oligo library where each oligo has common 5′ and 3′ ends that allow for PCR amplification and plasmid integration. The 5′ end sequence is the PCR primer, ENO2-Lib-F1 plus the sequence, AACTAATA, and the 3′ end sequence is the complementary to the FACS-uORF-YFP-R primer (*see* Table 1, and Fig. 2)
4. 1.5 mL microcentrifuge tubes.
5. Vortexer.
6. 0.2 mL PCR tube.
7. DNase/RNase-free distilled water (nuclease free).
8. Herculase II Fusion DNA polymerase with dNTPs Combo (Agilent).
9. Custom DNA oligo primers (*see* Table 1).

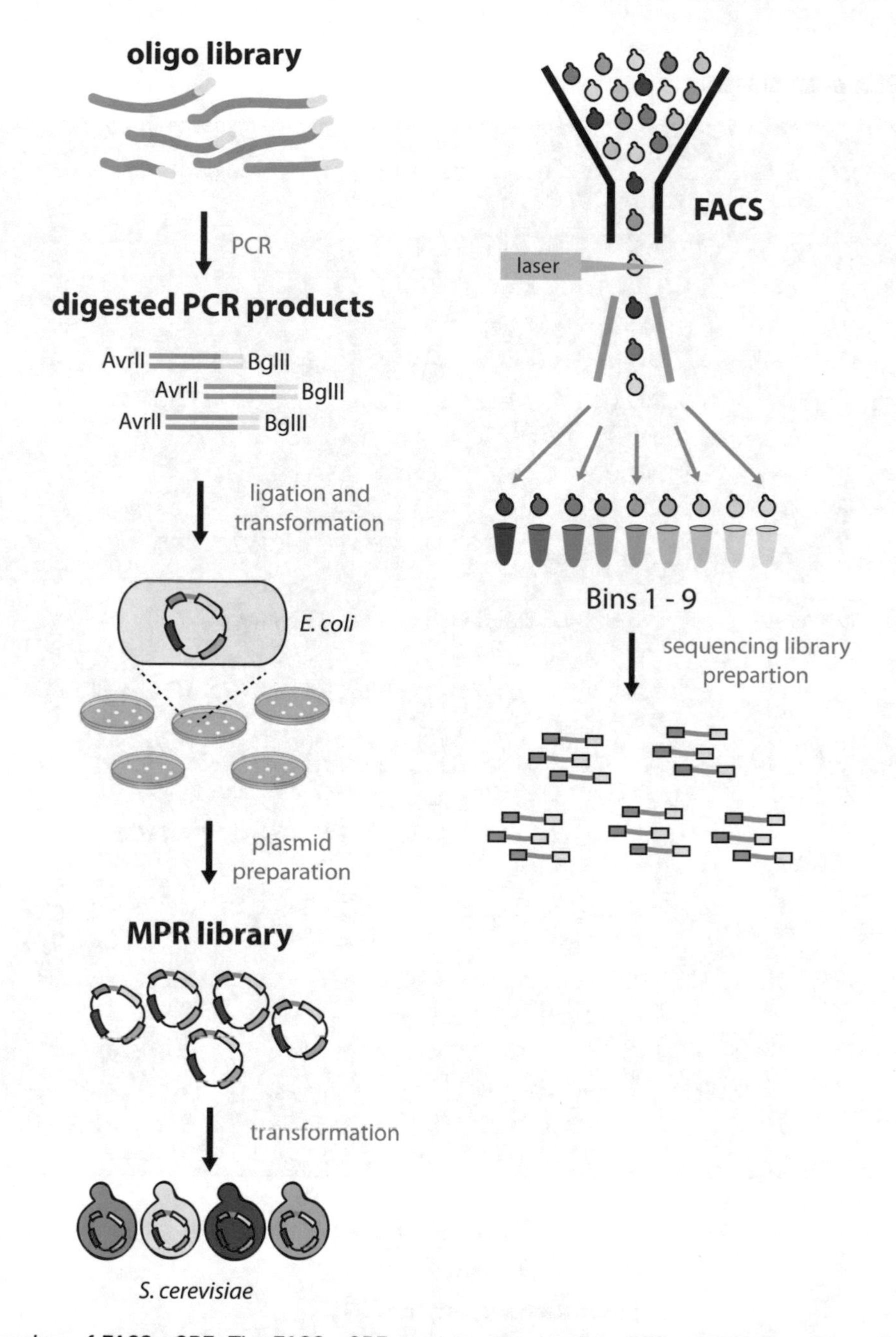

Fig. 1 Overview of FACS-uORF. The FACS-uORF assay system involves PCR amplification of an oligo library containing designer 5′ UTR templates for wild-type (uORF containing) and start-codon mutant (e.g., AUG→AAG) sequence pairs. The resulting products are cut and pasted into a plasmid background to generate a Massively Parallel Reporter library. Yeast transformed with this library are fluorescently sorted using FACS. After overnight outgrowth, reporter plasmids are recovered from each expression bin and used to generate high-throughput sequencing libraries for further analysis

Table 1
Custom DNA oligo primers

Primer Name	Sequence (5′–3′)
Eno2_lib_F1	AGTTTCTTTCATAACACCAAGC
FACS-uORF-YFP-R	AATTCTTCACCTTTAGATCTCAT
Eno2_lib_AvrII_F2	TTCCCTAGGCTCAATCTTTTATTTTTATTTTATTTTTCTTTTCTTAG TTTCTTTCATAACACCAAGC
FuORF_2_DNA_N0_F	TTCAGACGTGTGCTCTTCCGATCTAGTTTCTTTCATAACACCAAGC
FuORF_2_DNA_N1_F	TTCAGACGTGTGCTCTTCCGATCTNAGTTTCTTTCA TAACACCAAGC
FuORF_2_DNA_N2_F	TTCAGACGTGTGCTCTTCCGATCTNNAGTTTCTTTCA TAACACCAAGC
FuORF_2_DNA_N3_F	TTCAGACGTGTGCTCTTCCGATCTNNNAGTTTCTTTCA TAACACCAAGC
FuORF_2_DNA_N4_F	TTCAGACGTGTGCTCTTCCGATCTNNNNAGTTTCTTTCA TAACACCAAGC
FuORF_2_DNA_N5_F	TTCAGACGTGTGCTCTTCCGATCTNNNNNAGTTTCTTTCA TAACACCAAGC
FuORF_2_DNA_N6_F	TTCAGACGTGTGCTCTTCCGATCTNNNNNNAGTTTCTTTCA TAACACCAAGC
FuORF_2_DNA_N7_F	TTCAGACGTGTGCTCTTCCGATCTNNNNNNNAGTTTCTTTCA TAACACCAAGC
FuORF_2_DNA_N0_R	GCGACCACCGAGATCTACACTCTTTCCCTACACGACGCTC TTCCGATCTTCTTCACCTTTAGATCTCAT
FuORF_2_DNA_N2_R	GCGACCACCGAGATCTACACTCTTTCCCTACACGACGCTC TTCCGATCTNNTCTTCACCTTTAGATCTCAT
FuORF_2_DNA_N4_R	GCGACCACCGAGATCTACACTCTTTCCCTACACGACGCTC TTCCGATCTNNNNTCTTCACCTTTAGATCTCAT
FuORF_2_DNA_N6_R	GCGACCACCGAGATCTACACTCTTTCCCTACACGACGCTC TTCCGATCTNNNNNNTCTTCACCTTTAGATCTCAT

10. Plasmid vectors (*pGM-ENO2-YFP-mCherry, pGM-YFP, pGM-mCherry*).
11. Thermocycler, any model.
12. DNA Clean & Concentrator™—5 (Zymo Research) or similar PCR column clean up kit.
13. Tabletop centrifuge capable of accommodating 1.5 mL ultracentrifuge tubes.
14. D1000 ScreenTape and sample buffer (Agilent).
15. TapeStation Instrument (Agilent Technologies, model 4400 or similar).

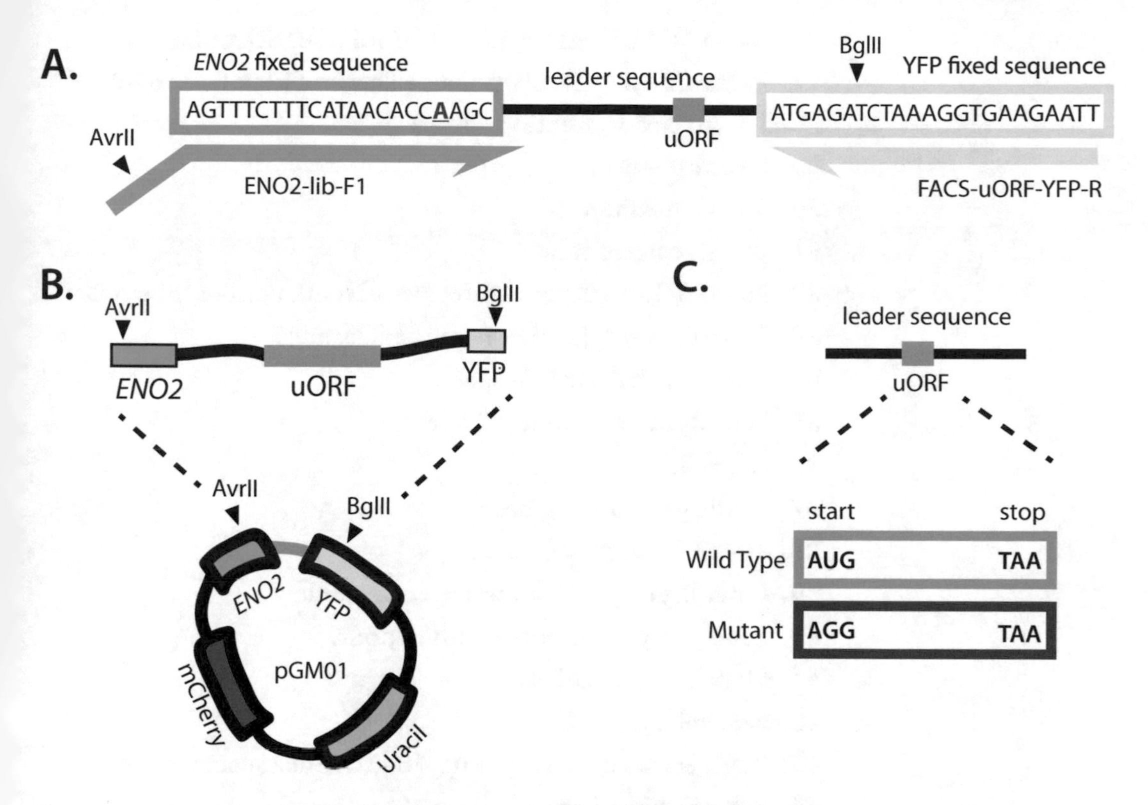

Fig. 2 MPR Assay Design. (**a**) Upstream open reading frames (orange) are present in their native leader sequence (black). The 5′ end of the oligo contains the 3′ end of the ENO2 promotor sequence (blue), which initiates transcription at the bold and underlined "A" (red). The 3′ end of the oligo contains the 5′ end of the yellow fluorescent protein (YFP) sequence (yellow). PCR primers ENO2-lib-F1 and FACS-uORF-YFP-R anneal to the oligo library and contain AvrII and BglII restriction sites. (**b**) The vector, pGM01 contains an AvrII site in the ENO2 promotor (blue) and a BglII site in the downstream YFP gene (yellow). The digested oligo library is cloned into the vector using the AvrII and BglII sites so that the effect of the transcript leader containing the uORF on translation is measured by the levels of YFP. (**c**) The start codons of the uORFs are mutated to a nonfunctional start codon (AAG or AGG) for the mutant uORF, such that each uORF containing construct has a matched nonfunctional mutant uORF construct

16. Q5® High-Fidelity DNA polymerase.
17. dNTP mix, 10 mM each.
18. NanoDrop™ 2000 Spectrophotometer or similar.
19. FastDigest XmaJI (Cuts AvrII sites (isoschizomer)) (ThermoFisher Scientific).
20. FastDigest BglII (ThermoFisher Scientific).
21. Phenol–chloroform–isoamyl alcohol (25:24:1, v/v).
22. Shrimp alkaline phosphatase (rSAP, New England Biolabs).
23. AMPure XP beads (Beckman Coulter).
24. Magnetic rack or plate for 0.2 mL tubes.
25. Ethanol, absolute 200 proof.

26. Rapid DNA Ligation Kit (ThermoFisher Scientific).
27. ElectroMAX™ DH10B Cells (ThermoFisher Scientific).
28. 1.1 cm electroporation cuvette.
29. Electroporator.
30. S.O.C. medium.
31. 16 mL culture tube.
32. 30–37 °C shaking incubator with 16 mL culture tube holders.
33. 100 mL and 1 L Erlenmeyer flask clamps.
34. Luria Broth Agar (LB agar).
35. Petri dishes 100 mm × 15 mm.
36. Ampicillin.
37. Sterile glass plating beads.
38. 30–37 °C plate incubator.
39. Luria Broth (LB), autoclaved and sterile.
40. Laboratory wash bottle with a spout.
41. 50 mL conical tubes.
42. General use scale.
43. Refrigerated centrifuge with 50 mL tube capacity rotor.
44. Plasmid Maxi Prep Kit.
45. Yeast extract-peptone-dextrose growth medium agar (YEPD agar) plates, sterile.
46. Yeast extract-peptone-dextrose growth medium (YEPD broth), sterile.
47. Frozen-EZ Yeast Transformation II Kit (Zymo Research).
48. *S. cerevisiae* strain BY4741 (MATa his3Δ1 leu2Δ0 met15Δ0 ura3Δ0).
49. 100 mL and 1 L glass Erlenmeyer flasks.
50. Autoclavable glass storage bottles (1 L).
51. 1 mL spectrophotometer cuvette.
52. Spectrophotometer with 1.0 mL cuvette capacity.
53. SD-URA plates and broth (yeast nitrogen base without amino acids (Sigma-Aldrich), yeast Synthetic drop-out medium supplements without uracil (Sigma-Aldrich), dextrose, bacteriological agar).
54. Sterile serological pipettes (5, 10, 25, 50 mL).
55. Pipet-Aid or similar pipette controller.
56. Bunsen burner and striker.
57. Sterile cryovial (1.8 mL) or similar cryogenic tubes.
58. Sterile 30% glycerol.

59. 5 mL sterile culture tubes.
60. FACSVantage Digital Cell Sorter, or similar.
61. Tecan Microplate reader, or similar.
62. White 96 well flat bottom plates compatible with plate reader.
63. Zymoprep Yeast Plasmid Miniprep II kit (Zymo Research).
64. NEBNext™ dual index primers (New England Biolabs, E6440).
65. Qubit™ dsDNA HS Assay Kit (Invitrogen).
66. 0.5 mL optically clear individual PCR tubes
67. Qubit™ Fluorometer, any model (Thermo Fisher Scientific).

3 Methods

3.1 Oligo Library PCR Amplification

1. Resuspend 10 pmoles of the custom-designed oligo pool (**item 3** in Subheading 2) by adding 100 μL of TE to the lyophilized oligos. Vortex to mix.
2. Set up the following reaction in a 1.5 mL ultracentrifuge tube. Herculase polymerase is specifically used to maintain library complexity when amplifying from a dilute oligo pool:

Reagent	Volume
Nuclease-free water	255.6 μL
Oligo pool	30 μL
5× Herculase buffer	80 μL
dNTPs (25 mM each)	6.4 μL
Eno2_lib_F1 (10 μM)	10 μL
FACS-uORF-YFP-R (10 μM)	10 μL
Herculase II fusion polymerase	8 μL

3. Mix the reaction by setting the pipettor to 150 μL and pipetting up and down several times.
4. Aliquot the reaction into eight 0.2 mL PCR tubes with 50 μL in each tube.
5. Put the reaction in a thermocycler and run the program.
   (a) 95 °C for 1 min.
   (b) 10 cycles of 95 °C for 20 s, 55 °C for 20 s, and 68 °C for 20 s.
   (c) 68 °C for 4 min.
6. Combine the PCR into one 1.5 mL ultracentrifuge tube and purify over a DNA Clean & Concentrator™ column (or similar

PCR clean up column kit) following the manufacturer's instructions. Elute the PCR product in 20 μL of nuclease-free water. The PCR product can be saved at −20 °C.

7. Visualize 1 μL of PCR product on a D1000 DNA ScreenTape on a TapeStation (or similar) to confirm that the correct sized products are present. The PCRs add 90 base pairs to the designed insert size.
8. Set up the following reaction in a 1.5 mL ultracentrifuge tube.

Reagent	Volume
Nuclease-free water	259 μL
Oligo pool PCR product (Subheading 3.1, **step 6**)	9 μL
5× Q5 buffer	80 μL
dNTPs (10 mM)	8 μL
Eno2-lib-AvrII-F2 (10 μM)	20 μL
FACS-uORF-YFP-R (10 μM)	20 μL
Q5® polymerase	4 μL

9. Aliquot 50 μL of the reaction in to eight 0.2 mL PCR tubes. Place the tubes in a thermocycler and run the program.
   (a) 98 °C for 30 s.
   (b) 10 cycles of 98 °C for 10 s, 55 °C for 20 s, and 72 °C for 30 s.
   (c) 68 °C for 4 min.
10. Combine the PCR into one 1.5 mL ultracentrifuge tube and purify over a DNA Clean & Concentrator™ column following the manufacturer's instructions. Elute the PCR product in 50 μL of nuclease-free water.
11. Quantify the PCR product using a Nanodrop. The typical yield is ~7.5 μg. The purified PCR product can be saved at −20 °C.

3.2 Restriction Digest—Digest the Oligo PCR Products and Vector

1. Digest the PCR products and vector in preparation for ligation. Set up the following reactions in one 0.2 mL PCR tube for each reaction.

Reagent	Amount
Oligo library PCR product (Subheading 3.1, **step 11**)	1 μg
FD 10× buffer	15 μL
FD BglII	5 μL
FD XmaJI (AvrII)	5 μL
Nuclease-free water	To 150 μL

Reagent	Amount
pGM-ENO2-YFP-mCherry	5 μg
FD 10× buffer	15 μL
FD BglII	5 μL
FD XmaJI (AvrII)	5 μL
Nuclease-free water	To 150 μL

2. Mix each restriction digest reaction thoroughly with a pipettor and incubate in a thermocycler for 1 h at 37 °C.
3. Transfer the restriction digests to 1.5 mL microcentrifuge tubes. Add 150 μL of TE to each restriction digest.
4. In a fume hood, add 300 μL of phenol–chloroform–isoamyl alcohol (25:24:1, v/v) to each reaction. Vortex for 30 s.
5. Centrifuge the sample at top speed, or ~14,000 × *g* in a tabletop microcentrifuge for 2 min at room temperature.
6. In a fume hood, remove 250 μL the supernatant to a fresh 1.5 mL tube, without disturbing the bottom phenol layer.
7. Purify the supernatant over a DNA Clean & Concentrator™ column following the manufacturer's instructions and elute the cut PCR product and vector in 15 μL and 43 μL of nuclease-free water respectively.
8. Quantify the digested PCR product using the NanoDrop. The typical yield for the PCR product is ~750 ng. Proceed immediately to Subheading 3.3, saving the PCR product on ice.

3.3 Vector Phosphatase Treatment

1. Set up the following reaction in a 0.2 mL PCR tube for the vector.

Reagent	Amount
pGM-ENO2-YFP-mCherry (Subheading 3.2, **step** 7)	43 μL
10X CutSmart buffer	5 μL
rSAP (New England Biolabs)	3 μL

2. Place the above reaction in a thermocycler and incubate at 37 °C for 30 min. Heat inactivate at 65 °C for 5 min.
3. Remove the phosphatase reaction from the thermocycler and add 50 μL of nuclease-free water.
4. Vortex the AMPure XP beads until they are resuspended and add 100 μL of beads to the phosphatase reaction. Briefly vortex the reaction and incubate for 5 min at room temperature.

5. Place the tube on the magnet and incubate for 5 min at room temperature.
6. With the tube still on the magnet, carefully remove the supernatant and discard the supernatant.
7. Wash the beads by adding 150 μL of freshly prepared 80% ethanol without disturbing the pellet. Incubate for 30 s and remove the ethanol.
8. Let the pellet air-dry for around 3 min. Using a pipettor, pipet any remaining ethanol from the tube.
9. Remove the tube from the magnet and add 20 μL of nuclease-free water to the beads and resuspend by pipetting up and down. Incubate at room temperature for 2 min.
10. Place the tube on the magnet and incubate for 2 min or until the supernatant is clear. Remove the supernatant to a fresh 1.5 mL microcentrifuge tube.
11. Quantify the digested and phosphatased vector using a NanoDrop. The typical yield is 1.5 μg. Proceed immediately to the ligation (Subheading 3.4).

3.4 Ligation

1. Set up the following reaction in a 0.2 mL PCR tube. For a negative control, set up a duplicate ligation reaction adding nuclease-free water instead of digested PCR product.

Reagent	Amount
pGM-ENO2-YFP-mCherry (Subheading 3.3, **step 11**)	200 ng
XmaJI/BglII digested PCR product (Subheading 3.2, **step 8**)	30 ng
5× rapid ligation buffer	8 μL
T4 DNA ligase (5 U/μL)	2 μL
Nuclease-free water	To 40 μL

2. Incubate both reactions at 22 °C for 10 min Proceed to the next step or save the reactions at −20 °C overnight.
3. Purify the ligation reactions over a DNA Clean & Concentrator™ column following the manufacturer's instructions. Elute in 10 μL of nuclease-free water. Proceed to Subheading 3.5 or save the ligations overnight at −20 °C.

3.5 E. coli Transformation

Note: A small-scale initial transformation is performed to estimate the transformation efficiency and yield. Once these are established, a larger-scale transformation is performed to recover enough colonies to maintain library diversity.

1. Prechill one 0.1 cm electroporation cuvette per reaction on ice.
2. In a cold 1.5 mL microcentrifuge tube, add 1.5 μL of the ligation to 30 μL of competent cells. Gently mix pipetting up and down 2–3 times with a pipettor set to 20 μL.
3. Transfer the cells and ligation mix to the electroporation cuvettes.
4. Electroporate each reaction at 2.0 kV, 200 Ω, and 25 μF. Immediately add 1.5 mL of room temperature S.O.C medium and transfer the cells and liquid to a 16 mL culture tube.
5. Place the cells in a shaking incubator set to 37 °C and incubate shaking at 225 rpm for 1 h. During this time, prewarm the necessary number of LB-AMP (LB agar medium + 100 μg/mL ampicillin) plates in a 37 °C plate incubator. *See* **Note 1** for LB-AMP plate preparation.
6. To determine the ligation and transformation efficiency of the reaction, use sterile glass plating beads to plate 10, 25, 50, 100, and 200 μL of cells on 5 LB-AMP plates. Leave the glass plating beads in the dish lids and incubate overnight at 37 °C with the agar side up.
7. Choose a plate that contains at least a few hundred colonies and count the colonies on the plate. Extrapolate the best volume for plating so that there will be about 2000–5000 colonies on a single plate. This volume can range from anywhere from 10 to 200 μL of transformed *E. coli* per plate. The negative control may have a few colonies (less than 50) on the 200 μL plates. These false positives are likely due to a very low rate of incomplete digestion or religation and does signify any technical issues of consequence.
8. Once the ligation and transformation efficiencies are determined for the library construct, repeat **steps 1–7** in this section until the desired number of colonies is reached (*see* **Note 2**). LB-AMP plates containing colonies can be saved at 4 °C for a few days if multiple rounds of transformation are required to obtain enough colonies.

3.6 Harvesting Colonies and Extracting Massively Parallel Reporter (MPR) Plasmid DNA

1. Fill a lab wash bottle with LB.
2. Open the lids to at most 4 plates with 4000–5000 colonies each and squirt enough LB onto the plates so that the glass beads can freely roll on the plates. Be sure to not add too much LB so that liquid overflows out of the dish.
3. Put the lids back on the plates and stack them on top of each other. Gently roll the beads across the colonies several times so that all the colonies on each plate are resuspended in the LB.
4. Weigh a 50 mL conical (with the lid on) and record the weight. Open the 50 mL conical and place the lid to one side.

5. For each plate, gently tilt the dish so that the liquid flows to one side. Using a 1 mL pipettor, remove the liquid from the dish and pipet it into the 50 mL conical. Repeat this step for each of the four dishes.
6. Repeat **steps 2–5** in this section as necessary so that all the colonies from all of the plates have been collected. Colonies are collected from four plates at a time because the liquid will soak into the LB agar after a few minutes. If the liquid has soaked into the plate before the colonies are collected, more liquid can be added to the dish prior to collection.
7. Once all of the colonies have been collected, screw the lid back onto the 50 mL conical and centrifuge at 6000 × *g* for 15 min at 4 °C.
8. Carefully decant the clear liquid taking care to not disturb the cell pellet.
9. Weigh the 50 mL conical again with the cell pellet to determine the weight of the cell pellet. If desired, the cell pellet can be stored at −20 °C at this step.
10. Extract the MPR plasmid DNA using a QIAGEN Maxi prep kit according to the manufacturer's instructions. Be sure to not overload the QIAGEN-tip 500 (*see* **Note 3**).
11. Measure the concentration of the extracted MPR library using a NanoDrop. The typical yield is approximately 500 μg. Store the library at −20 °C.
12. If desired, the sequence diversity of the MPR plasmid library can be verified. Continue with Subheading 3.11, **steps 2–19** to create a high-throughput sequencing library from the plasmid library (*see* **Note 4** for modifications to the protocol).

3.7 Making Competent Yeast

1. Streak BY4741 (*see* **Note 5**) on a YEPD agar plate. Incubate at 30 °C for 48 h in a plate incubator.
2. Pick a single colony and inoculate 5 mL of YEPD in a 16 mL culture tube. Incubate overnight shaking at 300 rpm at 30 °C.
3. In a sterile 100 mL Erlenmeyer flask, start a 50 mL culture in YEPD at an OD_{600} of 0.1 by diluting the overnight culture into fresh YEPD. Incubate the culture shaking at 300 rpm at 30 °C for 4–6 h until mid-log phase (OD_{600} of 0.6–0.8).
4. Pellet the cells by transferring the culture to two 50 mL conical tubes and centrifuging for 4 min at 500 × *g* at room temperature.
5. Carefully decant the clear supernatant and resuspend the pellet in 10 mL EZ 1 solution (Frozen-EZ Yeast Transformation II Kit).

6. Pellet the cells by centrifuging for 4 min at 500 × *g* at room temperature. Remove the supernatant by carefully decanting. Remove any excess liquid using a P200 pipettor.
7. Add 500 μL of EZ 2 solution to each of the pellets. Resuspend the cells and combine the cells into one 1.5 mL microcentrifuge tube.
8. Aliquot 400 μL of cells into two 1.5 mL microcentrifuge tubes. Aliquot the remaining 200 μL as 50 μL aliquots in among 1.5 mL microcentrifuge tubes.
9. Slowly freeze the cells by placing them inside a Styrofoam box with a lid. Put the Styrofoam box in a −80 °C freezer and allow the cells to freeze slowly overnight.

3.8 Transforming Saccharomyces yeast with the pGM-YFP, and pGM-mCherry Controls

1. Thaw two 50 μL aliquots of cells on ice. Add 200 ng of *pGM-YFP* to one aliquot and 200 ng of *pGM-mCherry* to the other aliquot. Add 500 μL of EZ 3 solution and mix the transformations thoroughly. Incubate the cells shaking at 300 rpm at 30 °C for 2 h.
2. Plate 10 μL of each transformation on SD-URA plates. Incubate for 2 days at 30 °C. Discard the remaining transformation mixtures.
3. Yeast transformed with the mCherry only vector should appear pink in color. The YFP only transformed yeast should not be pink. If desired to further check for positive transformants, examine the plates using a blue transilluminator with an amber filter. Positive colonies should appear to be glowing red (mCherry) and yellow (YFP).
4. Prepare glycerol stocks by inoculating 5 mL of SD-URA- liquid media in a 16 mL culture tube from one positive colony from each transformation. Grow overnight shaking at 300 rpm at 30 °C.
5. Transfer each culture to a 15 mL conical tube. Centrifuge the conical tubes at 500 × *g* for 5 min at room temperature. Decant or pipet off the media leaving the cell pellet behind.
6. Resuspend the pellets in 800 μL of SD-URA media and add 800 μL of 30% glycerol. Transfer the cells to 1.8 mL cryovials and store the glycerol stocks at −80 °C.

3.9 Transforming Saccharomyces yeast with the MPR Library

1. Thaw one 400 μL aliquot of cells on ice. Add 2 μg of MPR plasmid library and 4 mL of EZ 3 solution and mix thoroughly. Incubate the cells shaking at 30 °C for 2 h.
2. Add 40 mL SD-URA liquid media to a 100 mL Erlenmeyer flask. Add the 4 mL of MPR transformation mixture. Incubate shaking overnight at 30 °C.

3. To assess the transformation efficiency, plate 10 μL of the MPR transformation on a SD-URA plate. Incubate the plate for 2 days in a plate incubator at 30 °C. Typically, 10 μL of transformation yields approximately 500 colonies.
4. Add 200 mL of URA- media to a 1 L flask and add the previous night's culture to the fresh media and incubate overnight shaking at 30 °C.
5. Prepare glycerol stocks. Aliquot the overnight culture into five 50 mL conical tubes with 49 mL of culture in each tube.
6. Centrifuge the conical tubes at 500 × *g* for 5 min at room temperature. Decant or pipet off the media leaving the cell pellet behind.
7. For each cell pellet, resuspend the pellet in 3 mL of SD-URA liquid media. Add 4 mL of 30% glycerol and pipet up and down several times to mix. The volume of the pellet usually is around 1 mL, if this is not the case adjust the amount of glycerol so that the final concentration of glycerol is 15%.
8. Aliquot 1.6 mL of the yeast culture and glycerol mix to 1.8 mL cryovial (22–25 vials). Store the glycerol stocks at −80 °C.

3.10 FACS

1. Thaw one aliquot of the yeast MPRA library (Subheading 3.9, **step 8**). Add the thawed yeast to 200 mL of liquid SD-URA media in a 1 L Erlenmeyer flask. Incubate the culture overnight shaking at 300 rpm at 30 °C. Go directly to step 2 without waiting for this overnight culture.
2. Start cultures of the YFP and mCherry only controls (from glycerol stocks, Subheading 3.8, **step 6**) in 5 mL of SD-URA- liquid media in 16 mL culture tubes. In addition, start a culture of the untransformed BY4741 ("cells only" control) from a glycerol stock in 5 mL of YEPD liquid media in a 16 mL culture tube. Incubate the cultures overnight shaking at 300 rpm at 30 °C.
3. The next day, restart the MPR transformed yeast cells in 50 mL of SD-URA liquid media in an Erlenmeyer flask at an OD_{600} of 0.1. Grown shaking 30 °C to an OD_{600} of 0.7. Restart the YFP only and mCherry only controls by pipetting 50 μL of the overnight culture into a fresh aliquot of 5 mL of SD-URA liquid media in a 16 mL culture tube. Restart the cells only control by pipetting 50 μL of the overnight culture into a fresh aliquot of 5 mL YEPD liquid media in a 16 mL culture tube. Grow all the cultures shaking at 300 rpm at 30 °C until the MPR transformed yeast grow to an OD_{600} of 0.7 (approximately 5 h).
4. Save some unsorted "Bin 0" cells for plasmid DNA extraction. Aliquot 1 mL of the MPR culture into a 1.5 mL

microcentrifuge tube. Centrifuge the cells for 1 min at 12,000 × *g* at room temperature. Pipet off the supernatant and save the pellet at −20 °C.

5. Transfer the cells only, YFP only, and mCherry only controls into 5 mL culture tubes or tubes compatible with the FACS machine being used. Place the tubes on ice.
6. Concentrate the MPR transformed yeast by transferring the culture to four 15 mL conical tubes and centrifuge them at 500 × *g* for 5 min in a refrigerated centrifuge. Pipet off the supernatant leaving around 2 mL at the bottom of the tube. Resuspend the pellets in the remaining liquid and transfer the remaining cultures into 5 mL culture tubes and place the tubes on ice.
7. Create a gate using the nonfluorescent yeast control, then set fluorescence compensation levels using the YFP-only and mCherry-only controls.
8. Place the MPR yeast on the sorter while in a chiller. If no chiller is available, a plastic bag filled with ice, wrapped around the tube and secured with tape is sufficient. Calculate the YFP-to-mCherry ratios using the available FACS software and divide the cells into nine bins (Bin 1–Bin 9), based on this ratio (*see* Fig. 3). Gate the cells that fall in the mid-range for mCherry values (excluding negative mCherry cells). Record how many events are counted for each of the gated bins. For each bin, deposit 100,000 cells into culture tubes containing 5 mL of SD-URA liquid media. Grow each bin's culture overnight shaking at 30 °C at 300 rpm.
9. The next morning, measure the ratio of YFP to mCherry for each bin using the Tecan fluorometer (or other plate reader). Pipet 100 μL of each culture into three separate wells (300 μL total) into a white 96 well plate compatible with the plate reader being used. For a negative control, pipet 3 wells containing 100 μL of SD-URA- media only.
10. Measure the ratio of YFP to mCherry in the plate reader with the lid off at room temperature. For YFP, set the excitation to 512 ± 5 nm and emission to 532 ± 5 nm. For mCherry, set the excitation to 586 ± 5 nm and emission to 608 ± 5 nm. Before calculating the ratios, subtract the value for the negative control for YFP and mCherry from each of the corresponding values for YFP and mCherry for the bins. *See* Fig. 4 for typical replicates of YFP-to-mCherry ratios among bins.
11. Harvest the cells from the bins by pipetting 1 mL of cells from each bin into nine 1.5 mL microcentrifuge tubes. Centrifuge the cells in a microcentrifuge at room temperature for 1 min at top speed. Pipet off the supernatant. The cells can be frozen at

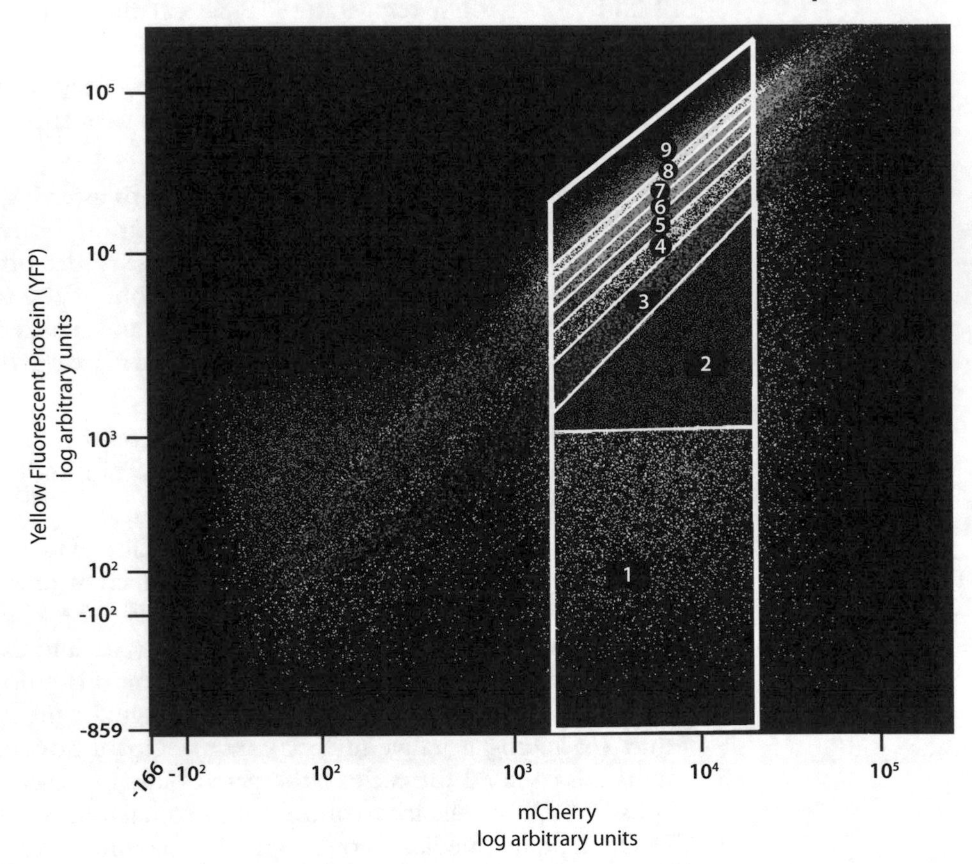

Fig. 3 FACS of *S. cerevisiae* transformed with the FACS-uORF MPRA library (YFP/mCherry sorting). The ratios of YFP to mCherry are calculated and divided into nine bins ranging from high YFP to mCherry values (pink mask) to low (no mask). The cells are also gated restricting the range of mCherry values (white box), avoiding large cells with high mCherry values and cells which may have lost the vector (low or negative mCherry values)

−20 °C or the plasmid DNA can be extracted at this point (Subheading 3.11, **step 1**).

3.11 Prepare Plasmid DNA Libraries

1. Extract the plasmid DNA from bin 0 (Subheading 3.10, **step 4**) and each sorted bin (Subheading 3.10, **step 11**) using the Zymoprep™ Yeast Plasmid Miniprep II kit according to the manufacturer's instructions. Measure the concentration of DNA using a NanoDrop.
2. Dilute each plasmid preparation to a concentration of 5 ng/μL. Set up the following PCR in 0.2 mL PCR tubes on ice for each plasmid prep. Include a negative control reaction where nuclease-free water is added to the reaction instead of plasmid template.

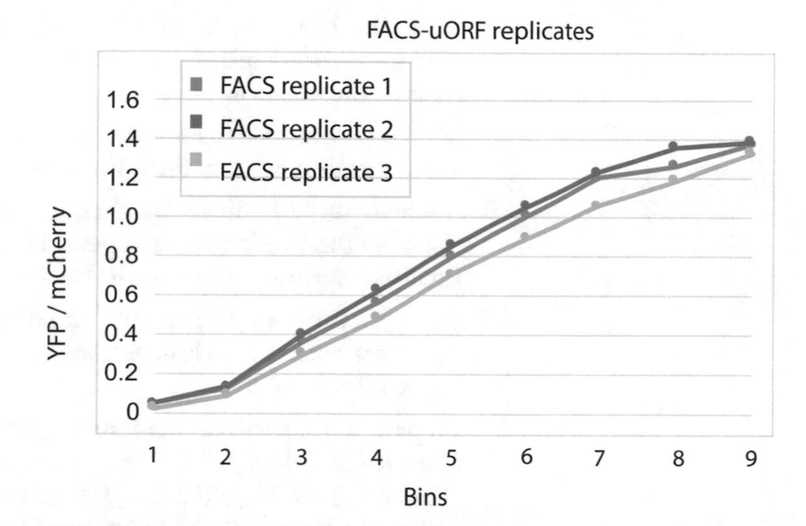

Fig. 4 Typical YFP–mCherry ratios from replicate FACS-uORF sorted cell bins after overnight growth. The ratios were read with a Tecan Fluorometer. These measurements are used to verify consistent sorting across replicates and can be used for further normalization and data analysis after high-throughput sequencing

Reagent	Volume
Nuclease-free water	14.5 μL
Plasmid template 5 ng/μL	2 μL
5× Q5 buffer	5 μL
dNTPs (10 mM)	1 μL
Eno2-lib-F1 (10 μM)	1 μL
FACS-uORF-YFP-R (10 μM)	1 μL
Q5® polymerase	1.5 μL

3. Run the program.
   (a) 98 °C for 30 s.
   (b) 8 cycles of 98 °C for 30 s, 55 °C for 30 s, and 72 °C for 20 s.
4. Purify each reaction (including the negative control) using AMPure XP beads. Vortex the beads until they are resuspended and add 37.5 μL of beads to each reaction. Briefly vortex the reactions and incubate for 5 min at room temperature.
5. Place the tubes on the magnet and incubate for 5 min at room temperature.
6. With the tubes still on the magnet, carefully remove the supernatant and discard the supernatants.

7. Wash the beads by adding 150 μL of freshly prepared 80% ethanol without disturbing the pellets. Incubate for 30 s and remove the ethanol.
8. Let the pellets air-dry for around 3 min. Using a pipettor, pipet any remaining ethanol from the tube.
9. Remove the tubes from the magnet and 10 μL of nuclease-free water to the beads and resuspend by pipetting up and down. Incubate at room temperature for 2 min.
10. Place the tubes on the magnet and incubate for 2 min or until the supernatants are clear. Remove the supernatants to a fresh 0.2 mL PCR tubes.
11. Prepare a pool of primers such that all 7 forward primers (FuORF_2_DNA_N0-N7_F) and all 4 reverse primers (FuORF_2_DNA_N0,2,4,6_R) are mixed together at a final concentration of 10 μM in nuclease-free water.
12. Set up the following PCR in 0.2 mL PCR tubes on ice for each purified PCR reaction from **step 10**. Include the purified negative control reaction from **step 10** as the negative control reaction for this step.

Reagent	Volume
Nuclease-free water	23 μL
Purified PCR product (**step 10**)	10 μL
5× Q5 buffer	10 μL
dNTPs (10 mM)	2 μL
10 μM F primer pool (**step 11**)	2 μL
10 μM R primer pool (**step 11**)	2 μL
Q5® polymerase	1 μL

13. Run the program.
    (a) 98 °C for 30 s.
    (b) 6 cycles of 98 °C for 30 s, 55 °C for 30 s, and 72 °C for 20 s.
14. Purify each reaction using AMPure XP beads (**steps 4–10** in this section) with the following exceptions. For **step 4**, add 75 μL of vortexed beads, and for **step 9**, add 50 μL of nuclease-free water to elute the sample.
15. Set up the following PCR in 0.2 mL PCR tubes on ice for each purified PCR reaction, and the purified negative control reaction from the previous step [14]. Use a different indexed primer pair for each sample. The remaining PCR products from **step 14** can be saved at −20 °C.

Reagent	Volume

(continued)

Reagent	Volume
Nuclease-free water	15.5 μL
Purified PCR product (**step 14**)	2 μL
5× Q5 buffer	5 μL
dNTPs (10 mM)	1 μL
NEBNext dual index primers (10 μM)	1 μL
Q5® polymerase	0.5 μL

16. Run the program.
    (a) 98 °C for 30 s.
    (b) 15 cycles of 98 °C for 10 s, 64 °C for 30 s, and 72 °C for 20 s.
17. Purify each reaction using AMPure XP beads (**steps 4–10** in this section) with the following exception, for **step 9**, add 15 μL of nuclease-free water to elute the sample.
18. Assess the quality of the libraries on a DNA ScreenTape (D1000 or similar). Load 1 μL of sample onto the TapeStation (*see* Fig. 5).

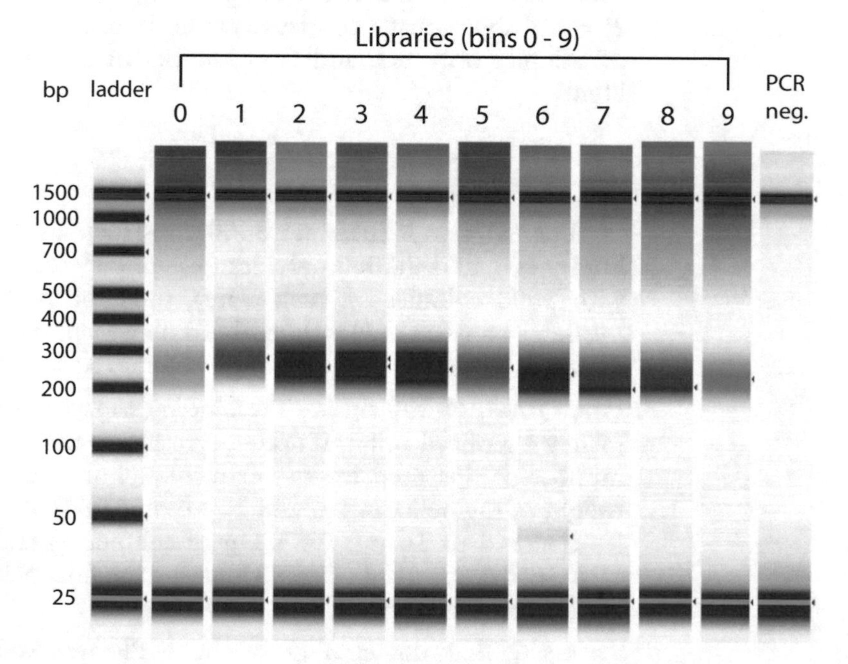

Fig. 5 TapeStation DNA ScreenTape gel image showing FACS-uORF libraries made from bin 0 (unsorted), and bins 1–9. The DNA ladder (leftmost lane) ranges from 25 bp to 1500 bp. Two control markers (in green and purple) at 25 and 1500 bp in every lane are used as internal sizing controls. A PCR negative (rightmost lane) carried through during the entire library preparation procedure (see methods) and does not contain amplified product. The typical library size shown ranges from approximately 200 bp to 350 bp. Occasionally, a minimal amount of small sized PCR products are present due to the amplification of primer dimers (bin 6 at ~40 bp). These small products do not interfere with library sequencing

19. Quantify the concentration using the Qubit™ dsDNA HS Assay Kit according to the manufacturer's instructions. Typically, 1–2 μL of library is sufficient for accurate quantification. At this point, the libraries can be pooled and run on an Illumina sequencer.

4 Notes

1. It is very important that the LB-AMP plates are dry and warm before use. If the plates are too wet, the bacteria will not adhere properly to the agar and the number of colonies obtained will be drastically reduced. After pouring the LB-AMP plates, leave the lids ajar so that the lids do not collect moisture. After the agar has solidified, turn the plates upside down and leave them at room temperature for 1–2 days. Do not stack the plates on top of each other as this interferes with drying. Use the plates immediately, do no store them in a cold room or refrigerator as this adds moisture to the plates.
2. The number of colonies needed to obtain sufficient sequence diversity depends on the number of sequence constructs contained in the library. The probability of missing a single construct can be calculated using the equation below, where $P =$ probability that a construct variant is missed, $n =$ number of colonies obtained, and C = number of constructs in the library.

$$P = \left(1 - \frac{1}{C}\right)^n$$

   For example if there are 5000 constructs in the plasmid library (C), and 30,000 colonies (n) are collected (10 plates with 3000 colonies on each plate), then the probability of missing a construct (P) is 0.0025. To obtain enough sequence diversity, be sure to collect enough colonies so that $P < 0.01$.
3. One QIAGEN-500 tip has the binding capacity for approximately 1 g of cell pellet. If necessary, scale up the volumes and number of tips used for the extraction. For example, if the weight of the pellet is 2 g, add 20 mL of QIAGEN Buffer P1 (as opposed to 10 mL for a 1 g of cell pellet) and use two QIAGEN tips instead of one. Combine the final plasmid prep elution into one tube.
4. To verify that the cloning of the MPR was successful, a sequencing library of the plasmid DNA library can be made and sequenced using the protocol outlined in Subheading 3.11, **steps 2–19** with the following modifications. For the template for the first PCR, dilute the plasmid library to a

concentration of 1 ng/μL and use 1 μL of the diluted library for the PCR. For the second PCR, perform 5 cycles of amplification (as opposed to 6 cycles). For the third PCR perform 12 cycles of PCR (as opposed to 15). The library can be sequenced on a MiSeq using a Nano kit, which yields about one million (2 × 150 bp) reads.

5. We use the *S. cerevisiae* strain BY4741 which has the genotype, MATa his3Δ1 leu2Δ0 met15Δ0 ura3Δ0. The deletion of the uracil gene allows for the selection of the dual fluorescent plasmid in media lacking uracil. In addition, this strain is nonflocculating, which allows for the individual cells to be sorted during FACS. Other strains of nonflocculating yeast can be used, as long as they have the *URA3* gene completely deleted (transposon insertion strains are somewhat leaky in our hands).

References

1. Barbosa C, Peixeiro I, Romão L (2013) Gene expression regulation by upstream open reading frames and human disease. PLoS Genet 9:e1003529
2. Zhang H, Wang Y, Lu J (2019) Function and evolution of upstream ORFs in eukaryotes. Trends Biochem Sci 44:782–794
3. Young SK, Wek RC (2016) Upstream open Reading frames differentially regulate gene-specific translation in the integrated stress response. J Biol Chem 291:16927–16935
4. Lin Y, May GE, Kready H et al (2019) Impacts of uORF codon identity and position on translation regulation. Nucleic Acids Res 2:1–10
5. Hinnebusch AG, Ivanov IP, Sonenberg N (2016) Translational control by 5-′-untranslated regions of eukaryotic mRNAs. Science 352:1413–1416
6. McGillivray P, Ault R, Pawashe M et al (2018) A comprehensive catalog of predicted functional upstream open reading frames in humans. Nucleic Acids Res 46:3326–3338
7. Melnikov A, Murugan A, Zhang X et al (2012) Systematic dissection and optimization of inducible enhancers in human cells using a massively parallel reporter assay. Nat Biotechnol 30:271–277
8. Sharon E, Kalma Y, Sharp A et al (2012) Inferring gene regulatory logic from high-throughput measurements of thousands of systematically designed promoters. Nat Biotechnol 30:521–530
9. Soemedi R, Cygan KJ, Rhine CL et al (2017) Pathogenic variants that alter protein code often disrupt splicing. Nat Genet 49:848–855
10. Wong MS, Kinney JB, Krainer AR (2018) Quantitative activity profile and context dependence of all human 5′ splice sites. Mol Cell 71:1012–1026.e3
11. Adamson SI, Zhan L, Graveley BR (2018) Vex-seq: high-throughput identification of the impact of genetic variation on pre-mRNA splicing efficiency. Genome Biol 19:1–12
12. Dvir S, Velten L, Sharon E et al (2013) Deciphering the rules by which 5′-UTR sequences affect protein expression in yeast. Proc Natl Acad Sci U S A 110:E2792–E2801
13. Weingarten-Gabbay S, Elias-Kirma S, Nir R et al (2016) Comparative genetics: systematic discovery of cap-independent translation sequences in human and viral genomes. Science 351:aad4939
14. Cuperus JT, Groves B, Kuchina A et al (2017) Deep learning of the regulatory grammar of yeast 5′ untranslated regions from 500,000 random sequences. Genome Res 27:2015–2024
15. Sample PJ, Wang B, Reid DW et al (2019) Human 5′ UTR design and variant effect prediction from a massively parallel translation assay. Nat Biotechnol 37:803–809

Part V

RNA Modifications

Chapter 19

m⁶A RNA Immunoprecipitation Followed by High-Throughput Sequencing to Map N^6-Methyladenosine

Devi Prasad Bhattarai and Francesca Aguilo

Abstract

N^6-methyladenosine (m^6A) is the most abundant internal modification on messenger RNAs (mRNAs) and long noncoding RNAs (lncRNAs) in eukaryotes. It influences gene expression by regulating RNA processing, nuclear export, mRNA decay, and translation. Hence, m^6A controls fundamental cellular processes, and dysregulated deposition of m^6A has been acknowledged to play a role in a broad range of human diseases, including cancer. m^6A RNA immunoprecipitation followed by high-throughput sequencing (MeRIP-seq or m^6A-seq) is a powerful technique to map m^6A in a transcriptome-wide level. After immunoprecipitation of fragmented polyadenylated (poly(A)$^+$) rich RNA by using specific anti-m^6A antibodies, both the immunoprecipitated RNA fragments together with the input control are subjected to massively parallel sequencing. The generation of such comprehensive methylation profiles of signal enrichment relative to input control is necessary in order to better comprehend the pathogenesis behind aberrant m^6A deposition.

Key words MeRIP-seq or m^6A-seq, N^6-Methyladenosine, Epitranscriptomics, METTL3

1 Introduction

The coding and noncoding transcriptome is subjected to extensive chemical modification constituting a new layer of gene expression control. To date, at least 170 distinct RNA modifications have been identified [1], among which N^6-methyladenosine (m^6A) is the most prevalent and reversible internal modification in eukaryotic messenger RNAs (mRNAs) and long noncoding RNAs (lncRNAs) [2] m^6A methylation is added cotranscriptionally [3–6] at the RRACH motif sequence (R = A/G, H = A/C/U) by the core m^6A methyltransferase complex (writers) which includes the methyltransferase-like 3 (METTL3) and the methyltransferase-like 14 (METTL14) [7, 8]. The heterodimer interacts with other writer proteins for the proper deposition of the m^6A mark [9]. The presence of m^6A demethylases (erasers), namely the Fat mass and obesity-associated protein (FTO) and α-ketoglutarate-dependent

Erik Dassi (ed.), *Post-Transcriptional Gene Regulation*, Methods in Molecular Biology, vol. 2404,
https://doi.org/10.1007/978-1-0716-1851-6_19,

dioxygenase alkB homolog 5 (ALKBH5), suggested that the effects of m^6A can potentially be reversed, and that cellular demethylation pathways might be a mechanism of dynamic m^6A regulation [10, 11]. The m^6A mark is deciphered by a group of RNA binding proteins (RBPs; readers) that ultimately dictate the RNA fate, such as RNA processing [12–14], nuclear export [15, 16], mRNA decay [17] and translation [18]. Hence, m^6A modulates fundamental cellular processes including pluripotency [19–21] and differentiation [22, 23], regulation of the circadian clock [24], and fertility [25], among others. Not surprisingly, dysregulated deposition of m^6A has been reported in a broad range of human diseases, including cancer and neurological and metabolic disorders [26–28].

Although m^6A was first discovered in poly(A)$^+$ RNA in the 1970s [29–34] it is only recently that transcriptome-wide profiling studies, mostly performed by m^6A RNA immunoprecipitation followed by high-throughput sequencing (MeRIP-seq or m^6A-seq), have revealed that m^6A is pervasive throughout the transcriptome and it is predominantly located at long internal exons, near stop codons, and along 3′ untranslated regions (3′UTRs) [35, 36]. In addition, it has been shown that despite the strong consensus, only a small fraction of RACH sites is detectably methylated in vivo, arguing that the sequence motif is not sufficient to determine the distribution of m^6A. m^6A-seq has been widely used to map m^6A in distinct organisms, cell lines, and experimental conditions [35, 36]. In this chapter, we have detailed the protocol for MeRIP-seq which is the most widely used method to enable transcriptome-wide studies of m^6A (Fig. 1). The immunoprecipitation of m^6A-modified RNA fragments is carried out on fragmented poly(A)$^+$ RNA using anti-m^6A antibodies and both m^6A enriched RNA and non-immunoprecipitated control input are subjected to massive parallel sequencing. Captured m^6A-RNA fragments are detected as peaks relative to the background mRNA abundance which is assessed with the input reads. The generation of such comprehensive methylation profiles is necessary in order to better understand the role of m^6A in RNA metabolism and in turn, its downstream effects in physiology and disease.

2 Materials

Prepare all the buffers and solutions in DEPC-treated double distilled water (ddH_2O). Use RNase/DNase-free pipette tips and glassware to prepare the reagents.

2.1 Buffers

1. DEPC treated ddH_2O: add 1 mL DEPC in 1000 mL miliQ H_2O, Mix and incubate at RT for 4 h, autoclave.

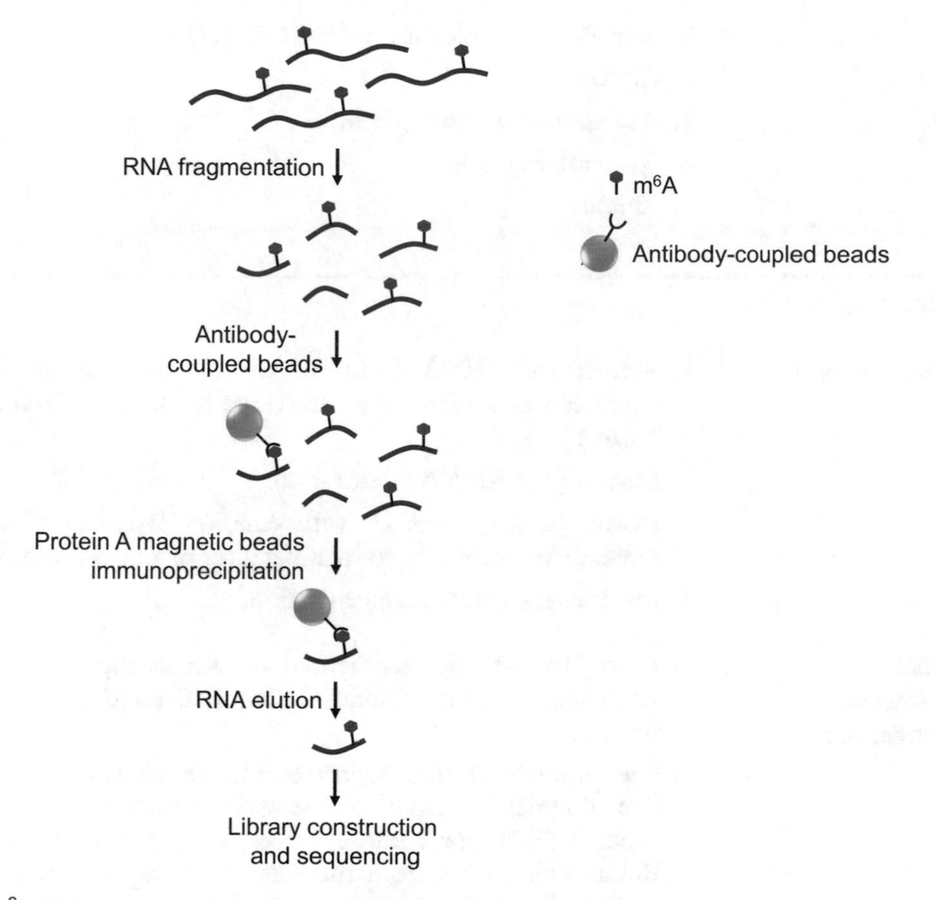

Fig. 1 m^{6}A-RNA immunoprecipitation followed by sequencing (MeRIP-seq)

2. 1 M Tris–HCl (pH 7.4) (100 mL): weigh 12.11 g Tris powder, dissolve in 90 mL DEPC H_2O, adjust pH to 7.4 with HCl, then add DEPC H_2O to make the volume 100 mL.
3. 5 M NaCl (100 mL): dissolve 29.22 g NaCl in 80 mL DEPC H_2O, then add DEPC H_2O to make the final volume 100 mL.
4. 10% NP-40 (50 mL): dilute 5 mL NP-40 solution in 45 mL DEPC H_2O, mix thoroughly on the shaker at RT.
5. Immunoprecipitation (IPP) buffer: 10 mM Tris–HCl pH 7.4, 150 mM NaCl, 0.1% NP-40.
6. PK buffer: 100 mM Tris–HCl pH 7.4, 50 mM NaCl, 10 mM EDTA.

2.2 m^{6}A MeRIP

1. Dynabeads Protein A for Immunoprecipitation.
2. Dynabeads® mRNA Purification Kit (Ambion).
3. RNA Fragmentation Reagents.
4. Antibody-N^6-methyladenosine (Synaptic Systems GmbH, 202,003).

5. Acid Phenol: Chloroform (pH 4.3–4.7).
6. Glycogen.
7. 3 M sodium acetate, pH 5.5.
8. Nuclease-free water.
9. RNasin.

3 Methods

3.1 Prepare RNA

1. Extract total RNA from human cell lines using Qiagen plus RNA extraction kit as described by the manufacturer (*see* **Note 1**).
2. Measure the RNA concentration.
3. Isolate poly(A)$^+$ mRNA with Ambion Dynabeads® mRNA direct™ kit according to manufacturer protocol (*see* **Note 2**).
4. Measure the mRNA concentration.

3.2 RNA Fragmentation and Purification

1. Keep 100–200 ng fragmented mRNA as input control for RNA-seq. The input library is required for determining the background.
2. 5 μg of mRNA is divided into 5 separate tubes. For 50–100 nt size, fragmentation kit can be used as follows. Incubate the tubes at 90 °C for 1 min in a preheated thermal cycler block. Remove the tubes from the block and stop the reaction by adding the appropriate amount of stop reagent provided in the kit. Pool the five tubes (total volume = ~50 μL). Vortex, spin down, and place the tubes on ice (*see* **Note 3**).
3. Add 5 μL of 3 M NaOAc (pH 5.2) (1/10 volume), 190 μL 100% ethanol (2.5 volume), and 2 μL glycogen as a carrier. Mix and incubate overnight at −80 °C (*see* **Note 4**).
4. Precipitate mRNA by centrifuging the tubes at 12,000 × *g* for 30 min at 4 °C. Discard the supernatant without disrupting the pellet. Wash the RNA pellet with 1 mL of 70% ethanol (vol/-vol) and centrifuge at 12,000 × *g* for 10 min at 4 °C.
5. Discard supernatant and spin at 12,000 × *g* for 1 min to remove the remaining ethanol, taking care not to disrupt RNA pellet. Let the pellet air dry for 10 min on ice. Resuspend the RNA pellet with 20 μL nuclease free water.
6. Dilute 0.5 μL fragmented RNA with 0.5 μL nuclease free water to measure the RNA concentration with NanoDrop spectrophotometer. Validate RNA size distribution on 1.5% agarose gel ~30 min (*see* **Note 5**).

3.3 Preparation of Anti-m^{6}A Antibody Conjugated to Dynabeads® Protein A for Immunoprecipitation

1. Use large-orifice tips or cut off the end of a regular micropipette tip for transferring 25 μL beads into low-binding microcentrifuge tubes (*see* **Note 6**). Wash the beads three times with 1 mL 1× IPP buffer for 3 min with gentle rotation at room temperature.
2. Discard the supernatant. Resuspend the beads in 500 μL 1× IPP buffer with 0.5 μL RNasin and add 5 μg anti-m^{6}A antibody (*see* **Note 7**). Incubate with protein A beads and anti-m^{6}A antibody for 1 h with gentle rotation at room temperature.

3.4 Immunoprecipitation

1. Denature the fragmented mRNA (~5 μg) at 75 °C for 5 min, and cooled it on ice for 5 min.
2. Add the denatured mRNA into the protein A beads-anti-m^{6}A antibody-containing tube and incubate for 4 h with gentle rotation at 4 °C.

3.5 Elution of m^{6}A-Containing mRNA

1. To check binding efficiency, take 150 μL of supernatant and extract RNA with Trizol and chloroform method. Any remaining supernatant can be stored in −80 °C.
2. Wash the beads three times with 1 mL of 1× IPP buffer for 3 min with gentle rotation at 4 °C.
3. Add 300 μL of PK buffer and 30 μL proteinase K to the beads and incubate with shaking at 1100 rpm at 55 °C for 30 min and collect the supernatant in a new tube.
4. Repeat **step 3**.
5. Combine the two eluates and add 660 μL of acid phenol (vol/vol) (pH 4.3–4.7). Vortex thoroughly and incubate on ice for 5 min.
6. Centrifuge at 12,000 × *g* for 15 min at 4 °C and transfer the supernatant into two new 1.5 mL tubes (330 μL/tube).
7. Add 33 μL 3 M NaOAc (pH 5.5) (1/10 volume), 1190 μL 100% ethanol (2.5 volume) and 2 μL glycogen (*see* **Note 8**). Vortex to mix thoroughly and incubate overnight at −80 °C.
8. Centrifuge the tubes for 30 min at 12,000 × *g* at 4 °C. Discard the supernatant. Wash the pellet with 1 mL cold 70% ethanol (vol/vol) and centrifuge at 12,000 × *g* for 10 min at 4 °C. Aspirate the supernatant and let the pellet air-dry. Resuspend the pellet in 15 μL of RNase-free water.
9. The samples can be used for direct library construction or put into 95% ethanol for longer preservation.

4 Notes

1. A large number of reagents, protocols, and kits are suitable for RNA extraction. However, we use column-based purification kits because they retrieve high yield and good quality RNA. The purity, integrity, and quality of the RNA is essential. Avoid salt and DNA contamination as they can interfere with downstream analysis. Perform DNase treatment and elute the RNA with molecular biology–grade water.
2. An additional round of poly(A)$^+$ mRNA purification with Ambion Dynabeads® mRNA direct™ is recommended in order to reduce ribosomal RNA contamination. After poly (A)$^+$ purification, we usually obtain 5 μg of mRNA from 300 μg of total RNA.
3. RNA fragmentation is optimized in order to obtain 50–100 nt fragments. Small volumes (10 μL) in several batches (no more than five tubes) facilitates size reproductivity.
4. RNA precipitation can be carried out at −20 °C for 2 h, but we highly recommend −80 °C overnight.
5. Alternatively, the size of the fragmented RNA can be assessed with an Agilent RNA 6000 Pico kit on an Agilent 2100 Bioanalyzer according to the manufacturer's instructions.
6. Dynabeads® Protein A, which has the binding capacity of 8 μg human IgG/mg beads contains 30 mg Dynabeads®/mL.
7. Antibody-N^6-methyladenosine (Synaptic Systems GmbH, 202,003).
8. Alternate to the Glycogen RNA precipitation during IP RNA recovery, can be used RNeasy mini kit with RNeasy spin column to reduce chances of free m^{6}A.

Acknowledgments

This research was supported by grants from the Knut and Alice Wallenberg Foundation, Umeå University, Västerbotten County Council, Swedish Research Council (2017-01636), Kempe Foundation (SMK-1766), and Cancerfonden (19 0337 Pj).

References

1. Boccaletto P, MacHnicka MA, Purta E, Pitkowski P, Baginski B, Wirecki TK, De Crécy-Lagard V, Ross R, Limbach PA, Kotter A, Helm M, Bujnicki JM (2018) MODOMICS: a database of RNA modification pathways. 2017 update. Nucleic Acids Res 46:D303–D307. https://doi.org/10.1093/nar/gkx1030
2. Roundtree IA, Evans ME, Pan T, He C (2017) Dynamic RNA modifications in gene expression regulation. Cell 169:1187–1200. https://doi.org/10.1016/j.cell.2017.05.045

3. Huang H, Weng H, Zhou K, Wu T, Zhao BS, Sun M, Chen Z, Deng X, Xiao G, Auer F, Klemm L, Wu H, Zuo Z, Qin X, Dong Y, Zhou Y, Qin H, Tao S, Du J, Liu J, Lu Z, Yin H, Mesquita A, Yuan CL, Hu YC, Sun W, Su R, Dong L, Shen C, Li C, Qing Y, Jiang X, Wu X, Sun M, Guan JL, Qu L, Wei M, Müschen M, Huang G, He C, Yang J, Chen J (2019) Histone H3 trimethylation at lysine 36 guides m6A RNA modification co-transcriptionally. Nature 567:414–419. https://doi.org/10.1038/s41586-019-1016-7
4. Ke S, Pandya-Jones A, Saito Y, Fak JJ, Vågbø CB, Geula S, Hanna JH, Black DL, Darnell JE, Darnell RB (2017) m6A mRNA modifications are deposited in nascent pre-mRNA and are not required for splicing but do specify cytoplasmic turnover. Genes Dev 31:990–1006. https://doi.org/10.1101/gad.301036.117
5. Knuckles P, Carl SH, Musheev M, Niehrs C, Wenger A, Bühler M (2017) RNA fate determination through cotranscriptional adenosine methylation and microprocessor binding. Nat Struct Mol Biol 24(7):561–569. https://doi.org/10.1038/nsmb.3419
6. Louloupi A, Ntini E, Conrad T, Ørom UAV (2018) Transient *N*-6-methyladenosine transcriptome sequencing reveals a regulatory role of m6A in splicing efficiency. Cell Rep 23:3429–3437. https://doi.org/10.1016/j.celrep.2018.05.077
7. Liu J, Yue Y, Han D, Wang X, Fu Y, Zhang L, Jia G, Yu M, Lu Z, Deng X, Dai Q, Chen W, He C (2014) A METTL3-METTL14 complex mediates mammalian nuclear RNA N6-adenosine methylation. Nat Chem Biol 10:93–95. https://doi.org/10.1038/nchembio.1432
8. Wang P, Doxtader KA, Nam Y (2016) Structural basis for cooperative function of Mettl3 and Mettl14 methyltransferases. Mol Cell 63:306–317. https://doi.org/10.1016/j.molcel.2016.05.041
9. Lence T, Paolantoni C, Worpenberg L, Roignant JY (2019) Mechanistic insights into m 6 a RNA enzymes. Biochim Biophys Acta Gene Regul Mech 1862:222–229. https://doi.org/10.1016/j.bbagrm.2018.10.014
10. Jia G, Fu Y, Zhao X, Dai Q, Zheng G, Yang Y, Yi C, Lindahl T, Pan T, Yang Y-G, He C (2011) N6-Methyladenosine in nuclear RNA is a major substrate of the obesity-associated FTO. Nat Chem Biol 7:885–887. https://doi.org/10.1038/nchembio.687
11. Zheng G, Dahl JA, Niu Y, Fedorcsak P, Huang CM, Li CJ, Vågbø CB, Shi Y, Wang WL, Song SH, Lu Z, Bosmans RPG, Dai Q, Hao YJ, Yang X, Zhao WM, Tong WM, Wang XJ, Bogdan F, Furu K, Fu Y, Jia G, Zhao X, Liu J, Krokan HE, Klungland A, Yang YG, He C (2013) ALKBH5 is a mammalian RNA demethylase that impacts RNA metabolism and mouse fertility. Mol Cell 49:18–29. https://doi.org/10.1016/j.molcel.2012.10.015
12. Du H, Zhao Y, He J, Zhang Y, Xi H, Liu M, Ma J, Wu L (2016) YTHDF2 destabilizes m6A-containing RNA through direct recruitment of the CCR4–NOT deadenylase complex. Nat Commun 7:12626. https://doi.org/10.1038/ncomms12626
13. Shi H, Wang X, Lu Z, Zhao BS, Ma H, Hsu PJ, He C (2017) YTHDF3 facilitates translation and decay of N6-methyladenosine-modified RNA. Cell Res 27(3):315–328. https://doi.org/10.1038/cr.2017.15
14. Li A, Chen Y-S, Ping X-L, Yang X, Xiao W, Yang Y, Sun H-Y, Zhu Q, Baidya P, Wang X, Bhattarai DP, Zhao Y-L, Sun B-F, Yang Y-G (2017) Cytoplasmic m6A reader YTHDF3 promotes mRNA translation. Cell Res 1:1–4. https://doi.org/10.1038/cr.2017.10
15. Lesbirel S, Wilson SA (2018) The m6A-methylase complex and mRNA export. Biochim Biophys Acta Gene Regul Mech 1862(3):319–328. https://doi.org/10.1016/j.bbagrm.2018.09.008
16. Roundtree IA, Luo G-Z, Zhang Z, Wang X, Zhou T, Cui Y, Sha J, Huang X, Guerrero L, Xie P, He E, Bin Shen CH (2017) YTHDC1 mediates nuclear export of N6-methyladenosine methylated mRNAs. Elife 6:e31311
17. Wang X, Lu Z, Gomez A, Hon GC, Yue Y, Han D, Fu Y, Parisien M, Dai Q, Jia G, Ren B, Pan T, He C (2014) N 6-methyladenosine-dependent regulation of messenger RNA stability. Nature 505:117–120. https://doi.org/10.1038/nature12730
18. Wang X, Zhao BS, Roundtree IA, Lu Z, Han D, Ma H, Weng X, Chen K, Shi H, He C (2015) N6-methyladenosine modulates messenger RNA translation efficiency. Cell 161:1388–1399. https://doi.org/10.1016/j.cell.2015.05.014
19. Batista PJ, Molinie B, Wang J, Qu K, Zhang J, Li L, Bouley DM, Lujan E, Haddad B, Daneshvar K, Carter AC, Flynn RA, Zhou C, Lim KS, Dedon P, Wernig M, Mullen AC, Xing Y, Giallourakis CC, Chang HY (2014) M^{6}A RNA modification controls cell fate transition in mammalian embryonic stem cells. Cell Stem Cell 15:707–719. https://doi.org/10.1016/j.stem.2014.09.019

20. Geula S, Moshitch-Moshkovitz S, Dominissini D, Mansour AAF, Kol N, Salmon-Divon M, Hershkovitz V, Peer E, Mor N, Manor YS, Ben-Haim MS, Eyal E, Yunger S, Pinto Y, Jaitin DA, Viukov S, Rais Y, Krupalnik V, Chomsky E, Zerbib M, Maza I, Rechavi Y, Massarwa R, Hanna S, Amit I, Levanon EY, Amariglio N, Stern-Ginossar N, Novershtern N, Rechavi G, Hanna JH (2015) m6A mRNA methylation facilitates resolution of naïve pluripotency toward differentiation. Science 347:1002–1006. https://doi.org/10.1126/science.1261417

21. Aguilo F, Zhang F, Sancho A, Fidalgo M, Di Cecilia S, Vashisht A, Lee D, Chen C, Rengasamy M, Jahouh F, Roman A, Krig SR, Wang R, Zhang W, Wohlschlegel JA, Wang J, Walsh MJ (2015) Coordination of m6 a mRNA methylation and gene transcription by ZFP217 regulates pluripotency and reprogramming. Cell Stem Cell 17:689–704. https://doi.org/10.1016/j.stem.2015.09.005.Coordination

22. Malla S, Melguizo-Sanchis D, Aguilo F (2018) Steering pluripotency and differentiation with N6-methyladenosine RNA modification. Biochim Biophys Acta Gene Regul Mech 1862 (3):394–340. https://doi.org/10.1016/j.bbagrm.2018.10.013

23. Aguilo F, Walsh MJ (2017) ScienceDirect the N 6 -Methyladenosine RNA modification in pluripotency and reprogramming. Curr Opin Genet Dev 46:77–82. https://doi.org/10.1016/j.gde.2017.06.006

24. Fustin JM, Doi M, Yamaguchi Y, Hida H, Nishimura S, Yoshida M, Isagawa T, Morioka MS, Kakeya H, Manabe I, Okamura H (2013) RNA-methylation-dependent RNA processing controls the speed of the circadian clock. Cell 155:793–806. https://doi.org/10.1016/j.cell.2013.10.026

25. Tang C, Klukovich R, Peng H, Wang Z, Yu T, Zhang Y, Zheng H (2017) Splicing and stability of long 3 ′ -UTR mRNAs in male germ cells. Proc Natl Acad Sci U S A 115(2):E325–E333. https://doi.org/10.1073/pnas.1717794115

26. Lin S, Choe J, Du P, Triboulet R, Gregory RI, Lin S, Choe J, Du P, Triboulet R, Gregory RI (2016) The m 6 a methyltransferase METTL3 promotes translation in human cancer cells article the m 6 a methyltransferase METTL3 promotes translation in human cancer cells. Mol Cell:1–11. https://doi.org/10.1016/j.molcel.2016.03.021

27. Cui Q, Shi H, Ye P, Li L, Qu Q, Sun G, Sun G, Lu Z, Huang Y, Yang C-G, Riggs AD, He C, Shi Y (2017) M 6 a RNA methylation regulates the self-renewal and tumorigenesis of glioblastoma stem cells. Cell Rep 18:2622–2634. https://doi.org/10.1016/j.celrep.2017.02.059

28. Xie W, Ma LL, Xu YQ, Wang BH, Li SM (2019) METTL3 inhibits hepatic insulin sensitivity via N6-methyladenosine modification of Fasn mRNA and promoting fatty acid metabolism. Biochem Biophys Res Commun 518:120–126. https://doi.org/10.1016/j.bbrc.2019.08.018

29. Desrosiers R, Friderici K, Rottman F (1974) Identification of methylated nucleosides in messenger RNA from Novikoff hepatoma cells. Proc Natl Acad Sci U S A 71:3971–3975. https://doi.org/10.1073/pnas.71.10.3971

30. Perry RP, Kelley DE (1974) Existence of methylated messenger RNA in mouse L cells. Cell 1:37–42. https://doi.org/10.1016/0092-8674(74)90153-6

31. Lavi S, Shatkin AJ (1975) Methylated simian virus 40 specific RNA from nuclei and cytoplasm of infected BSC 1 cells. Proc Natl Acad Sci U S A 72:2012–2016. https://doi.org/10.1073/pnas.72.6.2012

32. Wei C-M, Gershowitz A, Moss B (1975) Methylated nucleotides block 5′ terminus of HeLa cell messenger RNA. Cell 4:379–386. https://doi.org/10.1016/0092-8674(75)90158-0

33. Wei CM, Moss B (1977) Nucleotide sequences at the N6-Methyladenosine sites of HeLa cell messenger ribonucleic acid. Biochemistry 16:1672–1676. https://doi.org/10.1021/bi00627a023

34. Schibler U, Kelley DE, Perry RP (1977) Comparison of methylated sequences in messenger RNA and heterogeneous nuclear RNA from mouse L cells. J Mol Biol 115:695–714. https://doi.org/10.1016/0022-2836(77)90110-3

35. Meyer KD, Saletore Y, Zumbo P, Elemento O, Mason CE, Jaffrey SR (2012) Comprehensive analysis of mRNA methylation reveals enrichment in 3′ UTRs and near stop codons. Cell 149:1635–1646. https://doi.org/10.1016/j.cell.2012.05.003

36. Dominissini D, Moshitch-Moshkovitz S, Schwartz S, Salmon-Divon M, Ungar L, Osenberg S, Cesarkas K, Jacob-Hirsch J, Amariglio N, Kupiec M, Sorek R, Rechavi G (2012) Topology of the human and mouse m6A RNA methylomes revealed by m6A-seq. Nature 485:201–206. https://doi.org/10.1038/nature11112

Chapter 20

Detecting m^{6}A with In Vitro DART-Seq

Matthew Tegowski, Huanyu Zhu, and Kate D. Meyer

Abstract

Recent studies have uncovered that cellular mRNAs contain a diverse epitranscriptome comprising chemically modified bases which play important roles in gene expression regulation. Among these is m^{6}A, which is a highly prevalent modification that contributes to several aspects of RNA regulation and cellular function. Traditional methods for m^{6}A profiling have used m^{6}A antibodies to immunoprecipitate methylated RNAs. Although powerful, such methods require high amounts of input material. Recently, we developed DART-seq, an antibody-free method for m^{6}A profiling from low-input RNA samples. DART-seq relies on deamination of cytidines that invariably follow m^{6}A sites and can be performed using a simple in vitro assay with only 50 ng of total RNA. Here, we describe the in vitro DART method and present a detailed protocol for highly sensitive m^{6}A profiling from any RNA sample of interest.

Key words DART-seq, m^{6}A, Epitranscriptome, In vitro deamination, Protein purification, RNA isolation

1 Introduction

N^6-methyladenosine (m^{6}A) is the most abundant internal mRNA modification and is found in thousands of cellular RNAs. Recent studies have revealed that m^{6}A plays diverse roles in regulating RNA function and gene expression and that it contributes to a wide range of physiological processes such as stem cell maintenance and learning and memory [1–4]. Therefore, the development of methods to profile m^{6}A has become increasingly important in order to understand the distribution and regulation of this mark across different cell and tissue types. Recently, we developed DART-seq (*d*eamination *a*djacent to *R*NA modification *t*argets), a novel method for global m^{6}A detection from low-input RNA samples [5]. This method utilizes a fusion protein consisting of the m^{6}A-binding YTH domain fused to the cytidine deaminase enzyme APOBEC1. When the APOBEC1-YTH protein encounters methylated RNA, it binds to the m^{6}A residue and converts adjacent cytidine residues that invariably follow m^{6}A sites into uridines.

Erik Dassi (ed.), *Post-Transcriptional Gene Regulation*, Methods in Molecular Biology, vol. 2404,
https://doi.org/10.1007/978-1-0716-1851-6_20,

The resulting C to U mutations can then be detected with next-generation sequencing or simple Sanger sequencing to identify m^6A sites. This in vitro DART assay is simple, fast, and highly sensitive.

Here, we demonstrate that m^6A residues in cellular RNA samples can be detected with in vitro DART assays using as little as 50 ng of total RNA. We describe a detailed protocol for purification of the APOBEC1-YTH protein as well as a companion protein, APOBEC1-YTH^{mut}, which has diminished m^6A binding ability and is employed as a negative control to increase m^6A detection stringency. We also present detailed assay conditions and provide important notes to improve assay performance.

2 Materials

2.1 Induction of DART Protein Expression in E. coli

1. pCMV-APOBEC1-YTH (Addgene #131636).
2. pCMV-APOBEC1-YTH^{mut} (Addgene #131637).
3. pET His6 MBP TEV LIC cloning vector (Addgene #29656).
4. Rosetta™ 2 (DE3) pLysS Singles™ Competent Cells.
5. 30% glycerol, autoclaved.
6. LB medium: 1% (w/v) tryptone, 1% (w/v) NaCl, 0.5% (w/v) yeast extract, 50 μg/mL kanamycin.
7. Autoinduction medium (1 L): 958 mL ZY medium, 20 mL M medium (50×), 20 mL 5052 medium (50×), 2 mL $MgSO_4$ (1 M), 100 μg/mL kanamycin.
8. ZY medium: 1% (w/v) tryptone, 0.5% (w/v) yeast extract. Autoclave and store at 4 °C.
9. M medium (50×): 1.25 M Na_2HPO_4, 1.25 M KH_2PO_4, 2.5 M NH_4Cl, 250 mM Na_2SO_4. Autoclave or sterile filter.
10. 5052 (50×): 25% (w/v) glycerol, 2.5% (w/v) glucose, 10% (w/v) α-lactose. Autoclave or sterile filter.
11. Liquid nitrogen.

2.2 Purification of DART Protein by Affinity Chromatography

1. Qproteome Bacterial Protein Prep Kit.
2. Ni-NTA Agarose Beads.
3. Poly-Prep Chromatography column.
4. PBS: 137 mM NaCl, 2.7 mM KCl, 10 mM Na_2HPO_4, 1.8 mM KH_2PO_4, pH 7.4.
5. Equilibration buffer: 50 mM Na_2HPO_4, 100 mM NaCl.
6. Wash buffer 1: 10 mM imidazole, 5 mM β-ME dissolved in PBS.

7. Wash buffer 2: 25 mM imidazole, 5 mM β-ME dissolved in PBS.
8. Wash buffer 3: 40 mM imidazole, 5 mM β-ME dissolved in PBS.
9. Wash buffer 4: 60 mM imidazole, 5 mM β-ME dissolved in PBS.
10. Elution buffer 1: 10 mM Tris–HCl (pH: 7.4), 100 mM NaCl, 250 mM imidazole, 5 mM β-ME.
11. Elution buffer 2: 10 mM Tris–HCl (pH: 7.4), 100 mM NaCl, 300 mM imidazole, 5 mM β-ME.
12. Elution buffer 3: 10 mM Tris–HCl (pH: 7.4), 100 mM NaCl, 500 mM imidazole, 5 mM β-ME.

2.3 Dialysis of Purified DART Protein

1. Slide-A-Lyzer Dialysis Cassette (10,000 MWCO).
2. Dialysis Buffer: 10 mM Tris–HCl (pH: 7.4), 100 mM NaCl, 1 mM DTT.
3. 18- or 21-gauge needles.
4. 5 mL syringes.

2.4 Assessing Protein Quality and Purity by Coomassie and Immunoblot

1. PVDF membranes.
2. Whatman filter paper.
3. 2× sample buffer: 50% (v/v) 4× NuPAGE™ LDS sample buffer, 20% (v/v) DTT (1 M), 30% (v/v) water. Store at −20 °C.
4. MES SDS running buffer (pH: 7.3): 50 mM MES, 50 mM Tris base, 0.1% (w/v) SDS, 1 mM EDTA.
5. NuPAGE™ 4–12%, Bis–Tris 15-well gel.
6. Transfer buffer (pH: 8.3): 25 mM Tris base, 192 mM glycine, 20% (v/v) methanol.
7. PBS: 137 mM NaCl, 2.7 mM KCl, 10 mM Na_2HPO_4, 1.8 mM KH_2PO_4, pH 7.4.
8. PBST: PBS with 0.05% (v/v) Tween 20.
9. Blocking solution: 5% (w/v) milk in PBST. Store at 4 °C.
10. Primary antibody solution: 1:1000 dilution of Anti-HA Antibody—Rabbit monoclonal (Cell Signaling) in 5% (w/v) BSA in PBST, 0.05% (w/v) sodium azide. Store at 4 °C.
11. Secondary antibody solution: 1:10,000 dilution of HRP anti-rabbit antibody in PBST. Dilute antibody in PBST immediately before use.

2.5 Long-Term Storage of APOBEC1-YTH Purified Proteins

1. Glycerol—100%.
2. Dialysis Buffer: 10 mM Tris–HCl (pH: 7.4), 100 mM NaCl, 1 mM DTT.
3. Liquid nitrogen.

2.6 In Vitro DART Assay

1. DART buffer (10×): 100 mM Tris–HCl (pH 7.4), 500 mM KCl, 1 μM $ZnCl_2$.
2. RNaseOUT™ Recombinant Ribonuclease Inhibitor.
3. Purified APOBEC1-YTH protein.
4. RNeasy Micro Kit.
5. iScript Reverse Transcription Supermix.
6. CloneAmp™ HiFi PCR Premix.
7. 6× purple gel loading dye.
8. Agarose.
9. QIAquick Gel Extraction Kit.
10. TAE gel running buffer (50×): 40 mM Tris base, 2 mM EDTA, 20 mM glacial acetic acid.

3 Methods

3.1 Induction of DART Protein Expression in E. coli

1. Clone APOBEC1-YTH into the pET-His6-MBP vector so that the 6×His tag and the maltose binding protein (MBP) solubility tag are added to the N-terminal portion of the protein (*see* **Note 1**). Transform 3 μL into 50 μL of the pLysS bacterial strain.
2. Pick a colony and culture in 3 mL of LB medium with 50 μg/mL kanamycin for 16–18 h (*see* **Note 2**).
3. Make a glycerol stock by adding 500 μL of cultured cells and 500 μL of sterile 30% glycerol to a 1.5 mL tube. Mix well and quickly store at −80 °C.
4. Add a small stab from the glycerol stock to 30 mL of LB medium with 50 μg/mL kanamycin.
5. Culture 16–18 h at 37 °C with shaking (200 rpm).
6. Prepare 250 mL of autoinduction medium in a 1 L Erlenmeyer flask (*see* **Note 3**) and add 100 μg/mL kanamycin.
7. Add 6 mL of the bacterial culture to the autoinduction medium.
8. Culture for 8–10 h at 37 °C with shaking (250 rpm).
9. Culture for 16–18 h at 18 °C with shaking (250 rpm).
10. Spin cultures at >5000 × *g* for 10 min at 4 °C to pellet the bacterial cells.

11. Decant the supernatant, removing as much supernatant as possible without disturbing the pellet.
12. Carefully submerge the container with the cell pellet into liquid nitrogen for 10–15 s, or until the pellet has frozen solid.
13. The cell pellets may be stored at −80 °C at this stage. Alternatively, you may proceed directly to protein purification.

3.2 Purification of DART Protein by Affinity Chromatography

1. Thaw the cell pellets on ice or at 4 °C. It is important at this step to thaw the pellet without excessive warming.
2. Prepare the complete cell lysis buffer from the Qiagen Qproteome Bacterial Protein Prep Kit (*see* **Note 4**). Add 10 mL of complete lysis buffer and pipet up/down to resuspend the pellet in the lysis buffer. It will take some time to completely resuspend the pellet, and the solution may be quite viscous. Do not vortex to resuspend the pellet. Once resuspended, transfer to a 50 mL conical tube.
3. If multiple flasks for each sample were prepared, you can pool the lysates here.
4. Incubate on ice for 30 min. Gently swirl or shake the lysate every 5–10 min.
5. At the end of the incubation, spin the samples at 12,500 × *g* for 60 min at 4 °C to pellet the insoluble material. Carefully remove the supernatant and place in a new tube. The supernatant should be yellow in color but cleared of insoluble material. Additional spin time may be required if the lysate is cloudy. It is important to obtain a clear lysate at this step.
6. Take a 50 μL sample of the lysate for SDS-PAGE analysis (Fig. 1).
7. Carefully pipet 750 μL of resuspended Ni-NTA bead slurry into a chromatography column and allow the liquid to drain by gravity flow (*see* **Note 5**).
8. Wash the beads on the column with 6 mL of equilibration buffer. Allow the liquid to drain (*see* **Note 6**).
9. Pipet the lysate directly into the column and allow it to flow through the Ni-NTA beads. Collect the flow-through in a tube below the column. After the supernatant has drained, take a 50 μL sample of the flow-through for SDS-PAGE analysis (Fig. 1).
10. Apply 10 mL of wash buffer 1. When the buffer has finished draining, collect a 50 μL sample of the wash buffer for SDS-PAGE analysis (Fig. 1).
11. Apply 6 mL of wash buffer 2. When the buffer has finished draining, collect a 50 μL sample of the wash buffer for SDS-PAGE analysis (Fig. 1).

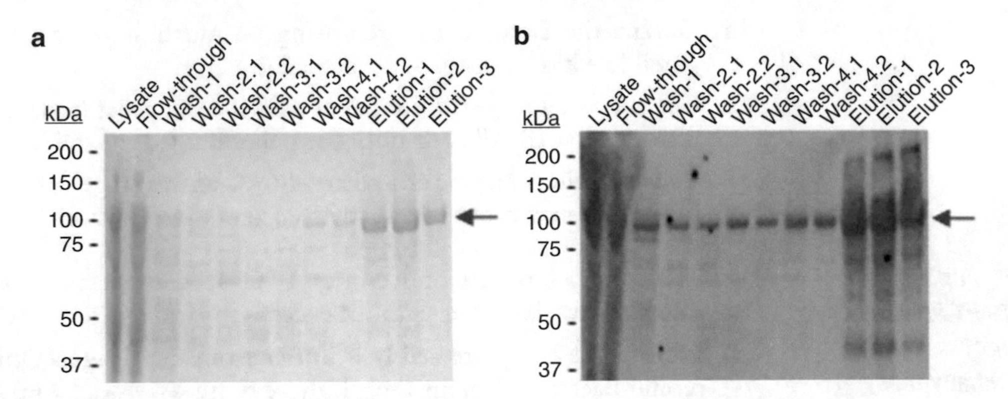

Fig. 1 SDS–PAGE analysis of APOBEC1-YTH protein purification. Red arrow labels the position of APOBEC1-YTH (96 kDa). (**a**) Coomassie blue stain of washes and elutions during the protein purification process. (**b**) Corresponding HA-tag western blot of panel (**a**)

12. Repeat **step 11** for a total of two washes.
13. Apply 6 mL of wash buffer 3. When the buffer has finished draining, collect a 50 μL sample of the wash buffer for SDS-PAGE analysis (Fig. 1).
14. Repeat **step 13** for a total of two washes.
15. Apply 6 mL of wash buffer 4. When the buffer has finished draining, collect a 50 μL sample of the wash buffer for SDS-PAGE analysis (Fig. 1).
16. Apply 10 mL of wash buffer 4. When the buffer has finished draining, collect a 50 μL sample of the wash buffer for SDS-PAGE analysis (*see* Fig. 1).
17. Apply 1.5 mL of elution buffer 1. When the buffer has finished draining, collect a 50 μL sample of the eluate for SDS-PAGE analysis (Fig. 1).
18. Apply 1.5 mL of elution buffer 2. When the buffer has finished draining, collect a 50 μL sample of the eluate for SDS-PAGE analysis (Fig. 1).
19. Apply 1.5 mL of elution buffer 3. When the buffer has finished draining, collect a 50 μL sample of the eluate for SDS-PAGE analysis (Fig. 1).
20. At this point, the eluate may be stored at 4 °C overnight (*see* **Note 7**), however it is recommended to continue with dialysis (*see* **Note 8**).

3.3 Dialysis of Purified DART Protein

1. Prepare 1 L of dialysis buffer in a 1 L beaker (*see* **Note 9**).
2. Place the Slide-a-Lyzer dialysis cassette in the beaker of dialysis buffer for 1–2 min to hydrate the membrane (*see* **Note 10**).
3. Using an 18 G or 21 G needle attached to a 5 mL syringe, aspirate the sample. Carefully insert the needle into the dialysis

cassette without touching the membrane and slowly inject the sample into the cassette. As you remove the needle, aspirate out the excess air so that there is as little remaining air as possible inside the dialysis cassette.

4. Place the dialysis cassette into the beaker of dialysis buffer. Place on a stir plate at 4 °C for 2 h at a low stir setting (*see* **Note 11**).
5. After 2 h, remove the cassette from the beaker and pour out the dialysis buffer. Add 1 L of fresh dialysis buffer and incubate for at least 2 h at 4 °C with stirring.
6. After the incubation, set a 5 mL syringe plunger at 2 mL and attach an 18 G or 21 G needle. Remove the dialysis cassette and carefully insert the needle without touching the membrane and inject 2 mL of air into the cassette. Then, invert the cassette so that the sample pools at the bottom of the cassette where the needle is inserted and draw up the plunger to completely remove all of the sample.
7. Take a 50 μL sample of the dialyzed purified protein for SDS-PAGE analysis (Fig. 1).

3.4 Assessing Protein Quality and Purity by Coomassie Stain and Immunoblot

1. Collect all 50 μL aliquots and add 50 μL of 2× sample buffer to each tube. Mix gently by pipetting (*see* **Note 12**).
2. Heat samples at $\geq$ 98 °C for 5 min and centrifuge briefly.
3. Load 10 μL of each sample into a well in two separate SDS-PAGE gels (*see* Fig. 1).
4. Perform electrophoresis at 150 V for ~1.5 h or until the dye reaches the bottom of the gel.
5. Remove the gels from their cassettes and rinse in 1× PBS.
6. Place one gel in 50–100 mL Coomassie stain solution for 1 h with gentle rocking at room temperature.
7. After 1 h, pour off the Coomassie stain solution and add 20 mL of Coomassie destain solution for 1 h with gentle rocking at room temperature.
8. After 1 h, pour off the Coomassie destain solution and add 20 mL of Coomassie destain solution. Cover the top of the container with Parafilm and keep on rocker at room temperature for at least 16 h.
9. Place the other gel in transfer buffer for 5 min.
10. To assemble transfer apparatus, open transfer cassette holder and place a sponge (soaked in 1× transfer buffer) on the bottom of the transfer cassette.
11. Place two pieces of filter paper (soaked in transfer buffer) on the sponge.

12. Wet the PVDF membrane in transfer buffer for 30 s and place on top of the filter papers. Use a roller to eliminate air bubbles.
13. Place the gel on the PVDF membrane and use a roller to eliminate air bubbles.
14. Place two filter papers (soaked in transfer buffer) on the gel and use a roller to eliminate air bubbles.
15. Place a sponge (soaked in transfer buffer) on the top of the assembly and close the cassette.
16. Insert the cassette into the transfer apparatus, fill the transfer tank with 1× transfer buffer, and run at 100 V for 60 min.
17. After the transfer is complete, soak the membrane in blocking solution for 1 h.
18. Place the membrane in 5 mL of primary antibody solution. Incubate with gentle rocking for at least 16 h at 4 °C.
19. Wash three times with PBST for 5 min at room temperature with rocking.
20. Place the membrane in 10 mL of secondary antibody solution. Incubate with gentle rocking for 1 h at room temperature.
21. Wash three times with PBST for 5 min at room temperature with rocking.
22. Mix 500 μL of ECL reagent A and 500 μL of ECL reagent B immediately before use.
23. Remove gel from PBST and apply 1 mL of the mixed ECL reagent to the entire membrane and incubate at room temperature for 1 min.
24. Pick up the membrane with forceps and allow the ECL reagent to drain off the membrane.
25. Immediately bring the membrane to a gel imaging system or expose using X-ray film (Fig. 1).

3.5 Long-Term Storage of APOBEC1-YTH Purified Proteins

1. Add 1.67 μg of purified protein to a clean 0.6 mL tube.
2. Add dialysis buffer for a final volume of 16 μL.
3. Add 4 μL of 100% glycerol for a final volume of 20 μL.
4. Mix thoroughly by pipetting.
5. Close tube and place in liquid nitrogen.
6. Repeat to store all of the purified protein prep.
7. Store tubes long term at −80 °C (*see* **Note 13**).

3.6 In Vitro DART Assay

1. Combine the following reagents in a 1.5 mL tube, adding the protein last (*see* **Notes 14** and **15**).
   (a) 5 μL of 10× DART buffer.
   (b) 1 μL RNaseOUT.

   (c) 50 ng total RNA.
   (d) 3 μL purified protein in 20% glycerol (250 ng).
   (e) Ultrapure water to 50 μL.
2. Gently mix by pipetting.
3. Incubate at 37 °C for 4 h (*see* **Note 16**).
4. After incubation, isolate RNA using RNeasy Micro Kit according to the manufacturer's instructions (*see* **Note 17**). Elute with 14 μL of RNase-free water.
5. The RNA can be stored at −80 °C after this step, or cDNA synthesis can be performed immediately following cleanup. The following steps describe targeted PCR amplification of a specific RNA region followed by Sanger sequencing to observe editing at specific loci. However, the purified RNA may also be used to prepare next-generation sequencing libraries for RNA-seq (*see* **Note 18**).
6. Add 7 μL of purified RNA to a clean PCR tube.
7. Add 9 μL of ultrapure water.
8. Add 4 μL of 5× iScript Supermix for RT (*see* **Note 19**).
9. Mix gently by pipetting and run the following program on a thermocycler:
   (a) 5 min at 25 °C.
   (b) 20 min at 46 °C.
   (c) 1 min at 95 °C.
   (d) Hold at 4 °C.
10. Dilute cDNA 1:5 with ultrapure water.
11. To generate an amplicon for Sanger sequencing, assemble the following reagents in a clean PCR tube (*see* **Note 20**):
   (a) 10 μL CloneAmp 2× mastermix.
   (b) 1 μL forward primer.
   (c) 1 μL reverse primer.
   (d) 1 μL diluted cDNA.
   (e) 7 μL ultrapure water.
12. Mix gently by pipetting and run the following program on a thermocycler:
   (a) 98 °C for 45 s.
   (b) 98 °C for 10 s.
   (c) 57 °C for 10 s.
   (d) 72 °C for 5 s.
   (e) Repeat **steps b–d** for a total of 30 cycles.

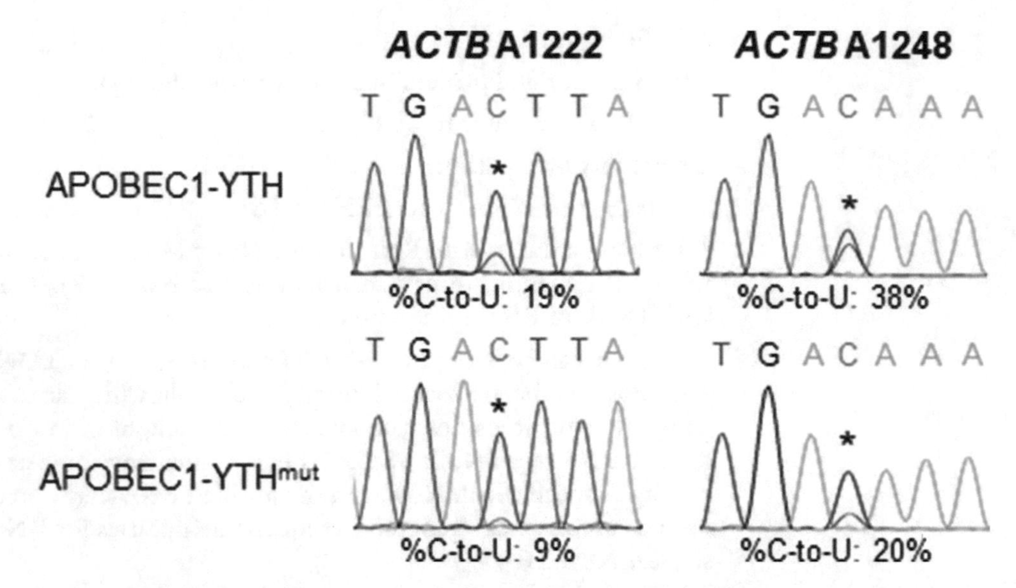

Fig. 2 In vitro DART was performed on 50 ng of total RNA from HEK293T cells, followed by RT-PCR to amplify the *ACTB* 3′UTR. Sanger sequencing chromatograms show C-to-U mutations in cytidine residues adjacent to two m^{6}A sites in *ACTB*: A1222 (left) and A1248 (right). C-to-U mutation rate is indicated for each site and indicates decreased editing using APOBEC1-YTH^{mut} protein compared to APOBEC1-YTH

   (f) 72 °C for 5 min.
   (g) Hold at 4 °C.

13. Add 4 μL of 6× purple gel loading dye and mix by pipetting.
14. Load samples into a 1% agarose gel and run until ladder is sufficiently separated.
15. Use a razorblade to excise desired products from the gel.
16. Extract DNA using Qiagen QIAquick Gel Extraction Kit according to manufacturer's protocol.
17. Perform Sanger sequencing using a primer of interest. We typically use one of the primers used for PCR in **step 11** above (Fig. 2).
18. Assess C-to-T transitions in Sanger sequencing data. We use EditR [6] to quantify % C-to-T (*see* **Note 21**).

4 Notes

1. Detailed cloning instructions can be found on the Addgene page for plasmid #29656.
2. Prepare all solutions using Milli-Q water. Prepare and store all reagents at room temperature (unless specified otherwise). Follow all chemical safety guidelines when preparing solutions. Follow all waste disposal regulations when disposing of waste materials.

3. Using beveled flasks can increase yield.
4. Ensure that the lysozyme in the lysis buffer is completely dissolved. This may require time and gentle warming followed by cooling the buffer again.
5. Keep all reagents on ice during protein purification or work in a cold room. Further, we use the Ni-NTA resin from GoldBio most successfully under the conditions used in this protocol. Substituting different resins with different binding capacities may require optimization to maximize yield and purity.
6. Do not allow the liquid to completely drain and the beads to dry. Always add the next wash/buffer right as the liquid level reaches the top of the beads.
7. Do not mix the elutions until you have confirmed their quality.
8. It is recommended to proceed with dialysis immediately after purification as there is a high concentration of imidazole in the elution buffers, and the protein is only stable for several weeks when stored at 4 °C.
9. Always make fresh dialysis buffer as it contains DTT.
10. Never touch the membrane of the dialysis cassette.
11. Each dialysis step should proceed for at least 2 h to allow for complete buffer exchange. However, the steps may proceed for 16–18 h if needed.
12. The Coomassie stained gel is for the assessment of purity, while the immunoblot is performed to confirm the expected band is indeed APOBEC1-YTH (this fusion protein contains an HA tag). It is highly recommended to perform this quality control step, as it may help with troubleshooting if a prep is unsuccessful.
13. Protein is still active when stored at −80 °C for more than 9 months.
14. It is highly recommended that RNA used for in vitro DART assays is first treated with DNase I, followed by ethanol precipitation.
15. Less than 50 ng of RNA may be used, if needed. In vitro DART assays using as little as 30 ng per reaction have been performed successfully in our hands as described. If using significantly more or less RNA, optimization of the RNA–protein ratio is recommended. For example, if using significantly more than 50 ng of RNA it is recommended to increase the protein concentration and/or time of incubation. If using less RNA, it is recommended to decrease the protein concentration and/or time of incubation.
16. Shaking during the 4-h incubation does not seem to increase or decrease editing efficiency.

17. Phenol–chloroform extraction or TRIzol extraction may also be used to isolate RNA after the DART assay. However, the RNeasy Micro Kit is recommended for its ease of use.
18. To make next-generation sequencing libraries, particularly if less than 50 ng of input RNA is used, it is recommended to use a template-switching approach, such as the Low Input/Single Cell Library Prep Kit from New England Biolabs.
19. Any cDNA synthesis system that can work with low input RNA can be used (e.g., SuperScript III).
20. Any high-fidelity polymerase can be used for amplicon generation. The annealing temperature and extension time should be adjusted for each primer pair and amplicon.
21. An alternate method of quantifying C-to-U editing is to use ImageJ to calculate the height of the C and T peak at the edited site and calculate % C-to-T = $T_{height}/(T_{height} + C_{height})$. This method does not consider the background in sequencing traces, nor the shape of the peaks, and is therefore not recommended.

References

1. Flamand MN, Meyer KD (2019) The epitranscriptome and synaptic plasticity. Curr Opin Neurobiol 59:41–48
2. Meyer KD, Jaffrey SR (2017) Rethinking m ^{6} a readers, writers, and erasers. Annu Rev Cell Dev Biol 33(1):319–342
3. Zaccara S, Ries RJ, Jaffrey SR (2019) Reading, writing and erasing mRNA methylation. Nat Rev Mol Cell Biol 20:608–624
4. Roundtree IA, Evans ME, Pan T, He C (2017) Dynamic RNA modifications in gene expression regulation. Cell 169(7):1187–1200
5. Meyer KD (2019) DART-seq: an antibody-free method for global m6A detection. Nat Methods 16:1275–1280
6. Kluesner MG, Nedveck DA, Lahr WS et al (2018) EditR: a method to quantify base editing from sanger sequencing. CRISPR J 1 (3):239–250

Chapter 21

Target-Specific Profiling of RNA m^{5}C Methylation Level Using Amplicon Sequencing

Tennille Sibbritt, Ulrike Schumann, Andrew Shafik, Marco Guarnacci, Susan J. Clark, and Thomas Preiss

Abstract

Mapping the position and quantifying the level of 5-methylcytosine (m^{5}C) as a modification in different types of cellular RNA is an important objective in the field of epitranscriptomics. Bisulfite conversion has long been the gold standard for the detection of m^{5}C in DNA, but it can also be applied to RNA. Here, we detail methods for bisulfite treatment of RNA, locus-specific PCR amplification, and detection of candidate sites by sequencing on the Illumina MiSeq platform.

Key words 5-methylcytosine, Epitranscriptomics, Bisulfite conversion, Next-generation sequencing, MiSeq

1 Introduction

Cellular RNAs can be richly modified with more than one hundred known chemically and structurally distinct nucleoside modifications [1–3]. The field of epitranscriptomics [4–6] has been greatly accelerated by the development of high-throughput mapping methods for RNA modifications, typically based on a next-generation sequencing (NGS) readout. Transcriptome-wide positions of 5-methylcytosine (m^{5}C) [7, 8], 5-hydroxymethylcytosine [9, 10], N^6-methyladenosine [11–13], N1-methyladenosine [14–16], and pseudouridine [17–19] have each been reported in this way. To detect m^{5}C in RNA, a range of methods have been developed, including the direct (meRIP [20]) or indirect immunoprecipitation of methylated RNA (aza-IP [21], miCLIP [22]). The bisulfite conversion approach in popular use for DNA methylation detection has also been successfully adapted to RNA [8, 23–

Tennille Sibbritt and Ulrike Schumann contributed equally to this work.

Erik Dassi (ed.), *Post-Transcriptional Gene Regulation*, Methods in Molecular Biology, vol. 2404,
https://doi.org/10.1007/978-1-0716-1851-6_21,

25]. Bisulfite conversion of nucleic acids takes advantage of the differential chemical reactivity of m^5C compared to unmethylated cytosines; unmethylated cytosines are quickly deaminated to uracil while m^5C deamination happens at a much slower rate. Therefore, using short time frames for the deamination reaction ensures that m^5C remains as a cytosine. However, rigorous bioinformatics analyses and candidate site filtering approaches are required to reduce false-positive site calling [26]. This has greatly accelerated m^5C research, leading to the identification of species-specific candidate sites in mammalian RNA [27], a putative reader protein [28], and an inverse correlation between m^5C content and mRNA translation [29].

We adapted an RNA bisulfite conversion method [23] for NGS-based transcriptome-wide readout and mapped a large number of novel candidate m^5C sites in a variety of RNA biotypes, including mRNA [8, 29]. Here, we detail our protocols for RNA bisulfite conversion and locus-specific, semiquantitative PCR-based detection of nonconverted sites.

Sequencing of PCR amplicons is conveniently done using the Illumina MiSeq platform, as this affords multiplexing of multiple distinct amplicons while still achieving ample read depth for estimating the proportion of m^5C at targeted positions. For instance, each of the 96 indexed adaptor primers available with the TruSeq® DNA Nano HT Kit could be assigned to a separate cellular RNA source material, and multiple RNA loci/PCR amplicons per sample could be included in the sequencing library. This multiplexing capability allows the generation of hundreds of independent quantitative measurements of the m^5C level in a single MiSeq run (Fig. 1).

2 Materials

Prepare all solutions using DNase and RNase-free H_2O and analytical grade reagents. Prepare and store all reagents at room temperature, unless indicated otherwise. Diligently follow all safety and waste disposal regulations when performing experiments.

Prepare and perform bisulfite conversion, cDNA synthesis, and PCR amplification experiments in a PCR and plasmid-free area.

2.1 *In Vitro* Transcription Components

1. pRL Renilla Luciferase Reporter Vectors (pRL-TK) (Promega).
2. MEGAScript™ T7 Kit (Invitrogen).
3. TURBO™ DNase (Invitrogen).
4. Phase Lock Gel Heavy (1.5 mL).
5. UltraPure™ phenol–water (3.75:1 v/v) (Invitrogen).

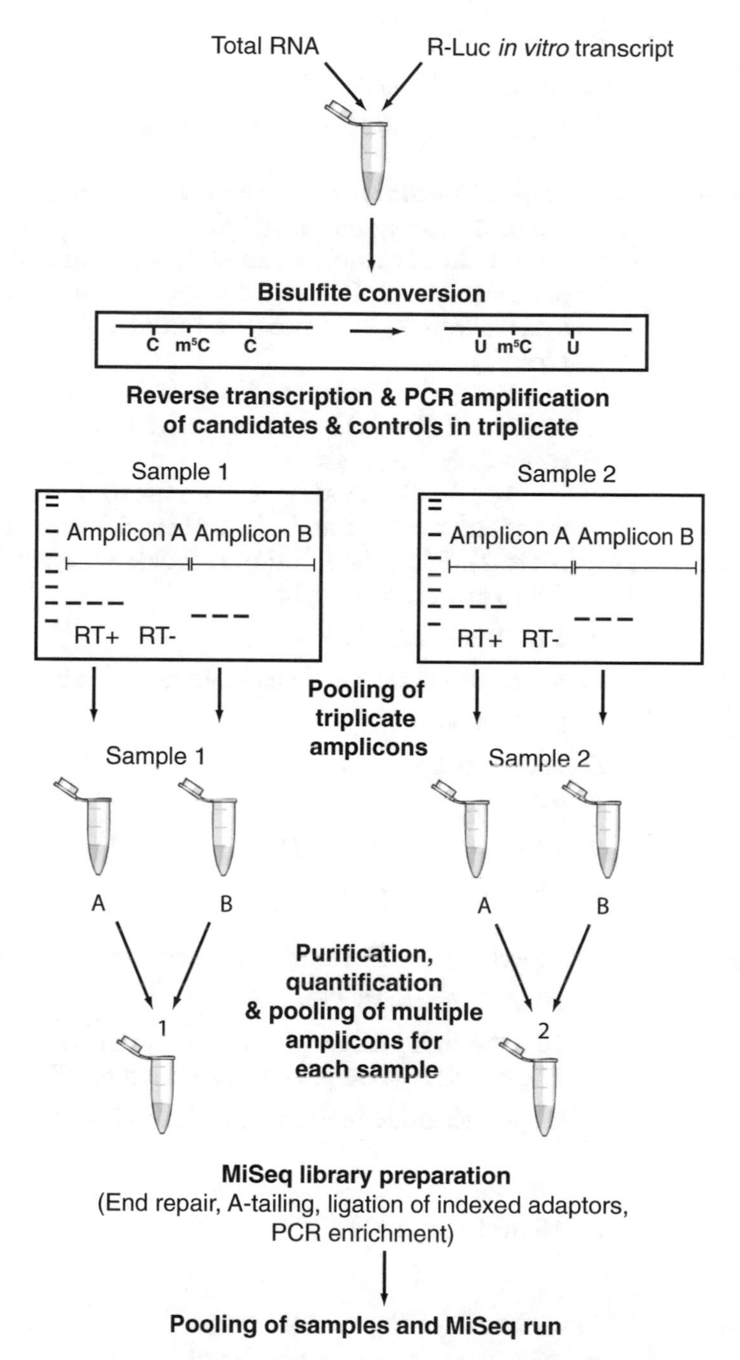

Fig. 1 Protocol overview, showing workflow and pooling strategy for effective sequencing. Total RNA spiked with the R-Luc in vitro transcript is bisulfite converted. The bisulfite converted RNA is reverse transcribed and candidates of interest, as well as the positive ($tRNA^{Asp(GUC)}$ and $tRNA^{Leu(CAA)}$) and negative (R-Luc) controls, are PCR amplified in triplicate for each sample to minimize PCR amplification bias. The triplicate amplicons are then pooled and subjected to purification. The purified amplicons are quantified and an equal amount of each amplicon is pooled for each sample. Following this, library preparation is performed using the TruSeq® Nano DNA Library Preparation Kit, which involves end repair and A-tailing of the amplicons, ligation of the indexed adaptors to the amplicons to enable multiplexing of samples, and enrichment of the libraries by PCR. Each library is then pooled, a PhiX Control library is spiked-in, and the libraries are subjected to sequencing on the MiSeq platform

6. Chloroform.
7. Glycogen (5 mg/mL).
8. Agilent RNA 6000 Nano Kit (Agilent).

2.2 Sodium Bisulfite Conversion Components

1. **Sodium bisulfite solution**: 40% (w/v) sodium metabisulfite, 0.6 mM hydroquinone, pH 5.1.

   0.6 M Hydroquinone: weigh 66 mg hydroquinone and place into a 1.5 mL tube. Add H_2O to 1 mL and cover in foil to protect from light. Place in an orbital shaker to dissolve (*see* **Note 1**).

   40% (w/v) sodium bisulfite: Dissolve 8 g sodium metabisulfite in 20 mL H_2O in a 50 mL falcon tube and vortex until completely dissolved.

   Add 20 μL 0.6 M hydroquinone to the 40% sodium bisulfite solution, vortex and adjust pH to 5.1 with 10 M NaOH (*see* **Note 2**). Filter the solution through a 0.2 μm filter. Cover in foil to protect from light.
2. 1 M Tris–HCl, pH 9.0.
3. Micro Bio-Spin™ P-6 Gel Columns, Tris buffer (Bio-Rad).
4. Mineral oil.
5. 100% ethanol.
6. 75% ethanol.
7. 3 M sodium acetate, pH 5.2.
8. Glycogen (5 mg/mL).

2.3 cDNA Synthesis Components

1. SuperScript™ III Reverse Transcriptase (Invitrogen).
2. 10 mM mixed dNTPs.
3. 20× random primer mix: 35 μM hexamers, 25 μM T12VN Oligonucleotide sequence for hexamers: NNNNNN

   Oligonucleotide sequence for T12VN: NVTTTTTTTTTTTT

2.4 PCR Amplification Components

1. Platinum™ Taq DNA Polymerase Kit (Invitrogen).
2. 10 mM mixed dNTPs.

2.5 Agarose Gel Electrophoresis and PCR Purification Components

1. SeaKem® LE Agarose (Lonza).
2. GelRed Nucleic Acid Stain (Millipore).
3. GeneRuler 1 kb Plus DNA Ladder (Thermo Scientific).
4. 1× TAE buffer: Make up 50× TAE buffer by combining 424 g Tris base, 57.1 mL acetic acid and 100 mL 0.5 M EDTA (pH 8.0) and make up to 1 L in H_2O. To make 1× TAE buffer, combine 40 mL 50× TAE buffer with 1.96 L H_2O.

5. Wizard® SV Gel and PCR Clean-Up System (Promega).
6. Qubit™ dsDNA BR Assay Kit (Molecular Probes).

2.6 Library Preparation Components

1. TruSeq® Nano DNA Library Preparation Kit (Illumina).
2. MinElute PCR Purification Kit (Qiagen).
3. Agencourt AMPure XP beads (Beckman Coulter).
4. Tween® 20 (Sigma-Aldrich).
5. Tris–HCl (pH 8.5) (*see* **Note 3**).
6. Tris–HCl (pH 8.0).
7. EBT buffer: 10 mM Tris–HCl (pH 8.5) and 0.1% Tween® 20. Add 19.8 mL H_2O to 0.2 mL of 1 M Tris–HCl (pH 8.5), then add 20 μL of Tween® 20 to the solution. Vortex solution thoroughly to ensure Tween® 20 is mixed throughout the solution.
8. Library Dilution Buffer: 10 mM Tris–HCl (pH 8.0) and 0.05% Tween® 20. Add 19.8 mL H_2O to 0.2 mL of 1 M Tris–HCl (pH 8.0) followed by 10 μL of Tween® 20. Vortex solution thoroughly to ensure Tween® 20 is mixed throughout the solution.
9. Kapa Library Quantification Kit Illumina® Platforms (Roche).
10. 0.2 M NaOH.
11. MiSeq Reagent Kit v2 (300 cycles) (Illumina) (*see* **Note 4**).
12. PhiX Control Library (Illumina).

3 Methods

Carry out all procedures at room temperature unless otherwise specified.

3.1 RNA Extraction and DNase Treatment

Total cellular RNA is extracted directly from PBS-washed adherent cells using 1 mL of TRIzol™ according to manufacturer's instructions (*see* **Note 5**). RNA is then treated with TURBO™ DNase as per the manufacturer's protocol. For transcriptome-wide bsRNA-seq, ribosomal RNA depletion (or poly-A enrichment) is essential prior to bisulfite conversion (*see* **Note 6**).

3.2 Generation of the Renilla Luciferase (R-Luc) In Vitro Transcript Spike-in Control (See Note 7)

1. Linearize the pRL-TK vector using *Bam*HI according to the MEGAScript™ T7 Kit protocol.
2. Perform in vitro transcription according to the MEGAScript™ T7 Kit protocol using 1 μg linearized DNA. An incubation period of 4 h at 37 °C with the kit components is sufficient.
3. Add 2 U of TURBO™ DNase and incubate at 37 °C for 30 min.

4. Transfer the reaction to a Phase Lock Gel Heavy (1.5 mL) tube and make the volume up to 100 μL with H_2O.
5. Add equal volumes of UltraPure™ phenol–water (3.75:1 v/v) and chloroform, shake vigorously for 15 s and centrifuge at 16,000 × *g* for 5 min.
6. Add the same volume of chloroform as in **step 5**, shake vigorously for 15 s and centrifuge again at 16,000 × *g* for 5 min.
7. Transfer the aqueous phase to a clean 1.5 mL tube. Add 1/10 volume of 3 M sodium acetate, 2.5 volumes of 100% ethanol and 1 μL glycogen (5 mg/mL), vortex and precipitate the RNA overnight at −20 °C.
8. Centrifuge the precipitate at 17,000 × *g* at 4 °C for 20 min and carefully discard the supernatant.
9. Add 1 mL of 75% ethanol to the RNA, vortex and centrifuge at 17,000 × *g* at 4 °C for 5 min.
10. Carefully discard the supernatant and allow the pellet to air dry for up to 5 min (*see* **Note 8**).
11. Resuspend the RNA in H_2O.
12. Treat 10 μg of in vitro transcript with 2 U of TURBO™ DNase according to the manufacturer's protocol to remove any residual template DNA.
13. Assess the size and integrity of the in vitro transcript using a RNA 6000 Nano Chip on the Agilent 2100 Bioanalyzer according to the manufacturer's protocol.

3.3 Bisulfite Conversion of RNA

1. Add 1/1000 of the R-Luc in vitro transcript to 2–4 μg DNase-treated RNA (*see* **Note 9**). The combined volume of the RNA sample and in vitro transcript should be <20 μL.
2. Denature RNA by heating to 75 °C for 5 min in a heat block.
3. Preheat sodium bisulfite solution to 75 °C, add 100 μL to the RNA, vortex thoroughly and briefly spin in a microcentrifuge.
4. Gently overlay the reaction mixture with 100 μL mineral oil (*see* **Note 10**). Cover the tube in aluminum foil to protect the reaction mixture from light.
5. Incubate at 75 °C for 4 h in a heat block.
6. About 20 min before the bisulfite conversion reaction is complete, prepare **two** Micro Bio-Spin™ P-6 Gel Columns **per reaction** by allowing the buffer to drain into a collection tube by gravity (*see* **Note 11**). Discard the flow through, place the column back into the collection tube and centrifuge at 1000 × *g* for 2 min. Transfer each column to a clean 1.5 mL tube.
7. Remove the bisulfite reaction mixture from the heat block and gently transfer the bottom layer containing the sodium

bisulfite/RNA mixture to the first Micro Bio-Spin™ P-6 Gel Column (*see* **Note 12**).

8. Centrifuge at 1000 × *g* for 4 min.
9. Carefully transfer the eluate to the second Micro Bio-Spin™ P-6 Gel Column and repeat **step 8**.
10. Discard the columns, add an equal volume of 1 M Tris–HCl (pH 9.0) to the second eluate, vortex, spin briefly and then overlay with 150 μL mineral oil. Cover the tube in aluminum foil to protect the reaction mixture from light.
11. Incubate at 75 °C for 1 h in a heat block. This step brings the RNA mixture to a neutral pH.
12. Carefully transfer the bottom layer containing the RNA to a clean 1.5 mL tube. Precipitate overnight and resuspend bisulfite-converted RNA in H_2O as described in **steps 7–11** in Subheading 3.2 (*see* **Notes 13** and **14**).

3.4 cDNA Synthesis

1. Combine 200–300 ng of bisulfite-converted RNA, 1 μL 20× random hexamers mix, 1 μL of 1 mM dNTP mix, and H_2O to a final volume of 13 μL and incubate at 65 °C for 5 min to denature the RNA.
2. Perform reverse transcription of the bisulfite-converted RNA according to the manufacturer's protocol. Use 1 μL of H_2O instead of SuperScript III™ reverse transcriptase for the reverse transcriptase (RT) negative control. It is essential to include RT negative controls for each sample, as the primers are not necessarily designed to span exon-exon junctions.
3. Once the reaction is complete, dilute the cDNA 1:10 in H_2O for bisulfite PCR amplification.

3.5 Bisulfite PCR Primer Design

1. For the amplification of the R Luc in vitro transcript, which does not contain m^5C sites, design the primers such that they avoid areas of bisulfite-converted cytosines (*see* **Note 15**) (Fig. 2a).
2. For the amplification of positive control transcripts (*see* **Note 16**) and transcripts containing candidate m^5C sites, design the primers such that they span regions containing converted cytosines to avoid preferentially amplifying nonconverted sequences (Fig. 2b).
3. Smaller amplicon sizes are desired to reduce amplification of nonconverted cytosines due to strong secondary structure (*see* **Note 17**). Amplicons should be designed such that they are 70–150 bp in length (*see* **Note 18**).
4. To ensure the designed primers uniquely amplify the region of interest, check primer specificity using the BiSearch webserver [30].

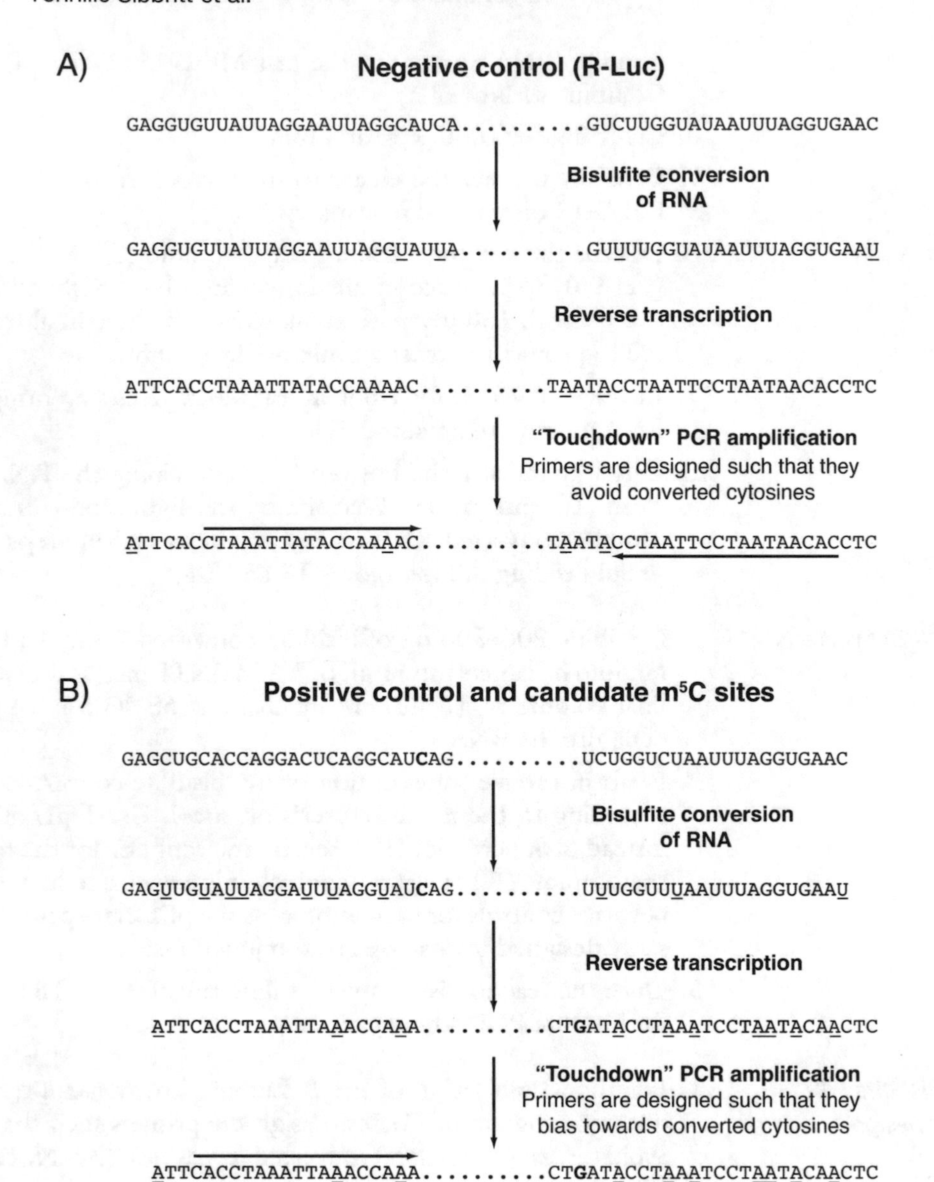

Fig. 2 Schematic demonstrating the bisulfite conversion of RNA, reverse transcription, primer design, and PCR amplification. (**a**) Primers designed for the R-Luc in vitro transcripts avoid areas of converted cytosines to prevent preferential amplification of converted sequences, which may falsely indicate efficient bisulfite conversion. (**b**) Primers designed for the tRNA controls and validation of candidate m^5C sites span areas containing converted cytosines to preferentially amplify converted sequences. Underlined bases represent converted (unmethylated) cytosines and bold bases represent m^5C. Primers were designed to amplify products that were 70–200 bp

3.6 PCR Amplification and Pooling of Amplicons

1. For a 25 μL PCR reaction, combine 0.1 μL Platinum® Taq DNA Polymerase, 1 μL diluted cDNA, 2.5 μL 10× buffer (without $MgCl_2$), 0.5 μL 10 mM dNTPs, 0.75 μL 50 mM $MgCl_2$, 0.25 μL 10 μM forward primer, 0.25 μL 10 μM reverse primer and 19.65 μL of H_2O. Perform PCR for each cDNA

sample in triplicate to reduce the potential for PCR amplification bias.

2. Mix gently, spin briefly and place into a thermal cycler.
3. Perform a touchdown PCR program (Table 1) (*see* **Note 19**).
4. Perform standard 2–3% agarose gel electrophoresis in 1× TAE of all amplicons along with 5 μL of GeneRuler 1 kb Plus DNA Ladder.
5. Pool the triplicate amplicons (*see* **Note 20**) and purify the products using the Wizard® SV Gel and PCR Clean-Up System according to the manufacturer's protocol (*see* **Note 21**).
6. Use 2 μL to quantify each set of amplicons using the Qubit™ dsDNA BR Assay Kit according to the manufacturer's protocol.
7. The MiSeq protocol enables the sequencing of multiple candidates on a single sequencing run. Pool 20–30 ng of each amplicon per sample into a single 1.5 mL tube.
8. Bring the volume to 65 μL using H_2O. If the volume is >65 μL, concentrate it to ~65 μL using a vacuum concentrator and bring to 65 μL in H_2O if required (*see* **Note 22**).

3.7 MiSeq Amplicon Sequencing Library Preparation

This protocol has been adapted from the TruSeq® Nano DNA Library Preparation Guide (*see* **Note 23**).

1. Add 40 μL of End Repair Mix to 60 μL of each pooled set of amplicons (*see* **Note 24**), gently pipette the entire volume up and down ten times and incubate at 30 °C for 30 min in a thermal cycler.

Table 1
"Touchdown" PCR cycling conditions for the amplification of candidates by Platinum® Taq DNA Polymerase

Stage		Temperature	Time
Initial denaturation		94 °C	2 min
Phase I			
Denaturation		94 °C	30 s
Annealing		Cycle from $T_{mhighest}$ + 5 °C to $T_{mlowest}$—5 °C	30 s
Extension		72 °C	15–30 s (*see* **Note 17**)
Phase II			
25–45 cycles	Denaturation	94 °C	30 s
	Annealing	$T_{mlowest}$	30 s
	Extension	72 °C	15–30 s
Final extension		72 °C	5 min
Hold		4 °C	Forever

2. Libraries containing amplicon sizes <100 nt are purified using the MinElute PCR Purification Kit, as per manufacturer's instructions (*see* **Note 25**). Libraries containing amplicon sizes >100 nt are purified using AMPure XP beads as detailed below (*see* **Note 26**).
3. Vortex AMPure XP beads to ensure even distribution. Dilute 136 μL beads in 24 μL H_2O, add to the End Repair reaction mixture, gently pipet the entire volume up and down ten times to mix and incubate at room temperature for 15 min.
4. Place the tubes on a magnetic rack at room temperature for 5 min then remove the supernatant.
5. Keep the tubes in the magnetic rack and wash the beads twice for 30 s by gently adding 200 μL of freshly made 80% ethanol to the tubes without disturbing the beads. Allow to air dry at room temperature for 15 min.
6. Remove the tubes from the magnetic rack and resuspend the beads in 18 μL Resuspension Buffer. Incubate at room temperature for 2 min.
7. Place the tubes on the magnetic rack for 5 min then transfer 17.5 μL of the cleared supernatant to a new 0.5 mL tube.
8. Add 12.5 μL of A-tailing Mix to 17.5 μL supernatant, pipet the entire volume of the reaction mixture up and down ten times to mix.
9. Incubate in a thermal cycler at 37 °C for 30 min, followed by 70 °C for 5 min and at 4 °C for 5 min. Centrifuge at 280 × *g* for 1 min.
10. Add 2.5 μL Resuspension Buffer, 2.5 μL Ligation Mix and 2.5 μL DNA Adapter to each 30 μL reaction mixture (*see* **Note 27**).
11. Pipette the entire volume up and down ten times and centrifuge at 280 × *g* for 1 min.
12. Incubate at 30 °C for 15 min in a thermal cycler, then cool to 4 °C.
13. Remove the reaction mixture from the thermal cycler and add 5 μL Stop Ligation Buffer.
14. Pipet the entire volume up and down ten times to mix and centrifuge at 280 × *g* for 1 min.
15. Purify the reaction mixture using 42.5 μL **undiluted** AMPure XP beads as described in **steps 3–7** of this section, resuspending the beads in 52.5 μL Resuspension Buffer (*see* **Note 28**).
16. Transfer 50 μL of the reaction mix to a new 0.5 mL tube.

Table 2
PCR cycling conditions for the enrichment of the MiSeq amplicon sequencing library

Stage		Temperature	Time
Initial denaturation		95 °C	3 min
8 cycles	Denaturation	95 °C	20 s
	Annealing	60 °C	15 s
	Extension	72 °C	30 s
Final extension		72 °C	5 min
Hold		4 °C	Hold

17. Purify the reaction mixture again using 50 μL **undiluted** AMPure XP as described in **steps 3–7** of this section, resuspending the beads in 27.5 μL Resuspension Buffer.
18. Transfer 25 μL of the cleared supernatant to a new 0.5 mL tube.
19. Add 5 μL PCR Primer Cocktail and 25 μL Enhanced PCR Mix to the 25 μL reaction mixture, pipette the entire volume up and down ten times to mix and centrifuge at 280 × *g* for 1 min.
20. Place the reaction mixture into a thermal cycler and perform the PCR program as outlined in Table 2.
21. Purify the enriched library using 50 μL **undiluted** AMPure XP beads, as described in **steps 3–7** of this section, resuspending the beads in 32.5 μL Resuspension Buffer.
22. Transfer 30 μL of the cleared supernatant to a new 0.5 mL tube. The library can now be stored at −20 °C for up to 7 days.

3.8 Validation and Quantification of the Libraries

1. Dilute 1 μL library in 5 μL H_2O and dilute 2 μL pooled amplicons prior to library preparation in 4 μL H_2O (*see* **Note 24**).
2. To validate that each library preparation was successful and to determine the average library size, perform standard 2–3% agarose gel electrophoresis in 1× TAE of all pooled amplicons and libraries along with 5 μL 1 kb Plus DNA ladder.
3. Quantify the libraries using the Qubit™ dsDNA BR Assay Kit, following the manufacturer's protocol.
4. Determine the concentration of each library in nM using the average fragment size determined from **step 2** and the concentration in ng/ μL determined from **step 3** of this section. The following formula may be used: $\frac{[\mathrm{DNA}](\mathrm{ng}/\mu\mathrm{L})}{\text{Average fragment size (bp)} \times 649} \times 10^6$.
5. Dilute the libraries to 50 nM with EBT buffer.
6. Perform the final quantification of each library by qPCR using the Kapa Library Quantification Kit Illumina® Platforms according to instructions with minor modifications. The manufacturer's protocol recommends performing twofold serial dilutions from 1:1000 to 1:8000; however, this results in

most samples being outside the range of the standard curve. Instead, perform twofold serial dilutions of the libraries from 1:1000 to 1:32000, to ensure they are within the range of the standard curve.

7. To calculate the concentration of each library from the qPCR, average the concentrations determined for each library based on the dilutions within the range of the standard curve.

3.9 Preparation of the Sample Sheet

1. Use the Illumina Experiment Manager based on the manufacturer's protocol to prepare the sample sheet (*see* **Note 29**).

3.10 Dilution of the Libraries and Loading of the Cartridge (See Note 30)

1. Dilute each library to 10 nM in EBT buffer, based on the concentrations determined by qPCR. From this point, keep the libraries on ice and store remaining libraries at −20 °C.
2. Pool 10 μL of each 10 nM library, and further dilute to 2 nM by adding 8 μL EBT buffer to 2 μL of the 10 nM pooled libraries.
3. Dilute the PhiX Control library to 2 nM by adding 8 μL EBT buffer to 2 μL of the 10 nM PhiX Control library (*see* **Note 31**).
4. Denature the pooled libraries and PhiX Control library separately by adding 10 μL of 0.2 M NaOH to 10 μL of the 2 nM libraries (*see* **Note 32**).
5. Vortex to mix and centrifuge at 280 × *g* for 1 min. Incubate at room temperature for 5 min.
6. Dilute the denatured pooled libraries and PhiX Control library separately to 20 pM by adding 980 μL prechilled HT1 to 20 μL denatured libraries.
7. Dilute the 20 pM pooled libraries and PhiX Control library separately to 10 pM by adding 500 μL prechilled HT1 to 500 μL 20 pM libraries.
8. Combine 300 μL of the 10 pM PhiX Control library with 700 μL of the 10 pM pooled libraries and vortex to mix (*see* **Notes 33** and **34**).
9. Load 600 μL of the final sample into the cartridge. Ensure air bubbles are removed by gently tapping the cartridge.
10. Perform the sequencing run according to the manufacturer's protocol.

3.11 Alignment of the MiSeq Data

Performing FastQC is recommended to analyze the quality of the sequencing reads. To trim the Illumina® adaptor sequences that were ligated to the ends of amplicons to facilitate sequencing of the 150 bp paired-end reads, Cutadapt [31] can be used. Sequencing reads can be aligned using the MeRanT tool within meRanTK [32] (*see* **Note 35**). As reference sequences for the alignment, pool all target RNA regions interrogated (including all positive and negative controls) into a single reference and perform C-to-T

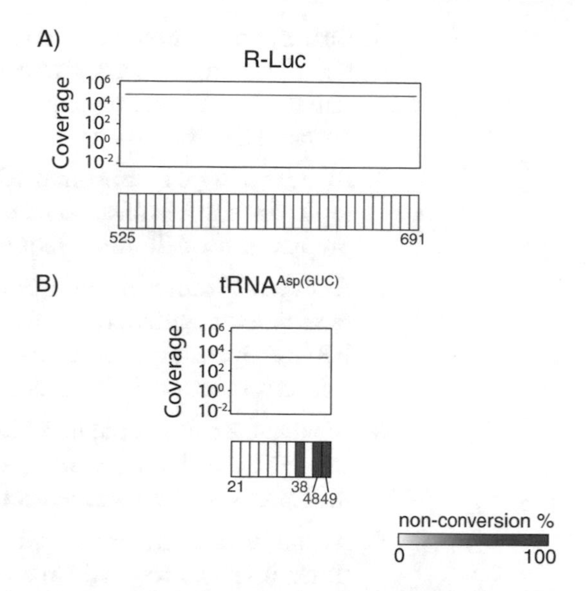

Fig. 3 Representative output of MiSeq amplicon sequencing showing a segment of the negative control R-Luc spike-in transcript and the positive control tRNA$^{Asp(GUC)}$. The top panel displays the coverage of the amplicon, which is relatively even across the amplicon. The heatmaps in the bottom panel display the cytosine nonconversion percentage. Numbers below the heatmaps represent the positions of cytosines relative to the start of the transcript. (**a**) A segment of the R-Luc in vitro transcript that does not exhibit nonconversion of cytosines, indicating the bisulfite conversion reaction was efficient. (**b**) A segment of endogenous HeLa cell tRNA$^{Asp(GUC)}$ displaying high levels of cytosine nonconversion (methylation) at the known m^5C sites (C38, C48, and C49)

conversion of the forward strand. The number of C and T calls at all C positions can be extracted from the aligned sequencing reads in order to determine the proportion of m^5C at a given cytosine (Fig. 3).

4 Notes

1. Dissolving hydroquinone in H_2O can be time consuming, particularly in cooler temperatures. It is recommended that this is prepared first.
2. 10 M NaOH is added dropwise to the sodium bisulfite solution while mixing. Slightly less than 1 mL is required to adjust the pH to 5.1.
3. 1 M Tris–HCl (pH 8.5) is equivalent to Buffer EB from Qiagen purification kits. This can be used for the preparation of EBT buffer.
4. The MiSeq Reagent Kit v2 (300 cycles) provides 2 × 150 bp reads.

5. Other source material, such as tissue samples, can also be used. Check manufacturer's instructions for suggested starting material to avoid overloading the TRIzol™ reagent, which will reduce RNA recovery.
6. At least 1 μg of ribosomal RNA-depleted RNA is required as input for the bisulfite conversion reaction to ensure successful preparation of Illumina sequencing libraries.
7. As in vitro transcripts do not carry m^5C sites, they are excellent targets for negative controls. Previously we have also used the ERCC transcript mix as spike-in negative control for transcriptome-wide bsRNA-seq.
8. Air-drying the sample in a biohazard hood is best. Ensure that the RNA pellet does not completely dry as this will cause difficulties in the resuspension.
9. As the in vitro transcript will most likely be at high concentrations, it is best to perform a serial dilution in H_2O. Dilute to a concentration such that 3–4 μL of the in vitro transcript is added to the RNA samples for accurate pipetting.
10. It is best to tilt the 1.5 mL tube at a 45° angle and then slowly pipette the mineral oil directly on top of the reaction mixture.
11. Emptying of the P-6 gel column takes ~5 min. If the gel column does not empty by gravity, place the lid back onto the column and remove again.
12. Slowly and gently pipette the reaction mixture onto the gel bed. Avoid disturbing the gel bed. Minimise the transfer of mineral oil, although there will be traces, which is unavoidable. The mineral oil will increase the A_{260}/A_{230} ratio to >2, however this does not hinder subsequent reactions.
13. ~500 ng is lost during this procedure, and we find that 13–15 μL of H_2O per 2 μg RNA used in the bisulfite conversion reaction results in concentrations of ~100 ng/μL.
14. For transcriptome-wide bsRNA-seq, confirmation of fragment size using the Agilent 2100 Bioanalyzer RNA 6000 Nano Chip according to the manufacturer's protocol is required. This is not required for preparation of RNA for locus-specific sequencing using Sanger sequencing or the MiSeq platform.
15. Inefficient bisulfite conversion may result in unconverted cytosines, so it is necessary to ensure that the primers are not biasing toward converted cytosines.
16. $tRNA^{Asp(GUC)}$ is known to contain m^5C sites at C38, C48, and C49, and $tRNA^{Leu(CAA)}$ is known to contain a m^5C site at C34. We have previously used these transcripts as positive controls for m^5C sites.

17. Longer amplicons increases the propensity of detecting non-converted cytosines in RNAs exhibiting strong secondary structure. We have previously experienced this for amplicons derived from ribosomal RNA.
18. The position and sequence environment of the candidate site will influence primer design and amplicon length, which may need to be optimized. We have previously experienced difficulty in designing primers for candidate sites within UTRs.
19. A touchdown PCR is performed to increase the specificity of the product. Initially an annealing temperature higher than the Tm is used, decreasing it by 1 °C to approximately 5 °C below the Tm. Then, a standard PCR protocol is implemented using the lowest annealing temperature from the first phase.
20. Sometimes not all triplicates are amplified and it may be necessary to optimize the PCR conditions or primers.
21. We recommend eluting the purified PCR products in 15–30 μL depending on the amount of amplified PCR product.
22. After purification of the amplicons with the Wizard® SV Gel and PCR Clean-Up System, residual ethanol may remain. We find that concentrating the pooled amplicons, even if there is <55 μL, and addition of H_2O to 55 μL is best to remove as much ethanol as possible.
23. As amplicons are short, fragmentation is not required. For transcriptome-wide bsRNA-seq, the TruSeq stranded total RNA Library preparation Kit may be used, omitting the rRNA depletion and RNA fragmentation steps.
24. The remaining 5 μL of pooled amplicons is kept for the validation of the libraries.
25. Agencourt AMPure XP beads exclude DNA fragments <100 nt, therefore purification of the DNA fragments between 70–100 nt is achieved using the MinElute PCR Purification Kit.
26. Remove Agencourt AMPure XP beads from the fridge at least 30 min prior to use.
27. The indexed adaptors allow multiplexing of samples on a single sequencing run. Ensure that you read the TruSeq® Library Preparation Guide if you have low sequence diversity libraries. Some indexed adaptors are not compatible with each other for low diversity libraries (i.e., <4 samples). This guide provides information on which indexed adaptors may be used for these libraries.

28. After the ligation of indexed adaptors, all DNA fragments should be >100 nt. Therefore, Agencourt AMPure XP beads can be used.
29. Most sequencing service providers will do this. If this is not the case, guidance on how to proceed is provided below. The sample sheet is required to enter in the sample names and adaptor indices used for each sample. We have previously selected "Other" as the category followed by "Fastq only" as the application for MiSeq amplicon sequencing. This generates fastq files only and also enables the deselection of adaptor trimming, allowing trimming and mapping to be performed separately. However, it is possible to select "Small Genome Sequencing" as the category and "Resequencing" as the application. This workflow includes adaptor trimming.
30. Most sequencing providers will perform library dilution and loading according to their experience. We provide guidance here should this not be the case.
31. The PhiX Control library is added to the pooled libraries as a control for the sequencing run.
32. Prepare fresh 0.2 M NaOH for the denaturation of libraries.
33. The percentage of PhiX Control library required depends on the sequence complexity of the libraries and may require optimization. We have previously found that up to 50% PhiX Control may be required to avoid cluster abortion.
34. The concentration of the library to be loaded into the cartridge can vary. We have previously loaded between 7 and 9 pM. Underloading can give cluster densities below the optimal range. Overloading can give cluster densities above the optimal range, reducing the quality of the data. The optimal cluster density is 700–1000 K/mm^2.
35. Alternatively, alignment can be performed using Bowtie2 within Bismark [33], allowing a single mismatch in the alignment, and candidate m^{5}C sites in overlapped paired-end reads are only counted once.

Acknowledgments

We thank Wenjia Qu for helpful suggestions for the MiSeq library preparation protocol. This work was supported by an NHMRC grant (APP1061551) and a senior research fellowship (514904) awarded to TP.

References

1. Cantara WA, Crain PF, Rozenski J, McCloskey JA, Harris KA, Zhang X, Vendeix FA, Fabris D, Agris PF (2011) The RNA modification database, RNAMDB: 2011 update. Nucleic Acids Res 39(Database issue):D195–D201. https://doi.org/10.1093/nar/gkq1028
2. Machnicka MA, Milanowska K, Osman Oglou O, Purta E, Kurkowska M, Olchowik A, Januszewski W, Kalinowski S, Dunin-Horkawicz S, Rother KM, Helm M, Bujnicki JM, Grosjean H (2013) MODOMICS: a database of RNA modification pathways--2013 update. Nucleic Acids Res 41 (Database issue):D262–D267. https://doi.org/10.1093/nar/gks1007
3. Milanowska K, Mikolajczak K, Lukasik A, Skorupski M, Balcer Z, Machnicka MA, Nowacka M, Rother KM, Bujnicki JM (2013) RNApathwaysDB--a database of RNA maturation and decay pathways. Nucleic Acids Res 41 (Database issue):D268–D272. https://doi.org/10.1093/nar/gks1052
4. Fu Y, He C (2012) Nucleic acid modifications with epigenetic significance. Curr Opin Chem Biol 16(5–6):516–524. https://doi.org/10.1016/j.cbpa.2012.10.002
5. Saletore Y, Meyer K, Korlach J, Vilfan ID, Jaffrey S, Mason CE (2012) The birth of the Epitranscriptome: deciphering the function of RNA modifications. Genome Biol 13(10):175. https://doi.org/10.1186/gb-2012-13-10-175
6. Sibbritt T, Patel HR, Preiss T (2013) Mapping and significance of the mRNA methylome. Wiley Interdiscip Rev RNA 4(4):397–422. https://doi.org/10.1002/wrna.1166
7. Amort T, Rieder D, Wille A, Khokhlova-Cubberley D, Riml C, Trixl L, Jia XY, Micura R, Lusser A (2017) Distinct 5-methylcytosine profiles in poly(A) RNA from mouse embryonic stem cells and brain. Genome Biol 18(1):1. https://doi.org/10.1186/s13059-016-1139-1
8. Squires JE, Patel HR, Nousch M, Sibbritt T, Humphreys DT, Parker BJ, Suter CM, Preiss T (2012) Widespread occurrence of 5-methylcytosine in human coding and non-coding RNA. Nucleic Acids Res 40 (11):5023–5033. https://doi.org/10.1093/nar/gks144
9. Fu L, Guerrero CR, Zhong N, Amato NJ, Liu Y, Liu S, Cai Q, Ji D, Jin SG, Niedernhofer LJ, Pfeifer GP, Xu GL, Wang Y (2014) Tet-mediated formation of 5-hydroxymethylcytosine in RNA. J Am Chem Soc 136(33):11582–11585. https://doi.org/10.1021/ja505305z
10. Huber SM, van Delft P, Mendil L, Bachman M, Smollett K, Werner F, Miska EA, Balasubramanian S (2015) Formation and abundance of 5-hydroxymethylcytosine in RNA. Chembiochem 16(5):752–755. https://doi.org/10.1002/cbic.201500013
11. Dominissini D, Moshitch-Moshkovitz S, Schwartz S, Salmon-Divon M, Ungar L, Osenberg S, Cesarkas K, Jacob-Hirsch J, Amariglio N, Kupiec M, Sorek R, Rechavi G (2012) Topology of the human and mouse m6A RNA methylomes revealed by m6A-seq. Nature 485(7397):201–206. https://doi.org/10.1038/nature11112
12. Meyer KD, Saletore Y, Zumbo P, Elemento O, Mason CE, Jaffrey SR (2012) Comprehensive analysis of mRNA methylation reveals enrichment in 3' UTRs and near stop codons. Cell 149(7):1635–1646. https://doi.org/10.1016/j.cell.2012.05.003
13. Schwartz S, Mumbach MR, Jovanovic M, Wang T, Maciag K, Bushkin GG, Mertins P, Ter-Ovanesyan D, Habib N, Cacchiarelli D, Sanjana NE, Freinkman E, Pacold ME, Satija R, Mikkelsen TS, Hacohen N, Zhang F, Carr SA, Lander ES, Regev A (2014) Perturbation of m6A writers reveals two distinct classes of mRNA methylation at internal and 5′ sites. Cell Rep 8(1):284–296. https://doi.org/10.1016/j.celrep.2014.05.048
14. Bar-Yaacov D, Frumkin I, Yashiro Y, Chujo T, Ishigami Y, Chemla Y, Blumberg A, Schlesinger O, Bieri P, Greber B, Ban N, Zarivach R, Alfonta L, Pilpel Y, Suzuki T, Mishmar D (2016) Mitochondrial 16S rRNA is methylated by tRNA methyltransferase TRMT61B in all vertebrates. PLoS Biol 14 (9):e1002557. https://doi.org/10.1371/journal.pbio.1002557
15. Dominissini D, Nachtergaele S, Moshitch-Moshkovitz S, Peer E, Kol N, Ben-Haim MS, Dai Q, Di Segni A, Salmon-Divon M, Clark WC, Zheng G, Pan T, Solomon O, Eyal E, Hershkovitz V, Han D, Dore LC, Amariglio N, Rechavi G, He C (2016) The dynamic N(1)-methyladenosine methylome in eukaryotic messenger RNA. Nature 530 (7591):441–446. https://doi.org/10.1038/nature16998
16. Hauenschild R, Tserovski L, Schmid K, Thuring K, Winz ML, Sharma S, Entian KD, Wacheul L, Lafontaine DL, Anderson J, Alfonzo J, Hildebrandt A, Jaschke A, Motorin Y, Helm M (2015) The reverse

transcription signature of N-1-methyladenosine in RNA-Seq is sequence dependent. Nucleic Acids Res 43(20):9950–9964. https://doi.org/10.1093/nar/gkv895

17. Carlile TM, Rojas-Duran MF, Zinshteyn B, Shin H, Bartoli KM, Gilbert WV (2014) Pseudouridine profiling reveals regulated mRNA pseudouridylation in yeast and human cells. Nature 515(7525):143–146. https://doi.org/10.1038/nature13802
18. Lovejoy AF, Riordan DP, Brown PO (2014) Transcriptome-wide mapping of pseudouridines: pseudouridine synthases modify specific mRNAs in *S. cerevisiae*. PLoS One 9(10): e110799. https://doi.org/10.1371/journal.pone.0110799
19. Schwartz S, Bernstein DA, Mumbach MR, Jovanovic M, Herbst RH, Leon-Ricardo BX, Engreitz JM, Guttman M, Satija R, Lander ES, Fink G, Regev A (2014) Transcriptome-wide mapping reveals widespread dynamic-regulated pseudouridylation of ncRNA and mRNA. Cell 159(1):148–162. https://doi.org/10.1016/j.cell.2014.08.028
20. Edelheit S, Schwartz S, Mumbach MR, Wurtzel O, Sorek R (2013) Transcriptome-wide mapping of 5-methylcytidine RNA modifications in bacteria, archaea, and yeast reveals m5C within archaeal mRNAs. PLoS Genet 9 (6):e1003602. https://doi.org/10.1371/journal.pgen.1003602
21. Khoddami V, Cairns BR (2013) Identification of direct targets and modified bases of RNA cytosine methyltransferases. Nat Biotechnol 31(5):458–464. https://doi.org/10.1038/nbt.2566
22. Hussain S, Sajini AA, Blanco S, Dietmann S, Lombard P, Sugimoto Y, Paramor M, Gleeson JG, Odom DT, Ule J, Frye M (2013) NSun2-mediated cytosine-5 methylation of vault non-coding RNA determines its processing into regulatory small RNAs. Cell Rep 4 (2):255–261. https://doi.org/10.1016/j.celrep.2013.06.029
23. Gu W, Hurto RL, Hopper AK, Grayhack EJ, Phizicky EM (2005) Depletion of *Saccharomyces cerevisiae* tRNA(His) guanylyltransferase Thg1p leads to uncharged tRNAHis with additional m(5)C. Mol Cell Biol 25 (18):8191–8201. https://doi.org/10.1128/MCB.25.18.8191-8201.2005
24. Pollex T, Hanna K, Schaefer M (2010) Detection of cytosine methylation in RNA using bisulfite sequencing. Cold Spring Harb Protoc 2010(10):pdb prot5505. https://doi.org/10.1101/pdb.prot5505
25. Schaefer M, Pollex T, Hanna K, Lyko F (2009) RNA cytosine methylation analysis by bisulfite sequencing. Nucleic Acids Res 37(2):e12. https://doi.org/10.1093/nar/gkn954
26. Legrand C, Tuorto F, Hartmann M, Liebers R, Jacob D, Helm M, Lyko F (2017) Statistically robust methylation calling for whole-transcriptome bisulfite sequencing reveals distinct methylation patterns for mouse RNAs. Genome Res 27(9):1589–1596. https://doi.org/10.1101/gr.210666.116
27. Huang T, Chen W, Liu J, Gu N, Zhang R (2019) Genome-wide identification of mRNA 5-methylcytosine in mammals. Nat Struct Mol Biol 26(5):380–388. https://doi.org/10.1038/s41594-019-0218-x
28. Yang Y, Wang L, Han X, Yang WL, Zhang M, Ma HL, Sun BF, Li A, Xia J, Chen J, Heng J, Wu B, Chen YS, Xu JW, Yang X, Yao H, Sun J, Lyu C, Wang HL, Huang Y, Sun YP, Zhao YL, Meng A, Ma J, Liu F, Yang YG (2019) RNA 5-methylcytosine facilitates the maternal-to-zygotic transition by preventing maternal mRNA decay. Mol Cell 75(6):1188–1202 e1111. https://doi.org/10.1016/j.molcel.2019.06.033
29. Schumann U, Zhang HN, Sibbritt T, Pan A, Horvath A, Gross S, Clark SJ, Yang L, Preiss T (2020) Multiple links between 5-methylcytosine content of mRNA and translation. BMC Biol 18(1):40. https://doi.org/10.1186/s12915-020-00769-5
30. Tusnady GE, Simon I, Varadi A, Aranyi T (2005) BiSearch: primer-design and search tool for PCR on bisulfite-treated genomes. Nucleic Acids Res 33(1):e9. https://doi.org/10.1093/nar/gni012
31. Martin M (2011) Cutadapt removes adapter sequences from high-throughput sequencing reads. EMBnet J 17(1):3. https://doi.org/10.14806/ej.17.1.200
32. Rieder D, Amort T, Kugler E, Lusser A, Trajanoski Z (2016) meRanTK: methylated RNA analysis ToolKit. Bioinformatics 32 (5):782–785. https://doi.org/10.1093/bioinformatics/btv647
33. Krueger F, Andrews SR (2011) Bismark: a flexible aligner and methylation caller for bisulfite-Seq applications. Bioinformatics 27 (11):1571–1572. https://doi.org/10.1093/bioinformatics/btr167

Chapter 22

Transcriptome-Wide Identification of 2′-*O*-Methylation Sites with RibOxi-Seq

Yinzhou Zhu, Christopher L. Holley, and Gordon G. Carmichael

Abstract

The ability to detect 2′-*O*-methylation sites (Nm) in high-throughput fashion is important, as increasing evidence points to a more diverse landscape for this RNA modification as well as the possibility of yet unidentified functions. Here we describe an optimized version of RibOxi-seq, which is built upon the original published method, that not only accurately profiles ribosomal RNA (rRNA) Nm sites with minimal RNA input but is also robust enough to identify mRNA intronic and exonic sites.

Key words 2′-*O*-methylation, Nm, Ribose methylation, rRNA modifications, mRNA modifications, RNA modifications

1 Introduction

RNA Nm modifications were discovered more than two decades ago, and since then most sites have been found on noncoding RNAs such as transfer RNAs (tRNAs), small nucleolar RNAs (snRNAs), and ribosomal RNAs (rRNAs) (reviewed in [1]). Although the exact functions of Nms have not been pinpointed, evidence suggests Nms can stabilize RNA alternative secondary structures by favoring C3'-*endo* sugar pucker conformation of the methylated ribose, as well as disrupting RNA–RNA interactions [2, 3]. In addition, both methylated ribonucleosides and fibrillarin (FBL), the methyltransferase involved in snoRNA guided 2′-*O*-methylation pathways, have been implicated in diseases [4, 5]. Furthermore, recent studies discovered that novel Nms can be found on mRNA transcripts and such sites within a gene alone can exert translational repression through disruption of mRNA–tRNA decoding [3, 6]. Thus, deciphering Nm landscape may provide

Supplementary Information The online version of this chapter (https://doi.org/10.1007/978-1-0716-1851-6_22) contains supplementary material, which is available to authorized users.

Erik Dassi (ed.), *Post-Transcriptional Gene Regulation*, Methods in Molecular Biology, vol. 2404,
https://doi.org/10.1007/978-1-0716-1851-6_22,

important new insights on potentially generalizable regulatory roles.

Currently there are three underlying strategies for high throughput profiling of Nm sites. The first is 2OMe-seq, which couples primer extension with next-generation sequencing (NGS), to find global RT stop sites. The basis of this technique is that under some conditions reverse transcriptase pauses at sites of 2′-*O*-methylation [7]. The second is RiboMeth-seq, which takes advantage of the resistance of Nm bases to either enzymatic or chemical cleavage. NGS libraries generated from RNAs extensively fragmented either enzymatically or chemically and sequenced at very high depth followed by alignment and summarization/visualization of aligned 3′-end counts allow the identification of base positions where read-3′-mapping is significantly lower than the rest of the positions. These positions often correspond to sites of 2′-*O*-methylation [8, 9]; The third approach is RibOxi/Nm-seq, which were developed independently but published almost simultaneously, and takes advantage of Nm's property of resistance to cleavage. Additional steps of oxidation, β-elimination, and dephosphorylation post-fragmentation allow the enrichment of Nm at 3′-end of fragmented RNAs. Owing to oxidation of nonmodified ends, only fragments with Nm ends in the final mixture can be ligated to linkers and subsequently converted to NGS libraries ([10, 11], Fig. 1).

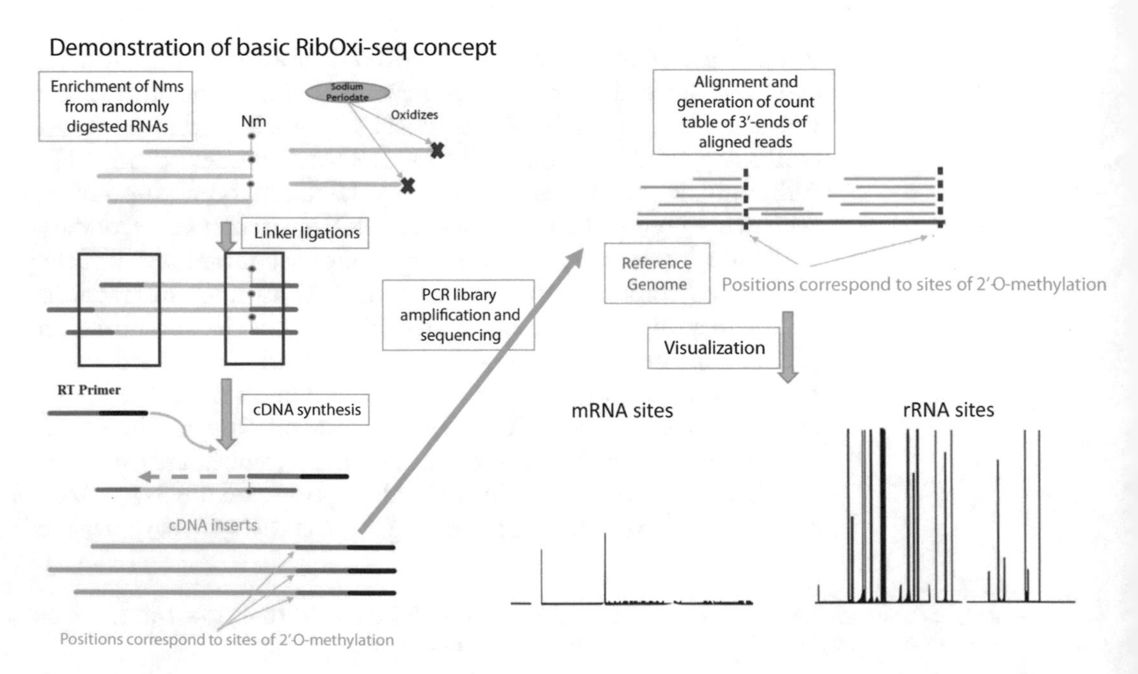

Fig. 1 Concept of RibOxi-seq chemistry and computational pipeline. RNA input is randomly digested, followed by enrichment of Nm ends through oxidation of unmethylated 3′-ends by sodium periodate. Libraries containing only nonoxidized fragments are made through the ligation method. Libraries are sequenced on Illumina platforms of choice. Reads are then aligned to a reference genome to allow 3′-end pile-up counting. Positions with significant 3′-end counts corresponds to sites protected by Nm

RibOxi-seq, as initially developed, was ideal for profiling rRNA Nm sites using modest input RNA material and sequencing depth while maintaining good accuracy (reviewed in [12]). Since then, the protocol has gone through further extensive optimization and testing. In our experience, the current protocol requires significantly less input RNA, as well as much lower sequencing depth for profiling rRNA sites. More importantly, the method now allows the detection of mRNA sites. The principle and outline of the protocol remain very similar to the original method, where major procedures include RNA preparation and fragmentation, RNA oxidation, elimination and dephosphorylation, 3′-DNA linker ligation, 5′-RNA linker ligation, cDNA synthesis, library amplification, library QC, and data processing/analysis. One of the improvements in this optimized protocol is that the 5′-RNA linker now contains a random sequence for PCR deduplication instead of the RT primer in the original design. In addition, the 3′-linker design provides the ability to bioinformatically remove RT mispriming events, which occurs in RT reactions with extremely low input ([13], Fig. 2).

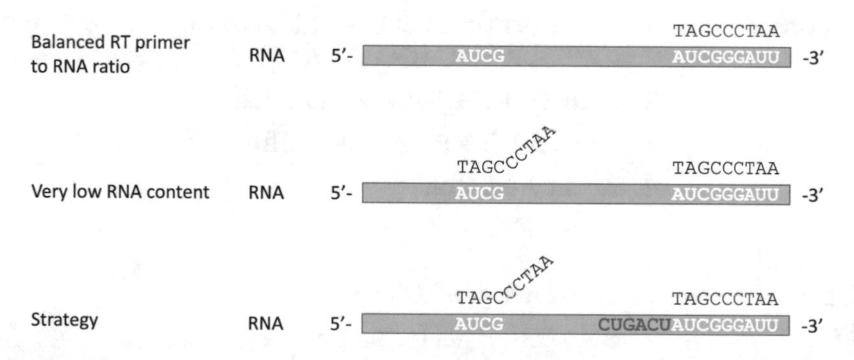

Fig. 2 Illustration of RT mispriming events and mitigation strategy. Under the condition where there is a reasonable amount of RNA material, the RT primer (shown as the floating sequence) mostly anneals to a designated target sequence during an RT reaction as shown in top panel. However, when RNA material is scarce, RT primer can prime with as few as 4 bases as shown in middle panel, drastically increasing nonspecific RT products. The strategy to mitigate the mispriming is to include a constant barcode during 3′-linker ligation as shown in the bottom panel. This does not prevent mispriming; however, after sequencing, the mispriming events can be filtered out by checking whether the reads have the barcode preceding the linker sequence

2 Materials

2.1 RNA Fragmentation

1. Benzonase nuclease: ≥99% purity.
2. 10× Benzonase buffer: 500 mM Tris–HCl pH 7.5, 200 mM NaCl, 20 mM $MgCl_2$, 1 mg/mL BSA.
3. Linear polyacrylamide (LPA): 10 μg/μL.
4. Acid-phenol–chloroform: with IAA, 125:24:1, pH 4.5.

2.2 RNA Oxidation, β-Elimination, and Dephosphorylation

1. Oxidation-elimination buffer: 2 M lysine hydrochloride pH 8.5.
2. Oxidation-only buffer: 4.375 mM sodium borate, 50 mM boric acid, pH 8.6.
3. Alkaline phosphatase.
4. Alkaline phosphatase buffer: 10× from phosphatase manufacturer.
5. T4 PNK: 1 Unit/μL.
6. 10× T4 PNK buffer: 700 mM Tris–HCl, 100 mM $MgCl_2$, 50 mM DTT, pH 6.0.
7. 10× T4 PNK buffer: 700 mM Tris–HCl, 100 mM $MgCl_2$, 50 mM DTT, pH 7.6.
8. 8 Buffer exchange columns.
9. RNA purification columns.
10. Sodium metaperiodate: ≥99%.

2.3 3′-DNA Linker Ligation

1. 5’ Preadenylated 3′ blocked DNA oligo: 5′−/5rApp/ATCACGCTGTAGGCACCATCAATGACAG/3SpC3/−3′, 10 μM.
2. NEB T4 RNA ligase 2 truncated KQ.
3. 10× NEB T4 RNA ligase buffer: without ATP.
4. RNase Inhibitor.
5. PEG 8000.

2.4 PAGE Gel Purification

1. TBE-Urea gel: 10%.
2. RNA PAGE gel loading dye: 2×.
3. SYBR Gold nucleic acid dye.
4. PAGE gel recovery kit.

2.5 5′-RNA Linker Ligation

1. 5′ blocked RNA oligo: 5′-/Biosg/ ACACUCUUUCCCUACACGACGCUCUUCCGAUCUNNNN-3′, 50 μM (*see* **Note 1**).
2. NEB T4 RNA ligase 1.
3. 10× NEB T4 RNA ligase buffer.

4. ATP: 10 mM.
5. DMSO: 100%.
6. RNase Inhibitor.

2.6 cDNA Synthesis

1. RT Primer: 5′- GTGACTGGAGTTCA GACGTGTGCTCTTCCGATCTGTCATTGATGGTGCC TACAG-3′, 10 μM.
2. NEB ProtoScript II RT kit.
3. dNTP: 10 mM.
4. RNase Inhibitor.
5. NaOH: 1 N solution.
6. Tris–HCl: 200 mM pH 7.5.
7. AMPure XP or equivalent SPRI beads.

2.7 Library Amplification

1. Illumina compatible I5 primer: 5′- AATGATACGGCGAC CACCGAGATCTACAC -(I5 index)- ACACTCTTTCCCTA CACGACGCTCTTCCGATCT-3′, 2.5 μM (*see* **Note 2**).
2. Illumina compatible I7 primer: 5′-CAAGCAGAAGACGGCA TACGAGAT -(I7 index)- GTGACTGGAGTTCA GACGTGTGCTCTTCCGATCT-3′, 2.5 μM (*see* **Note 2**).
3. High fidelity DNA polymerase: 2×.
4. Gel purification components: PAGE/agarose, DNA recovery reagents.

2.8 Library Quantification and Visualization

1. Broad range UV-spectrophotometer.
2. Qubit or equivalent.
3. Library size distribution QC equipment and reagents.

3 Methods

It is critical throughout the protocol to practice caution and avoid introducing RNase contamination. It is suggested that all incubation procedures that use a thermal cycler set lid temperature about 10 °C above block temperature. Since profiling rRNA sites is more efficient, it requires drastically less RNA input, thus the fragmentation reaction volume is significantly less. Thus, enzyme is further diluted to allow using larger volume while maintaining final units of enzyme per μL volume.

3.1 RNA Preparation and Fragmentation

To profile only rRNA sites, skip **step 2**. For the mRNA protocol (which also profiles rRNA sites at a lower depth), skip **step 4**.

1. Extract total RNA and prepare 1 μg total RNA in 9 μL water (or 600–800 μg mRNA in 500 μL H_2O for mRNA protocol).

2. [Skip if profiling rRNA only] Perform poly(A) enrichment and obtain ~3–7 μg of poly(A) enriched RNA in 44 μL (*see* **Note 3**).
3. Prepare 0.25 U/μL benzonase working solution by mixing 100 μL 10× benzonase buffer, 899 μL RNase-free H_2O, and 1 μL of 250 U/μL benzonase stock in a 1.5 mL Eppendorf tube.
4. [Skip if profiling mRNA] Further dilute benzonase working solution by mixing 100 μL of the 0.25 U/μL benzonase working solution, 40 μL 10× benzonase buffer, and 360 μL H_2O. Final benzonase concentration is now 0.05 U/μL.
5. Transfer 8 μL (44 μL for mRNA protocol) of the 1 μg RNA input to a 0.2 mL PCR tube and incubate at 95 °C for 3 min in a thermal cycler. Immediately place the tube on ice for 10 s.
6. Add 1 μL (5 μL for mRNA protocol) of 10× benzonase buffer and 1 μL of 0.05 U/μL (use 0.25 U/μL for mRNA protocol) benzonase working solution to the sample (final RNA concentration ~100 ng/μL), vortex to mix and immediately incubate on ice for 80 min (*see* **Note 4**).
7. In a clean Eppendorf tube, add 90 μL H_2O (50 μL for mRNA protocol) and 100 μL acid-phenol:chloroform. Transfer fragmentation reaction mixture from the previous step to the Eppendorf tube and vortex for 15–20 s at high intensity. Place at room temperature for 2 min.
8. Centrifuge at 18,000–20,000 × *g* for 5 min. Carefully transfer 90–95 μL aqueous phase to a new Eppendorf tube. Add 1 μL LPA to help precipitation. Ethanol precipitate RNA with 2.5 volumes of 100% EtOH, wash with 80% EtOH once, and resuspend into 33 μL H_2O.
9. Use 1 μL of the fragmented RNA for QC. In our experience, there should be a sharp peak between 25–200 nt skewing to the right (Fig. 3c).

3.2 RNA Oxidation, β-Elimination, and Dephosphorylation

1. Prepare 200 mM $NaIO_4$ solution by dissolving 42.78 mg of $NaIO_4$ powder in 1 mL nuclease free H_2O. Protect the solution from light and keep on ice and use only the same day.
2. Add 4 μL oxidation-elimination buffer and 4 μL $NaIO_4$ to the 32 μL fragmented RNA. Mix well by vortexing, briefly spin-down, and incubate at 37 °C with shaking for 45 min. Purify and resuspend RNA in 42 μL H_2O (*see* **Note 5**). If performing additional oxidation, β-elimination, and dephosphorylation cycles (a minimum of 3 for profiling mRNA sites), go to **step 3**, otherwise, continue from **step 4**.
3. Add 5 μL 10× alkaline phosphatase buffer, 1 μL RNase inhibitor, and 10 units of the phosphatase enzyme solution. Incubate according to manufacturer's instructions followed by

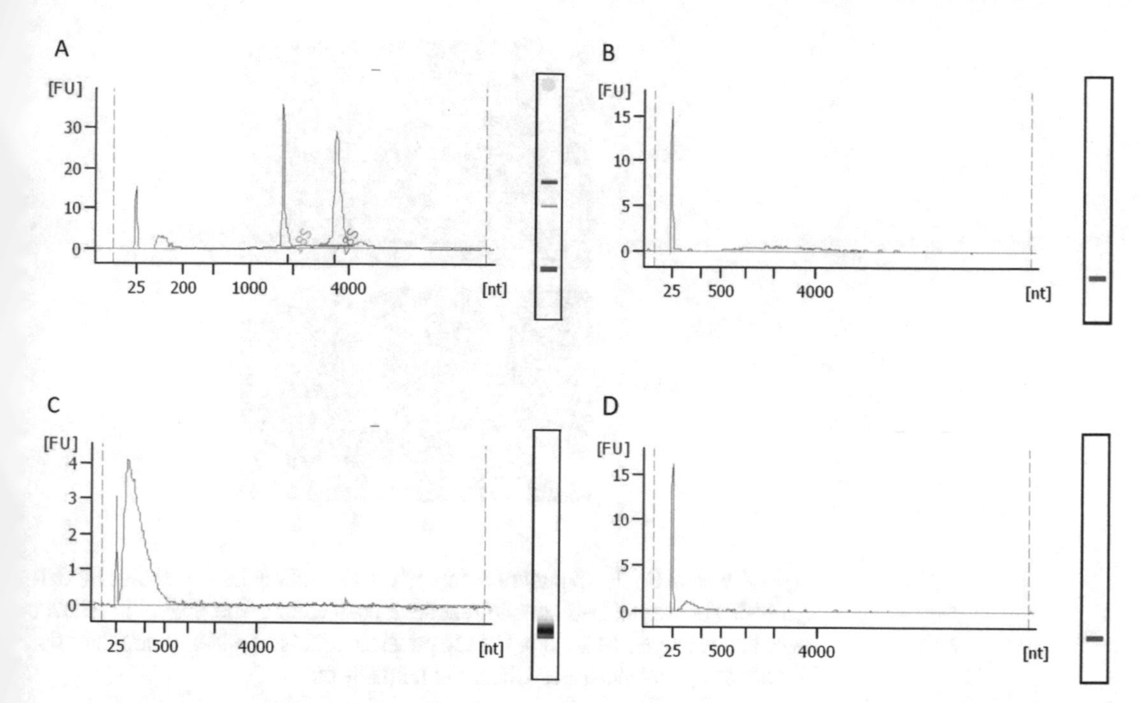

Fig. 3 Typical RNA fragment size-distributions during RibOxi-seq. All QC in this figure was done using Bioanalyzer Nano 6000 chips. (**a**) Typical size distributions of total RNA extracted from HEK293T cells with RNA integrity number of 10. (**b**) After poly(A) enrichment twice, there should not be any major rRNA peaks left. (**c**) The RNA size distribution after benzonase fragmentation. (**d**) Size distribution after 3′-linker ligation. A size shift is not observable, most likely due to the fact that only a very small portion of the RNA was protected from oxidation by Nm and ligated with linkers

inactivation. Purify and resuspend in 32 μL H_2O. Repeat from **step 1** to achieve desired number of oxidation, β-elimination, and dephosphorylation cycles (*see* **Note 6**).

4. Add 5 μL 10× T4 PNK buffer (pH 6.0), 20 units T4 PNK and 1 μL RNase inhibitor to purified RNA from **step 2**. Mix well by vortexing, briefly spin-down, and incubate at 37 °C for 3 h. Add 5 μL 10× T4 PNK buffer (pH 7.6), 20 more units T4 PNK, 10 μL 10 mM ATP, 33 μL H_2O to the reaction. Mix well by vortexing, briefly spin-down, and incubate at 37 °C for an additional 1 h. Inactivate the enzyme following the manufacturer's instructions.
5. Ethanol precipitate the reaction as detailed in previous steps into a new Eppendorf tube with help of 1 μL LPA. Thoroughly resuspend the pellet in 35 μL oxidation-only buffer. Add 5 μL $NaIO_4$ solution. Mix well by vortexing, briefly spin-down and incubate at 37 °C with shaking for 45 min.
6. Purify oxidized RNA and resuspend in 10 μL H_2O (*see* **Note 7**).

3.3 3′-DNA Linker Ligation

1. Transfer the oxidized RNA into a 0.2 mL PCR tube. Add 1 μL 3′ DNA linker, 1 μL RNase inhibitor, 8.5 μL PEG 8000, 2.5 μL

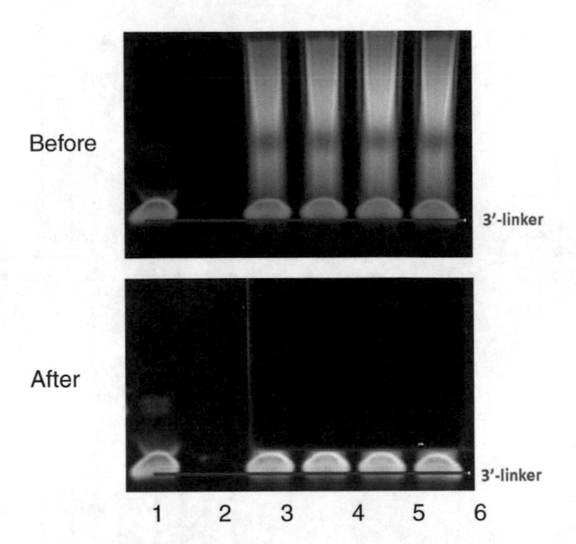

Fig. 4 A typical PAGE gel pattern before and after cutting. Lane 1 shows RibOxi-seq 3′-linker. Lanes 3–6 contain reactions postligation. The goal is to recover RNAs and exclude free linkers. The top panel illustrates a typical electrophoresis. The bottom panel illustrates where we normally cut

10× T4 RNA ligase buffer, and 2 μL (400 units) T4 RNA ligase 2 truncated KQ. Mix well, spin down, and incubate in a thermal cycler at 16 °C overnight for 18 h (*see* **Note 8**).

2. Cleanup and concentrate RNA to 15 μL. Perform gel purification to separate ligation products from free 3′ DNA linkers. Recover RNA with 11 μL H_2O (*see* **Note 9**, Fig. 4).

3.4 5′-RNA Linker Ligation

1. Thaw the 50 μM RNA linker, denature 1.3 μL of the linker in a 0.2 mL PCR tube in a thermal cycler at 72 °C for 2 min and return to ice.
2. Immediately after starting the denaturation of the RNA linker, prepare the ligation reaction by mixing 11 μL product from the previous step with 2 μL 100% DMSO, 2 μL T4 RNA ligase buffer, 2 μL 10 mM ATP, and 0.5 μL RNase inhibitor.
3. After placing denatured RNA linker on ice for 1 min, add 1 μL of the linker and 1.5 μL T4 RNA ligase I to the mixture in **step 2**. Mix well and spin down before incubating the reaction at 25 °C for 1 h. Inactivate the RNA ligase enzyme following the manufacturer's instructions.

3.5 cDNA Synthesis

1. Denature the RT primer by mixing 20 μL ligated product, 2 μL 10 mM dNTP, 2 μL H_2O, and 1 μL 10 μM RT primer in a 0.2 mL PCR tube and incubate under 65 °C in a thermal cycler for 5 min. During the 5 min, prepare reaction mix by mixing 8 μL Protoscript II buffer, 1 μL RNase inhibitor, 4 μL 10× DTT, and 2 μL Protoscript II enzyme mix (*see* **Note 10**).

2. Once the 5-min denaturation is finished, combine both mixtures, and mix well. In a thermal cycler, incubate the combined RT reaction at 50 °C for 1 h (*see* **Note 10**).
3. Hydrolyze RNA by adding 4.4 μL 1 N NaOH and incubating at 95 °C for 20 min in a thermal cycler. Immediately place on ice. Add 4.4 μL 1 M Tris–HCl pH 7.5 and mix well. Use AmpureXP beads or equivalent to clean up the reaction and elute with 20 μL H_2O.

3.6 Library Amplification and Library QC

1. Add 2.5 μL of Illumina compatible forward and reverse primers with indexes of choice to the cDNA. Combine with 25 μL 2× high fidelity DNA polymerase.
2. Use an appropriate thermal cycler program to allow 3 cycles of initial amplification with low annealing temperature to account for partial complementarity between cDNA ends and PCR primers. This should be then followed by an additional 15 cycles (18 cycles for mRNA protocol) at appropriately higher annealing temperature depending on the polymerase of choice (*see* **Note 11**).
3. Clean up the library using methods of choice to avoid PCR reaction buffer interfering with the subsequent gel electrophoresis. Gel-purify the library, using a commercial or custom DNA ladder for referencing sub 1 kb sizes (*see* **Note 12**).
4. Confirm size distribution using method of choice and quantify concentration with UV spectrophotometer followed by quantification with Qubit or equivalent (*see* **Note 12**).
5. Library is sequenced on an Illumina platform. 3–5 million reads (10–25 million reads for mRNA protocol) per sample for sequencing depth and 1 × 150 cycle configuration are recommended (*see* **Note 13**).

3.7 Data Processing and Analysis

A computational pipeline available at: https://github.com/yz201906/RibOxi-seq. It is under development to add more functions, but it can currently be used to process the sequencing reads, align them to a reference genome and generate count tables and visualization in one command line. Here we will go through the individual steps. The read output from the sequencer contains randomer used for deduplication at 5′-ends, a linker sequence at 3′-ends where the first six bases are used for mispriming removal, thus requiring additional processing before alignment and downstream steps (Fig. 5).

1. The non-barcoded portion of the linker needs to be trimmed first using *Cutadapt* or a similar package (Fig. 6a, blue portion). Since this part of the read is at the very 3′-end, which can have low quality and base-call errors especially with low complexity libraries, we allow for 20% mismatch rate. Any read

RNA structure after 3′ and 5′ linker ligations

5'-/Biosg/ACACUCUUUCCCUACACGACGCUCUUCCGAUCUNNNN------insert-------ATCACGCTGTAGGCACCATCAATGACAG/SpC3/- 3′

cDNA library structure

Randomer for deduplication 3′-linker sequence

3'-TGTGAGAAAGGGATGTGCTGCGAGAAGGCTAGANNNN----insert----TAGTGCGACATCCGTGGTAGTTACTGTCTAGCCTTCTCGTGTGCAGACTTGAGGTCAGTG-5'

Complementary to partial Illumina i5 PCR primer In-line barcode for filtering out RT mis-priming Identical to partial Illumina i7 PCR primer

dsDNA library structure

Forward	5'-Illumina_i5_adapterNNNN----insert----ATCACGCTGTAGGCACCATCAATGACIllumina_i7_adapter-3'
Reverse	3'-Illumina_i5_adapterNNNN----insert----TAGTGCGACATCCGTGGTAGTTACTGIllumina_i7_adapter-5'

Fig. 5 Pseudo-sequence demonstration of RNA, cDNA, and dsDNA library structures. The top panel shows RNA structure with both 3′ and 5′ linkers ligated. The blue portion functions as the RT primer annealing site. The middle panel shows cDNA structure, where yellow indicates sequence complementarity to a portion of the Illumina forward PCR primer and red marks the sequence identical to a portion of the Illumina reverse PCR primer. The bottom panel illustrates the structure of the final dsDNA library

Read:

A

```
@ Sequence_identifier+i7_index+i5_index
NNNN----INSERT----ATCACGCTGTAGGCACCATCAATGAC---MAYBE_READ-THROUGH_INTO_ILLUMINA_PCR_ADAPTER
+
_QUALITY_OF_THE_READ____________________________________________________________
```

B

```
@ Sequence_identifier+i7_index+i5_index
NNNN----INSERT----ATCACG
+
_QUALITY_OF_THE_READ____
```

C

```
@ Sequence_identifier_NNNN+i7_index+i5_index
INSERT----ATCACG
+
_OF_THE_READ____
```

D

```
@ Sequence_identifier_NNNN+i7_index+i5_index
INSERT
+
_OF_TH
```

Fig. 6 Pseudo-sequence demonstration of how sequencing reads are proccessed. The reads were generated with 1 × 150 cycle sequencing. (**a**) Illustration of a raw read before any processing. This is in the .fastq format where each read is represented by 4 lines. The first line contains read name and indexes. The second line contains actual sequences. The third line is for optional identifiers, but usually is left empty. The fourth line contains quality strings corresponding to each base in line 2. (**b**) Resulting read after removing the 3′-linker sequence and everything downstream of it. (**c**) Resulting read after moving the random sequence from line 2 to line 1. (**d**) The resulting read after final trimming to remove the barcode used for mispriming event filtering. Reads that do not contain this barcode sequence are discarded as mispriming events

without the linker match and subsequent trimming is discarded, Fig. 6b represents the sequence structure after this round of trimming.

2. A custom python script *move_umi.py*, which is part of the riboxi-pipeline, is used to move the randomer sequence to

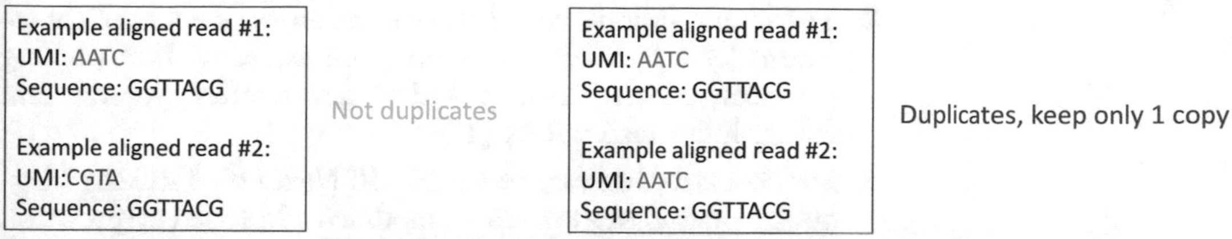

Example aligned read #1:
UMI: AATC
Sequence: GGTTACG

Example aligned read #2:
UMI:CGTA
Sequence: GGTTACG

Not duplicates

Example aligned read #1:
UMI: AATC
Sequence: GGTTACG

Example aligned read #2:
UMI: AATC
Sequence: GGTTACG

Duplicates, keep only 1 copy

Fig. 7 Demonstration of how PCR duplicates are removed. This process uses the random sequence that was moved to the read identifier line for deduplication as unique molecular identifiers (UMI). Each time an aligned read that is identical to another read that was previously seen is identified, UMIs of these two reads are compared to one another. If the UMIs are different, then both alignments are kept as shown on the left. On the other hand, if the UMIs are identical, then only one of the alignments is kept

the read-name/identifier line for each read. This allows PCR deduplication later on after alignment (Fig. 6c).

3. Final trimming occurs on the barcode region of the 3′-linker, by allowing 0 base mismatch and discarding untrimmed reads, thus ensuring thorough removal of RT mispriming events (Fig. 6d).
4. The processed reads are next aligned to a reference genome. The *.bam* output is the preferred format, which can then be directly indexed and deduplicated using *umi_tools* (Fig. 7).
5. The deduplicated alignment file can then be used with the *bedtools genomecov* function to generate a UCSC genome browser compatible file for visualization. In addition, after converting the deduplicated .bam file into .bed file, a 3′-end count table can be generated using *riboxi_bed_parsing.py* script from the riboxi_pipeline. The resulting table can be converted to *.RData* file that is loaded into the *riboxi_shinyapp* for further analysis and visualization.

4 Notes

1. Having a random UMI allows for distinguishing PCR duplicates vs. identical RNA fragments that physically exist. For example, a quadruple-mer can accommodate 9999 copies of identical fragments, which is not too common in random digestions. Thus, theoretically if 2 sequencing reads contain identical sequences but with different UMIs, then they should all be counted, while only 1 should be retained if the UMIs are the same. In our experience, a quadruple-mer is sufficient for deduplication for the mRNA protocol and low input and low depth rRNA protocol. We recommend 6–8 N bases for higher RNA input (>5 μg) and sequencing depth range (15–100 million clusters).

2. The i7 and i5 indexes portion can be any sequences of choice (normally a 6/8-base sequence). Be aware of best pooling practices, which can help with index sequence design and ensure index read quality [14].
3. For cost effectiveness, we used NEB Next polyA mRNA Magnetic Isolation Module with a modified and scaled up protocol. In our routine, we prepare 500 μL beads following original protocol, and use it for 600 μg HEK 293T total RNA that is in 500 μL H_2O. For the mRNA protocol, if is critical to have as little rRNA content as possible. Thus, we perform a second poly(A) selection immediately after elution reusing the magnetic beads for each sample (Wash beads once with wash buffer, then twice with RNA dilution buffer). You should see a significant difference in rRNA content for each sample between 1-round and 2-round poly(A) selection (Fig. 3b).
4. The digestion time requires some optimization especially for different cell lines/tissue samples. Ideally, a time gradient for digestion time should be performed. Under shorter durations, shift in peak sizes should be observed. Eventually, the size of the peak no longer shifts, and reduction in the height of the peak is observed. This is the incubation time that should be chosen.
5. Initially, we tried either ethanol precipitation or Zymo RNA-Clean concentrator-5/25 after each oxidation–elimination reaction; however, we always see white precipitate that becomes cloudy when vortexed, presumably from high salt concentration. Although we have not verified whether this impacts the subsequent dephosphorylation reaction, it is safer to perform buffer exchange at this step instead. We use Ambion Nucaway columns, which did not result in the same white precipitate.
6. The goal of using non-T4 PNK phosphatase is to speed up dephosphorylation steps if multiple oxidation-elimination and dephosphorylation cycles are to be performed. Most alkaline phosphatase can be used at this step. We have tested with Antarctic phosphatase from NEB for multiple rounds of oxidation, elimination and phosphorylation cycles followed by mass spectrometry. We were able to confirm that the procedure is robust against at least 6 nmole RNA material (Supplemental Fig. 1).
7. Alternatively, elute in 11 μL H_2O, and use 1 μL for concentration/distribution QC. It is normal to see significant loss of materials when compared to post-digestion concentration (Fig. 3d).
8. It is easiest to add PEG 8000 and oxidized RNA to the 0.2 mL tube first, then make a master mix for the rest of the reagents

and add appropriate amount to each sample when multiple samples are involved.

9. Here we use a 10% TBE-Urea gel in a Bio-Rad mini-protean system with 1× TBE buffer. The wells of the gel are flushed, and gel prerun at 180 mV for 45 min. Prior to electrophoresis, 2× loading dye is added to sample and subsequently incubated at 72 °C for 2 min and returned to ice for 1 min. Electrophoresis is performed at 180 mV for 40 min. Gels are stained with SYBR Gold. RNA recovery was performed with ZR small RNA PAGE recovery kit.

10. To reduce mispriming events, it is preferable to use a reverse transcriptase that can sustain activity at higher temperatures. We have only extensively tested the NEB ProtoScript II enzyme, where 48–50 °C temperature range is achievable. Other enzymes that could function at or above such temperature should also be suitable. We minimize RNA and primer spent at low temperature by adding the reaction mix to the RNA–primer mixture immediately after the 5-min denaturation is finished. While thermal cycler is set to 48 °C, we quickly vortex and spin down the combined reaction before placing the tubes into the thermal cycler.

11. For this step, we have extensively tested the NEB Q5 DNA polymerase 2× master mix. We have eventually landed on the annealing temperature for initial cycles at 52 °C, while setting 62 °C for the remaining cycles.

12. After library amplification, a cleanup should be performed either with ethanol precipitation, SPRI beads, buffer exchange or column purification. A gel-purification step is also necessary here due to unavoidable empty ligation products that have sizes around 200 bp (Fig. 8a). Either PAGE-gel purification or agarose gel (4%) purification is suitable, but we have opted for agarose since DNA recovery efficiency is much higher. The trade-off is that the desired library peak and empty ligation product peak usually form broader peaks that require more than 1 round of gel-purification (Fig. 8b).

13. Even if the experimental goal is to detect mRNA sites, the procedure is still very sensitive to rRNA sites which can act as internal technical control. In our experience, libraries prepared with 2× poly(A) enriched samples were still sufficient for rRNA sites profiling even at 1–2 M/sample depth (Supplemental Data 1, 2, Supplemental Figs. 2, and 3).

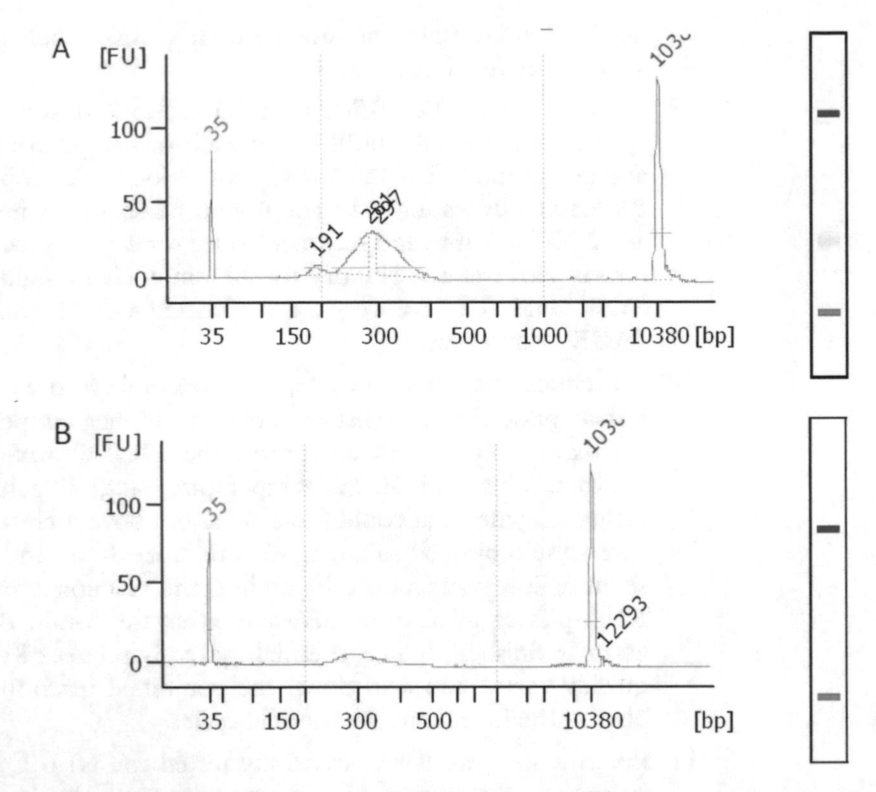

Fig. 8 Final library distribution QC. The QCs shown here were done using Bioanalyzer DNA 1000 high sensitivity chips. (**a**) RibOxi-seq library after 1 round of agarose gel purification (cutting above 220–250 bp). There is still an empty ligation product peak that is at 200 bp. (**b**) In our experience, a second round of the purification is required to completely remove the 200 bp empty ligations

Acknowledgments

The extension work of the original RibOxi-seq is supported NIH grant GM135383 and Duke University Department of Medicine Strong Start to C.L.H. GGC was supported by grant HD099975 from NICHD.

We thank Dr. Disa Elisabet Tehler and Dr. Anders Lund for helpful comments on the initial publication and current development of the method.

References

1. Ayadi L, Galvanin A, Pichot F, Marchand V, Motorin Y (2019) RNA ribose methylation (2′-O-methylation): occurrence, biosynthesis and biological functions. Biochim Biophys Acta Gene Regul Mech 1862(3):253–269
2. Abou Assi H, Rangadurai AK, Shi H, Liu B, Clay MC, Erharter K, Kreutz C, Holley CL, Al-Hashimi HM (2020) 2’-O-methylation can increase the abundance and lifetime of alternative RNA conformational states. Nucleic Acids Res 48(21):12365–12379
3. Choi J, Indrisiunaite G, DeMirci H, Ieong KW, Wang J, Petrov A, Prabhakar A, Rechavi G, Dominissini D, He C, Ehrenberg M, Puglisi JD (2018) 2′-O-methylation in mRNA disrupts tRNA decoding during translation elongation. Nat Struct Mol Biol 25(3):208–216

4. Dimitrova DG, Teysset L, Carré C (2019) RNA 2′-O-methylation (Nm) modification in human diseases. Genes (Basel) 10(2):117
5. Rajan KS, Zhu Y, Adler K, Doniger T, Cohen-Chalamish S, Srivastava A, Shalev-Benami M, Matzov D, Unger R, Tschudi C, Günzl A, Carmichael GG, Michaeli S (2020) The large repertoire of 2′-O-methylation guided by C/D snoRNAs on *Trypanosoma brucei* rRNA. RNA Biol 17(7):1018–1039
6. Elliott BA, Ho HT, Ranganathan SV, Vangaveti S, Ilkayeva O, Abou Assi H, Choi AK, Agris PF, Holley CL (2019) Modification of messenger RNA by 2'-O-methylation regulates gene expression in vivo. Nat Commun 10 (1):3401
7. Incarnato D, Anselmi F, Morandi E, Neri F, Maldotti M, Rapelli S, Parlato C, Basile G, Oliviero S (2017) High-throughput single-base resolution mapping of RNA 2′-O-methylated residues. Nucleic Acids Res 45 (3):1433–1441
8. Krogh N, Jansson MD, Häfner SJ, Tehler D, Birkedal U, Christensen-Dalsgaard M, Lund AH, Nielsen H (2016) Profiling of 2′-O-me in human rRNA reveals a subset of fractionally modified positions and provides evidence for ribosome heterogeneity. Nucleic Acids Res 44 (16):7884–7895
9. Marchand V, Blanloeil-Oillo F, Helm M, Motorin Y (2016) Illuminabased RiboMeth-Seq approach for mapping of 2′-O-Me residues in RNA. Nucleic Acids Research 44(16): e135–e135
10. Zhu Y, Pirnie SP, Carmichael GG (2017) High-throughput and site-specific identification of 2′-O-methylation sites using ribose oxidation sequencing (RibOxi-seq). RNA 23 (8):1303–1314
11. Dai Q, Moshitch-Moshkovitz S, Han D, Kol N, Amariglio N, Rechavi G, Dominissini D, He C (2018) Corrigendum: nm-seq maps 2′-O-methylation sites in human mRNA with base precision. Nat Methods 15 (3):226–227
12. Krogh N, Nielsen H (2019) Sequencing-based methods for detection and quantitation of ribose methylations in RNA. Methods 156:5–15
13. Gillen AE, Yamamoto TM, Kline E, Hesselberth JR, Kabos P (2016) Improvements to the HITS-CLIP protocol eliminate widespread mispriming artifacts. BMC Genomics 17:338
14. Illumina (2020) https://support.illumina.com/content/dam/illumina-support/documents/documentation/chemistry_documentation/experiment-design/index-adapters-pooling-guide-1000000041074-10.pdf
15. Lestrade L, Weber MJ (2006) snoRNA-LBME-db, a comprehensive database of human H/ACA and C/D box snoRNAs. Nucleic Acids Res 34:D158–D162
16. Yoshihama M, Nakao A, Kenmochi N (2013) snOPY: a small nucleolar RNA orthological gene database. BMC Res Notes 6:426

Correction to: Post-Transcriptional Gene Regulation

Erik Dassi

Correction to:
FM and Chapter 12 in: Erik Dassi (ed.), *Post-Transcriptional Gene Regulation, Methods in Molecular Biology, vol. 2404*,
https://doi.org/10.1007/978-1-0716-1851-6

The book was inadvertently published with incorrect affiliation for the authors "Katherine B. Henke, Rachel M. Miller, Rachel A. Knoener, Mark Scalf, Michele Spiniello, and Lloyd M. Smith" in the contributors' list in FM and in the metadata of chapter 12.

This error has now been corrected by updating the affiliation as "Department of Chemistry, University of Wisconsin-Madison, Madison, WI, USA" in the contributors' list in FM and in the metadata of chapter 12.

The updated online version of the book can be found at
https://doi.org/10.1007/978-1-0716-1851-6 and https://doi.org/10.1007/978-1-0716-1851-6_12

Erik Dassi (ed.), *Post-Transcriptional Gene Regulation*, Methods in Molecular Biology, vol. 2404,
https://doi.org/10.1007/978-1-0716-1851-6_23,

INDEX

Erik Dassi (ed.), *Post-Transcriptional Gene Regulation*, Methods in Molecular Biology, vol. 2404,
https://doi.org/10.1007/978-1-0716-1851-6,

Printed in the United States
by Baker & Taylor Publisher Services